RF-MEMS Technology for High-Performance Passives (Second Edition)

5G applications and prospects for 6G

RF-MEMS Technology for High-Performance Passives (Second Edition)

5G applications and prospects for 6G

Jacopo Iannacci
Center for Sensors and Devices (SD), Fondazione Bruno Kessler (FBK), Trento, Italy

IOP Publishing, Bristol, UK

Multimedia content is available for this book from http://iopscience.iop.org/book/978-0-7503-4199-8.

ISBN 978-0-7503-4199-8 (ebook)
ISBN 978-0-7503-4197-4 (print)
ISBN 978-0-7503-4200-1 (myPrint)
ISBN 978-0-7503-4198-1 (mobi)

DOI 10.1088/978-0-7503-4199-8

Version: 20220701

IOP ebooks

British Library Cataloguing-in-Publication Data: A catalogue record for this book is available from the British Library.

Published by IOP Publishing, wholly owned by The Institute of Physics, London

IOP Publishing, Temple Circus, Temple Way, Bristol, BS1 6HG, UK

US Office: IOP Publishing, Inc., 190 North Independence Mall West, Suite 601, Philadelphia, PA 19106, USA

This book is dedicated to the people below, without any relationship between the order of appearance and their relevance.

To Brando and Pietro and to their relentless energy and unconditional serenity, with the profoundly heartfelt wish of conserving such treasures to the largest and longest possible extents, across the extraordinary and endless journey unrolling ahead their limpid eyes.

To Moira, for making her way, our way, relentlessly, day after day, in spite of our adversities.

To Rossana and Pietro, with endless gratitude for their unconditional presence and for making everything possible.

To Luisa De Sanctis, in the awareness that wherever she is now, she is beautiful as always.

To Victor Manuel Caraballo, in the awareness that wherever he is now, he is dancing and smiling.

Jacopo Iannacci

Contents

Preface

The scientific area of microsystems, known as Micro-ElectroMechanical Systems (MEMS), has followed an evolutionary path intimately linked to that of semiconductors, albeit with relevant differentiation. MEMS and semiconductor technologies developed together, starting in the 1960s. While the manufacturing process of transistors was becoming more sophisticated, critical fabrication steps, like the patterning of buried conductive/piezoresistive thin-films and deposition of metals, began to be ventured to yield microstructures with mechanical properties. However, if transistors and semiconductors followed a relentless trend to miniaturisation (*More Moore*) across the last five decades and more, microsystems developed increasing and diversifying transduction principles, aiming at broad functionalities in the fields of sensors and actuators (*More than Moore*).

Within the scenario of microsystem technologies, RF-MEMS, i.e. MEMS for Radio Frequency passive components, started being investigated in the 1990s. Passive components for RF applications, such as waveguides, micro-relays, variable capacitors (varactors), as well as tunable filters, reconfigurable phase shifters, impedance tuners, and so on, were demonstrated in the literature. Their characteristics, in terms of low loss, high isolation and wide tunability/reconfigurability, were outstanding if compared both to standard RF/microwave components as well as to semiconductors.

Such remarkable performances triggered, in the early 2000s, inflated expectations around the massive absorption of RF-MEMS components in mass-market applications, especially referring to mobile handsets, which were growing particularly fast in those years, with the 2nd and 3rd generations of mobile communications (2G/3G). In actuality, market forecasts were systematically disappointed for more than a decade, and the foreseen revolution never took place.

More recently, starting from 2014, RF-MEMS commenced making their way into the market landscape, thanks to a need that other technologies were unable to address effectively. Modern 4th generation (4G) smartphones, due to the integration of more and more components, triggered a degradation trend in the quality of communication, which has been mitigated by the wide reconfigurability of MEMS-based RF passives. As a result, adaptive impedance tuners are now the first successful example of the exploitation of RF-MEMS technology in the consumer market segment. Besides, other MEMS components, such as high-performance switches, are starting to consolidate in applications like high-linearity RF Power Amplifiers (PAs) and RF Front Ends (RFFEs).

If this is the quite reassuring recent market snapshot of RF-MEMS, in addition the 5th generation (5G) of mobile communications and services has begun to be deployed in the last few years. Despite the fact that 5G is stepping in compliance with the 4G Long Term Evolution (4G-LTE) and according to a smooth fashion, it will pursue fundamental drivers of network densification and diversification. Compared to 4G-LTE, the full deployment of 5G in the years to come will demand an increase of data volume, up to 1000 times, of connected devices, from 10 to 100 times, as well as reduced End to End (E2E) latency, massive Machine-Type Communication (mMTC) and operation in the millimeter wave (mm-wave) range.

Further ahead, 6G is already starting to gain shape, capitalising on the visions and high-level Key Performance Indicators (KPIs) that are being elaborated in the scientific literature. Keeping as reference the year 2030, 6G is expected to take over 5G, marking a further 1000× increase in terms of data rate and speed, apart from largely increased reliability of communications, and, most of all, massively leveraging Artificial Intelligence (AI) to trigger resilient and self-evolutionary network paradigms.

These harshly stringent system requirements will push for very-high-performance, widely reconfigurable and frequency-agile passive components. RF-MEMS technology holds the fundamental features to be a Key-Enabling Technology (KET) for 5G and future 6G. From a different perspective, 5G/6G could be the killer applicative scenarios for a massive consolidation of RF-MEMS in the mass-market landscape.

The aim of this book is to outline the outstanding intrinsic potential of RF-MEMS technology with reference to future 5G/6G mobile communications and services. The way this objective is pursued is twofold. On one hand, the proper background concerning the characteristics of RF-MEMS is built, together with a definition of target specifications of interest for future telecommunications protocols and standards. On the other hand, practical insight around the design and development of RF-MEMS passive components is provided, by reviewing a few case studies of design concepts, also including multi-physics simulation approaches and techniques. In more detail, the book unfolds, chapter by chapter, as reported in the following.

Chapter 1 develops a comprehensive discussion on MEMS technologies. First, the inception of microsystems is analysed with reference to the evolution of semiconductor technologies, highlighting common features as well as their differentiation. The most diffused technology platforms for microsystems manufacturing are reported. Examples of consolidated (market) exploitations of MEMS sensors and actuators are also provided. Then, RF-MEMS are introduced, explaining their working principles and listing diverse actuation mechanisms. Subsequently, the main categories of RF-MEMS devices are reviewed, focusing both on simple components, as well as on complex high-order reconfigurable networks.

Chapter 2 places the evolution of the RF-MEMS market, since the early days of the technology, under the spotlight. First, fundamental market analysis concepts, like the hype curve and technology push/market pull scenarios, are introduced. Then, the boosted prospects around massive market penetration of RF-MEMS, counteracted, in fact, by fluctuations and disappointments for over a decade, will be studied in-depth. Two sets of reasons, namely intrinsic and extrinsic to RF-MEMS technology, are identified and discussed. Bearing in mind the lesson learned across multiple forecasted and then missed breakthroughs, a sound overview addressing the current state of the RF-MEMS market for mobile applications is provided.

Chapter 3 focuses on present and future mobile communication protocols. Initially, the working principles of mobile networks are introduced. The evolution of services and performance are reviewed, starting from the 1st generation (1G) of mobile communications, launched in the late 1970s, to the 4th generation (4G). Subsequently, the frame of the under deployment and future 5th generation (5G) communications is depicted. The expected performance and services offered to end users are discussed, along with a comprehensive recap of the current state of its roll-out. Concerning the further development of the paradigm, the main drivers of network virtualisation, densification and diversification pushed forth by 5G will also be debated, covering challenges and KETs to make this vision real within a few years. A brief outlook will also be developed on future 6G, with reference to the timeframe of the next decade, highlighting the potential limitations of the current Hardware-Software (HW-SW) systems development approaches in making the current visions turn into reality.

Chapter 4 deals with a few RF-MEMS design concepts, discussed through a quite practical and hands-on approach. First, capitalising on system-level requirements unrolled in the previous chapter, a few classes of passive components and a set of target specifications critical for 5G/6G will be reported. Then, examples of RF-MEMS designs will be studied in detail, with the aid both of electromechanical and electromagnetic multi-physics simulations. Leveraging the aforementioned tools, case study designs will be altered, showing how this influences the RF-MEMS characteristics and how they can be tailored with respect to the requirements.

Chapter 5 builds upon the hands-on modelling and simulation of RF-MEMS passive components. To this end, a purposefully developed software library of compact analytical models of basic MEMS structures will be reported and discussed in detail. The tool is based on the VerilogA programming language, and it can be exploited in any commercial or open source Integrated Circuits (ICs) development environment supporting Hardware Description Language (HDL) syntaxes. Apart from discussing the most relevant details of the implemented mathematical models, the chapter will also focus on practical exploitation examples of the mentioned tool, with reference to physical and experimentally characterised samples of RF-MEMS lumped components and complex networks.

Appendix A enriches the book's content, showing the dynamics of a few RF-MEMS devices by means of experimental videos (acquired through optical micro-scopy techniques) and simulated animations.

Appendix B completes the book, gathering practical information and hints, according to a user guide fashion, in order to introduce the reader to the practical use of the MEMS compact model library, previously reported in chapter 5, which is also made available for download along with the manuscript (available at https://iopscience.iop.org/book/978-0-7503-4199-8).

Acknowledgement

I want to express gratitude to the people below, without any relationship between the order of appearance and their relevance.

Dr John Navas, Senior Commissioning Manager at IOP Publishing, and all his colleagues at IOPP involved in the realisation and production of this book, for their availability and for building together a fruitful working relationship, based on reciprocal trust and esteem, which goes on since the early phase of the 1st edition's preparation, back in late 2016.

Dipl.-Ing. Harald Pötter, Dr-Ing. Ivan Ndip and all the colleagues with the RF & Smart Sensor Systems Department at Fraunhofer Institute for Reliability and Microintegration IZM (Berlin, Germany), for providing access to the High Frequency Laboratory equipment, necessary to perform the experimental characterisation discussed in chapter 4.

Prof. Dr Gerhard Wachutka, Prof. Dr Gabriele Schrag and Dr Thomas Kuenzig (the latter, now with Infineon Technologies) with the Institute for Physics of Electrotechnology at Munich University of Technology TUM (Munich, Germany), for providing access to the characterisation laboratory and optical measurement facilities, discussed in appendix A.

Author biography

Jacopo Iannacci

Jacopo Iannacci was born in Bologna, Italy, in 1977. He received the MSc (Laurea) degree in Electronics Engineering from the University of Bologna, Italy, in 2003, and a PhD in Information and Telecommunications Technology from the Advanced Research Center on Electronic Systems 'Ercole De Castro' (ARCES) at the University of Bologna, Italy, in 2007.

He received the Habilitation as Associate Professor in Electronics from the Italian Ministry of Education, University and Research (MIUR), in 2017, and the Habilitation as Full Professor in Electronics from the Italian Ministry of University and Research (MUR), in 2021.

He worked in 2005 and 2006 as visiting researcher at the DIMES Technology Center (currently Else Kooi Lab) of the Technical University of Delft (www.tudelft.nl), the Netherlands, focusing on the development of innovative packaging and integration technology solutions for RF-MEMS devices. In 2016, he visited as a seconded researcher the Fraunhofer Institute for Reliability and Microintegration IZM (www.izm.fraunhofer.de) in Berlin, Germany, to conduct high-frequency characterisation of RF-MEMS components jointly with the RF & Smart Sensor Systems Department at IZM. Since 2007, he is researcher (permanent staff) at the Center for Sensors and Devices of Fondazione Bruno Kessler (www.fbk.eu), in Trento, Italy.

His research interests and experience fall in the areas of Finite Element Method (FEM) multi-physics modelling, compact (analytical) modeling, design, optimisation, integration, packaging, experimental characterisation and testing for the reliability of MEMS and RF-MEMS devices and networks for sensors and actuators, Energy Harvesting (EH-MEMS) and telecommunication systems, with applications in the fields of 5G, Internet of Things (IoT), as well as future 6G, Tactile Internet (TI) and Super-IoT.

Dr Iannacci has authored more than 100 scientific contributions, including international journal papers, conference proceedings, books, book chapters and one patent. Among them, a few items are going to be mentioned here. The monograph *Practical guide to RF-MEMS*, published by Wiley-VCH (https://onlinelibrary.wiley.com/doi/book/10.1002/9783527680856), in 2013, was adopted in several courses as a reference textbook. The monograph *RF-MEMS Technology for High-Performance Passives—The challenge of 5G mobile applications*, published by IOP Publishing (https://iopscience.iop.org/book/978-0-7503-1545-6), in 2017, along with the articles 'RF-MEMS Technology for 5G: Series and Shunt Attenuator Modules Demonstrated up to 110 GHz' (https://ieeexplore.ieee.org/document/7556311) and 'RF-MEMS Technology for Future Mobile and High-Frequency Applications: Reconfigurable 8-Bit Power Attenuator Tested up to 110 GHz' (https://ieeexplore.ieee.org/document/7726036), published on the IEEE

Electron Device Letters (EDL), in 2016, were among the first contributions in the literature discussing the employment of RF-MEMS technology in the frame of 5G applications.

He was and is currently involved in several international conferences as symposium chair/co-chair, session chair, technical program committee member, international advisory board member, tutorial lecturer and invited speaker, among which the following few are mentioned: IEEE Sensors, SPIE Microtechnologies, ESSCIRC-ESSDERC, ESREF, MNDCS.

Jacopo Iannacci is Senior Member of the IEEE (www.ieee.org), and is currently Associate Editor of the Springer Microsystem Technologies (https://www.springer.com/journal/542) and of the Frontiers in Mechanical Engineering (https://www.frontiersin.org/journals/mechanical-engineering#).

Introduction

Driven by my passion for commercial aviation, I wrote the introduction to the 1st edition of this book, back in 2017, around the incident of the British Airways (BA) Flight 9. The text is entirely reported below, right at the end of this section. When the book was launched, after receiving the green light from Dr John Navas at IOP Publishing, on early December 2017, I decided to publish the BA Flight 9 story in a LinkedIn article, entitled 'A Lesson Transcending Science', available at the following link: https://www.linkedin.com/pulse/lesson-transcending-science-jacopo-iannacci.

Nearly four years later, precisely on September 21, 2021, I received an email of appreciation regarding the LinkedIn article. My reaction was twofold. I was happy to read the message, and I jumped on the chair when I reached its end. The sender was Andrew Townley-Freeman, son of Barry Townley-Freeman, Senior Engineer Officer of BA Flight 9.

Since then, I started exchanging messages with Andrew, led by my curiosity for aviation and not only. Moreover, at that time I was working on the 2nd edition of the book, so I asked Andrew if he was available to answer some questions I would have then included in this Introduction. As a matter of fact, Andrew kindly accepted, so I report below my questions and his answers.

Closing the circle of coincidences, I am writing these words in February 2022, just a few months before the 40th anniversary of the BA Flight 9 incident. This is a further motivation for which I want to express my gratitude to Andrew Townley-Freeman for contacting me, as well as to his father, Barry Townley-Freeman, to Roger Greaves and Eric Moody, Engineer Officer, First Officer and Captain, respectively, of BA Flight 9, for what they did that night of June 1982.

Jacopo Iannacci

The story of BA Flight 9—From the Introduction to the 1st edition

On the evening of June 24, 1982, British Airways Flight 9, operated by the Boeing 747 named 'City of Edinburgh', was *en route* from Kuala Lumpur to Perth, in a multiple stopover haul from London (Heathrow), to Auckland. At around 20:40 (Jakarta time) engine number 4 started malfunctioning and failed straight away. According to standard drills, the crew shut it down and armed the fire-extinguishing system. Shortly after, engine number 3 failed for no apparent reason, followed by engines 1 and 2. In a matter of two or three minutes, all four engines were down. It was the first time in the history of aviation that a Boeing 747 lost all four engines.

Starting from the altitude of the engines' failure, Flight 9 had approximately 23 minutes of gliding available before unavoidably touching down. Jakarta airport was within gliding distance, but a mountain in between had to be cleared; however, gaining altitude was not an option with no functioning engines. The only alternative was banking towards the open sea and trying an emergency ditching, even though, in the middle of the night and with no functioning engines at all, it was a very risky manoeuvre. On top of that, it would have been the first time that a Boeing 747 was trying to land on water.

After several minutes, just a steps before the point of no return in which the crew would have had to aim the Boeing 747 nose to the Indian Ocean, engine 4 came back to life. Two minutes later, all four engines were back on track. Flight 9 gained altitude, cleared the mountain and landed safely in Jakarta.

The reason for the failure remained unknown until scientific investigations were carried out. The cause was found to be a cloud of volcanic ash produced by an eruption of Mount Galunggung, in West Java. The particles engulfed the engines, killing their power. It was only when the aircraft significantly lost altitude and the engines cooled down, that the ash solidified and broke off from the fans and rotating parts, allowing them to resume normal operation.

In summary, it was the first time a Boeing 747 had lost all four engines. The crew had absolutely no idea of the reason for such a failure, and found themselves faced with a critical decision: on the one hand, an emergency landing in Jakarta, impossible without engines; on the other, an emergency ditching in the Indian Ocean, at night, with no thrust, attempting a landing that had never been done before with a Boeing 747.

What Captain Eric Moody, First Officer Roger Greaves and Engineer Officer Barry Townley-Freeman kept doing in those endless minutes, was one, and only one thing. They repeated the engines' in-flight starting procedure, time after time, and even with no response at all, they tried again, dozens of times, relentlessly, without caring about the unmistakeable ineffectiveness and uselessness of their actions; without paying too much attention to the hopelessness of the whole situation.

While studying averted and, unfortunately, in other circumstances, unavoidable aviation accidents, I unexpectedly found an invaluable source of cases from which important lessons can be learned. For instance, what the crew of Flight 9 did in a matter of less than 20 minutes, projected on the unrolling of a lifespan—of course without the pressure of imminent disaster—could be the key to success. Pushing forth and trying to do what one feels like doing, against unfavourable circumstances and adverse and sceptical people, as well as, on top of it all, a lack of confirmation, might be a way to reach the target, because, primarily, it is the proof of your own motivation, belief and passion.

In fact, immediately following landing in Jakarta, Moody, Greaves and Townley-Freeman, instead of having champagne (and celebrating) with the cabin crew and passengers, looked through the booklet containing all the flight procedures and emergency drills dozens of times, because they were concerned that the engines' failure could have been their fault.

Now, if someone would ask me to unfold the concept of commitment, I think that this example would sketch quite effectively the meaning that such a term bears.

A few questions to Andrew Townley-Freeman

JACOPO IANNACCI (JI): Andrew, as you recently wrote me, you were the person at home picking up the phone call of Berry from Jakarta. Which is the most vivid memory you have of that moment and of the rest of June 24?

ANDREW TOWNLEY-FREEMAN (ATF): I was getting ready to go to my part time job for the evening and the house phone rang—there were no mobiles then—and, to my surprise, it was my father. Probably because the telephone line wasn't that good, he was speaking very quickly. He said that they had had an incident and when the news breaks, don't worry, they are all ok! I had to go find mum, who was out shopping, and tell her, and also phone the homes of the Captain and First Officer. I didn't think much more about it until I got home from work and mum said that British Airways had called her, and there was a little more to the full story. The next morning, I was sent to the newsagent to buy every newspaper!

JI: Did the spread-out of the BA Flight 9 case, on media and among people, interfere on your daily life and on that of your family, in the weeks/months after it happened?

ATF: No, not at all. I think that in those days, news reporting was a lot less sensational and where this would be a massive story in this era, I don't really recall much being said after the first day or two.

JI: As at the time of this interview it is early 2022, we are very close to the 40 years anniversary of BA Flight 9. Wrapping together these four decades, what is the most relevant lesson you learned, in your personal and/or professional life, from what happened on June 24?

ATF: I don't know if this is driven by the events of that night, or just the sort of person that my father is in general, but what I learned (especially in my professional life) is that you stay calm and do things the right way. Panicking when things get tough is an unhelpful emotion and wastes time—you have to put that aside and solve the problem. I think that it has also instilled the value of team working.

Above all, we all know the difference between being professional and being unprofessional, and I strive to be professional every day. My father's approach to this guides me every day.

JI: Probably I can't help sticking to the idea that in June 1982 I was barely 5 years old, i.e. in that period of life when my father, police officers and airplane pilots were all heroes. If I think to this and to what I am today, the feeling that a lot of time passed, gets stronger and stronger. Differently, if I think that 40 years ago planes were flying as they are today, it seems like yesterday. Having said this, thinking of all the time gone, and reasoning in very general terms, is there one item that gives you the feeling that a huge time has passed, and something else that, on the contrary, makes you think that the journey from 1982 to 2022 was not so long, after all?

ATF: When the event happened, I was just 18 years old so, on the one hand, it's (almost) a lifetime ago. Things have changed, especially in flying. Flight Engineers are no longer needed on today's airplanes and if you look in a flight deck from those days to now, they are hugely different. Closer to your field, when dad called me from Jakarta, they had had to book a telephone line to call the UK and then they were able to call. Today, they would call on their mobile … in fact, the passengers would probably be live streaming the incident!

On the other side, there are things that haven't changed. When problems happen, we still need quick thinking humans to address them. If we think about Covid, the scientists that developed the vaccines have done a modern-day equivalent of the crew of BA009. They stayed calm, evaluated the information that they had available, used the processes that they were familiar with and adapted to the situation.

JI: Lastly, still recalling the common way of attributing the role of heroes to the piloting crew of BA Flight 9, is there a little thing, like a daily habit or anything else, you always observed in Barry, that makes him absolutely human?

ATF: This is difficult to answer because you are asking if there is something that makes the hero seem human, but for me, the human that I had known for 18 years became the hero – my hero. But, that wasn't a different role, it was an additional role. After the event, he returned from his role as a Flight Engineer to his role as a father and husband and really he kept the two as different worlds. Maybe that is the answer to your question—he remained the same person that he always was. Also, he feels that he didn't do anything particularly special—to him, he was just doing the job that he was trained to do.

Chapter 1

Introduction to MEMS and RF-MEMS: from the early days of microsystems to modern RF-MEMS passives

1.1 Introduction to semiconductor and microsystem technologies

Thinking of electronics and wondering about the intricacy of paths through which it has modified our habits, expectations and way of living in the last few decades, linking all these facts to the invention of the transistor (by John Bardeen, Walter Brattain and William Shockley at Bell Labs in 1947) seems quite spontaneous, both to people holding technical skills in semiconductor technologies, as well as to the general public. It is unequivocal that the transistor, as an elemental building block of any electronic circuit, was and is still today, the key element enabling the implementation of more complex and increasingly *smart* function/functionalities carried out by smaller, more integrated and less power-hungry devices.

Nonetheless, a quite critical consideration must be dragged into the spotlight before moving the discussion to the world of microsystems. In a rather effective attempt to reduce the complexity of a highly branched scenario, the transistor, according to a rather strict and classical definition, realises a limited set of main functions, as it can be exploited as a relay, i.e. an ON/OFF switch, as a varactor (i.e. variable capacitor), or as an amplifier, i.e. a device able to increase the amplitude of an electrical signal according to a certain proportionality law. None of these functions were enabled or *invented* by the transistor. The first electrically operated switch, or relay, is attributed to the American scientist Joseph Henry in 1835. Its development was driven by the advancement of telegraph technologies. On the other hand, the first thermionic valve for amplification purposes was invented by

doi:10.1088/978-0-7503-4199-8ch1

John Ambrose Fleming in 1904. Bearing in mind this scenario, it is straightforward that transistors and, more generally, semiconductor technologies, have been playing a key role in the development of electronic devices for decades, with no leverage on the novelty or complexity of the function implemented by a single device, with respect to its vacuum valve traditional counterpart. The actual key-enabling feature of semiconductor-based components is miniaturisation, closely linked to the ease of integration. Just to provide a simple visual interpretation of the latter concept, it is sufficient recalling that the computational capacity of a modern smartphone, 55 or 60 years ago, would have required a medium-size apartment full of thermionic valves, relays, wires and power cables, to be implemented. As a matter of fact, miniaturisation and integration enabled by semiconductor technologies, triggered a relentless trend in increasing the implemented complexity, counterbalanced by a moderate and, therefore, affordable spread in manufacturing and production costs, as well-framed by Moore's law [1]. As a matter of completeness, the latter states that with the advancement of technology, the number of transistors that can be integrated in a square inch of silicon doubles roughly every two years.

On the other hand, the development of microsystem technologies has followed a path that exhibits several factors in common with semiconductors, yet marking fundamental differences from such technologies for many other critical aspects.

Microsystems, which are universally referred to with the MEMS acronym (Micro-ElectroMechanical Systems), are millimetre/sub-millimetre devices, realising a certain transduction function between two (or more) distinct physical domains, among which the mechanical one is always involved. More simply, regardless of the specific function it is conceived for, a MEMS device always features tiny structural parts that move, bend, stretch, deform and/or contact together. These peculiarities make microsystem devices particularly suitable for the realisation of a very wide variety of micro-sized sensors and actuators.

Provided with these basic concepts, a few considerations around the differences and similarities of MEMS versus semiconductor technologies can now be developed. Commencing from the most obvious diversities, while semiconductor devices are active, i.e. able to amplify an electrical signal, MEMS are exclusively passive, i.e. can just attenuate an electrical signal. However, transistors do not feature any movable or deformable part, i.e. they do not exploit the mechanical/structural domain to realise transduction functions.

From a technological point of view, MEMS and semiconductors share most of the same micro-fabrication steps, as will be discussed later in more detail. Both feature selective deposition/removal of conductive/insulating thin-films by means of lithography, despite a few peculiar steps and sequences of fabrication that are typical of MEMS only.

Both MEMS and semiconductors pursue the concept of miniaturisation. However, if semiconductor devices, beyond down-scaling, implement in the electrical/electronic domain the multi-physical function of traditional components, MEMS often miniaturise classical objects, keeping their transduction across physical domains. To this regard, the example of the aforementioned relay is quite explanatory. The traditional electrically operated switch exploits the transduction between the electrical and mechanical domain to realise the ON/OFF function. The transistor (when exploited as a switch) realises such a function entirely in the electrical/electronic domain. In contrast, a MEMS switch commutes between the ON/OFF state by coupling the mechanical and electrical domains, likewise the traditional device, despite the former being typically two or even three orders of magnitude smaller compared to the latter.

Also, importantly, the concept of miniaturisation is inflected in a radically diverse fashion when referring to semiconductor and MEMS devices. In the first case, as mentioned earlier, the trend in down-scaling has been continuous for decades. In order to build a more circumstanced idea, in Complementary Metal Oxide Semiconductor (CMOS) technology, the reference geometrical feature characterising the transistor is the channel length. In the mid-1980s such a length was around 4 μm, in the mid-1990s it was roughly 600 nm, in 2010 it reached 30 nm, while nowadays it is well below 20 nm [2]. This trend is, broadly speaking, addressed by the turn of phrase '*More Moore*', indicating the substantial hold of validity of Moore's law.

The concept of miniaturisation *played* by microsystem technologies, is completely different with respect to the aforementioned sketched scenario. First, there is no such thing as a trend in evolving technologies and processes in order to make the same MEMS device smaller, from one year to the next. Instead, a strong driver exists in implementing more and more functionalities, possibly bringing them from the macro- to the micro-world. In other words, if the transistor was the same device over a number of decades, benefiting from being smaller and becoming, in turn, faster, less power consuming, more integrated and so on, the MEMS is a miniaturised object that benefits from implementing more and/or diverse sensing/actuating/transducing functions, by means of a device roughly of the same size. Because of these characteristics, microsystems, as well as other non-standard technologies not mentioned here for brevity, are generally labelled by the turn of phrase '*More than Moore*' [3], indicating that their evolution through time does not follow Moore's law, as they cannot be standardised according to a development trend exclusively built upon the continual shrinking of dimensions.

Eventually, from a different perspective, the concept of miniaturisation is radically dissimilar in quantitative terms, as well, when referring to semiconductors rather than microsystems. While CMOS transistors, as mentioned before, are framed today in the range of nanometres, a MEMS sensor/actuator can span from a few micrometres (in-plane dimensions), to hundreds of micrometres, or even to a few millimetres. Therefore, if a MEMS switch is nearly invisible to the naked eye when compared to a traditional relay, it is massive when placed beside a CMOS transistor.

In the following subsections, a few key considerations will be developed around the early days of MEMS, the most diffused micro-fabrication techniques and their market applications. Such concepts will help one to understand the core topic of this work, which will be introduced immediately after, i.e. MEMS for Radio Frequency applications, universally known as RF-MEMS.

1.1.1 The genesis of MEMS

As already discussed, the non-standard underlying peculiarity of microsystems with respect to semiconductors has emerged. Due to this reason, development of MEMS as a whole has not followed a well-established path, making it difficult to determine an exact point in time corresponding to the conception of microsystems.

From the point of view of technology, the key fabrication steps developed concomitant with the growth of semiconductor technologies starting in the 1950s. Nonetheless, the exploitation of such techniques aimed towards the manufacturing of microsystems commenced later, in the early 1970s. The advancement of silicon-based semiconductor technology motivated the scientific community to investigate, beside critical aspects related to the electrical/electronic characteristics, the mechanical properties of the materials involved in the manufacturing of semiconductors. In this regard, significant contributions can be found, for instance, in the valuable work of several authors concerning the mechanical properties of both bulk materials [4] and deposited thin-layers [5–7], dating from the mid-1950s to mid-1960s. Nevertheless, the exploitation of such techniques aimed towards the manufacturing of micro-devices with movable parts and membranes emerged later, in the period from the second half of the 1970s to the beginning of the 1980s.

Examples around how to exploit anisotropic etching to obtain a variety of 3D suspended structures from a silicon substrate are provided in [8]. Such techniques, together with those typically exploited for the fabrication of transistors and Integrated Circuits (ICs), led to the realisation of miniaturised pressure sensors [9], accelerometers [10, 11], switches [12, 13], and other devices for various applications, such as in the optical and biomedical fields. A remarkable article summarising the state-of-the-art microsystem technologies, and providing a comprehensive outlook around diverse applications, was authored by Petersen [14] at the beginning of the 1980s.

Nonetheless, it was with the further maturation of the surface micromachining fabrication technique [15] that the development of microsystems started to receive a significant boost, leading to the concepts of MEMS sensors and actuators as we know them today. A relevant contribution is represented by the work of Howe and Muller [16] in 1983, in which micro-cantilevers and double-supported beams were realised in polycrystalline silicon, and released suspended above the substrate, via silicon oxide utilised as a sacrificial layer. Since then, a wide variety of MEMS-based sensors, actuators, and various mechanisms, like gears and micro-motors, have been developed, tested, and reported in the literature [17–19].

In addition to what has just been outlined, another significant element of diversity emerges between semiconductor and microsystem technologies. The integration of transistors in silicon-based substrates on one side, and the novelty of the electrical/electronic properties of semiconductor materials, on the other hand, stimulated for decades the aggregation and strengthening of a distinct and unprecedented multi-disciplinary domain of science. Within it, classical physics and chemistry converged, mainly driven by the development of technology, together with mathematics and electrical techniques necessary for the functional understanding of novel devices.

This resulted in boosting the development of the electronics discipline, which today encompasses, at the same time, computer-based codes for designing circuits and systems, as well as quantum theory to describe the physical behaviour of state-of-the-art semiconductor devices.

On a different plane, microsystems did not prosper as a well-structured discipline in the beginning. Essentially, the possibility to realise mechanical micro-devices was seen in the early days as a sort of side-branch of the more and more standardised semiconductors manufacturing stream. As the small scientific community working in the MEMS field was mainly involved in the development of transistors and active devices, a consolidated background in mechanics and structural mechanics was essentially missing. However, the community of mechanical/structural engineers was neither particularly interested in, nor even fascinated by, the idea of bringing part of their expertise *down to* the micro-world, especially from the mid-1950s to the 1960s, when the massive development of modern aircraft was driving research. It was only much later, roughly from the first half of the 1990s, that MEMS started to emerge as a self-standing discipline, where basic knowledge of physics, chemistry, electronics and fabrication started to be blended together with structural mechanics, electro-mechanics and functional reliability. Such a statement is corroborated by the fact that in those years the first books, scientific journals and international conferences explicitly focused on microsystems started to emerge, supported by the growth of a sectorial community of researchers, designers, engineers and developers.

1.1.2 Micro-fabrication technology platforms

The manufacturing of semiconductor components, like transistors, consists in a sequence of steps in which different doses of dopant materials are selectively implanted/diffused within a substrate (typically silicon) to obtain (locally) certain electrical properties. The same implantation/diffusion, or alternatively the digging of deep trenches, can be performed to enhance isolation and reduce the cross-talk between adjacent devices. In addition, conductive and insulating layers are selectively deposited/grown, or deposited/grown everywhere and then selectively removed, in order to redistribute the electrical signals, from the intrinsic devices to the external world. The most important steps in the manufacturing of ICs are ion implantation, diffusion, epitaxial growth, chemical vapour deposition/physical vapour deposition and their variations, wet and dry-etching, sputtering, evaporation and electrodeposition of metals. The selection of the areas that have to undergo one or more of the previously listed steps, is achieved by means of lithography [20]. A typical cross-section view of CMOS transistors is depicted by the schematic in figure 1.1.

Looking at the cross section, a couple of considerations must be developed. First, the intrinsic device is *made into silicon*, where differently doped areas (with diverse electrical/electronic properties) are obtained through the aforementioned fabrication steps. Second, the transistors themselves are very small, while the metal and insulating layers stacked above them, necessary to redistribute the electrical signals from each transistor terminal to the external world, can be several times the size of

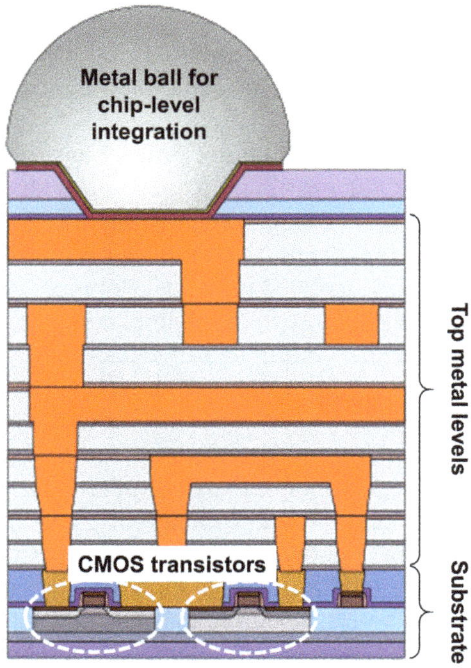

Figure 1.1. Schematic cross section of CMOS transistors. Image from [21] CC BY 3.0.

the semiconductor intrinsic devices. In figure 1.1 the stack is completed at the top by a metal ball, necessary for chip-level integration, i.e. for mounting/interfacing the CMOS circuit into a more complex system or subsystem.

Microsystems, as mentioned earlier, are manufactured relying (for the most part) on the fabrication steps commonly utilised in the processing of semiconductors, even though they are arranged according to different sequences. Despite the fact that MEMS technology is not highly standardised like CMOS, it is possible to identify two distinct processing streams, which are the most diffused, both in the research and commercial production of microsystems. For the sake of completeness, there also exist highly customised technology platforms strongly oriented to the manufacturing of MEMS. Such solutions enable one to achieve very high aspect ratios and extremely precise geometrical features. Nonetheless, their pronounced customisation makes them diverse from CMOS-like processes, resulting in significantly higher costs and more articulated issues in terms of integration with other (incompatible) technologies. Because of their suitability for niche applications, these solutions will not be discussed further in this work. However, it is worth mentioning that the Lithography, Electroplating, and Moulding process (in German, *Lithographie, Galvanoformung, Abformung—LIGA*), is one of the best-known highly customised technologies for the manufacturing of MEMS [22, 23].

Coming back to the most common aforementioned MEMS fabrication flows, they are substantially two: *surface micromachining* and *bulk micromachining*.

In surface micromachining processes, the substrate (silicon or other materials) is used as a sort of *ground floor* plane. All the selective depositions/removals of the conductive/insulating layers are performed above the substrate, through the techniques and steps mentioned above. From a conceptual point of view, a surface micromachining process is not crucially different from the above-CMOS stacking of layers depicted in figure 1.1. The substantial addition of a MEMS surface micromachining process is that membranes and movable parts need to be suspended in air. To do so, a temporary layer is necessary to support mechanically the micro-membranes during their manufacturing (e.g. through electrodeposition or sputtering). Afterwards, it has to be removed via an etching step, in order to release the so-called air gaps, i.e. membranes suspended above the substrate and, therefore, free to move. Such temporary support is referred to as a sacrificial layer, and it can be a photoresist material or a thin-film deposited during processing [24, 25]. Accordingly, it can be stated that in surface micromachining, MEMS devices are *made above silicon.*

A typical schematic cross section of a MEMS surface micromachining process is reported in figure 1.2(a), while a microphotograph of physical gold-based MEMS devices, manufactured with such a technology solution [26], is shown in figure 1.2(b).

In this example, the silicon substrate is 625 μm thick, the air gaps are around 3 μm, while the suspended gold membrane's thickness ranges between 2 and 5 μm. Each of the MEMS in figure 1.2(b) has in-plane dimensions of 2 mm by 0.7 mm.

In bulk micromachining processes, the substrate itself (typically, but not limited to, silicon) is exploited for the realisation of the structural parts of MEMS. By means of performing selective etching (removal) of substrate-specific regions, e.g. Tetramethylammonium Hydroxide (TMAH)-based wet-etching [27] or Deep Reactive Ion Etching (DRIE) dry-etching, thin and deformable membranes are released [28]. Still keeping the same idiomatic expression as above, it can be stated that in bulk micromachining, MEMS devices are *made of silicon.* In light of the discussion developed up to this point, bulk micromachining of MEMS deviates from standard semiconductor processes more than surface micromachining does.

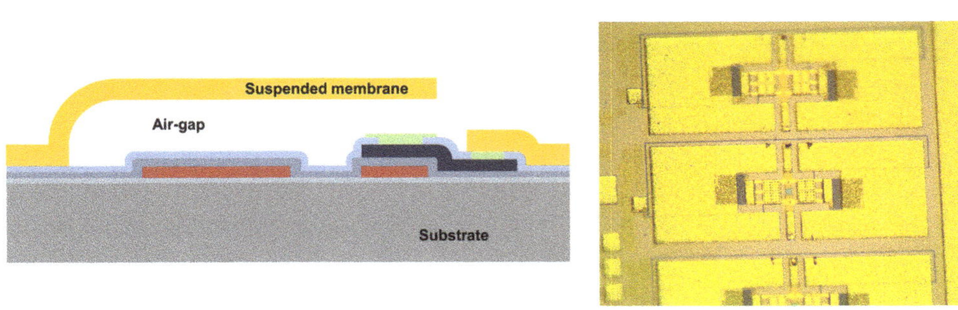

(a) (b)

Figure 1.2. (a) Schematic cross section of a typical surface micromachining MEMS process. (b) Microphotograph of physical MEMS devices realised by means of a surface micromachining process based on gold [26].

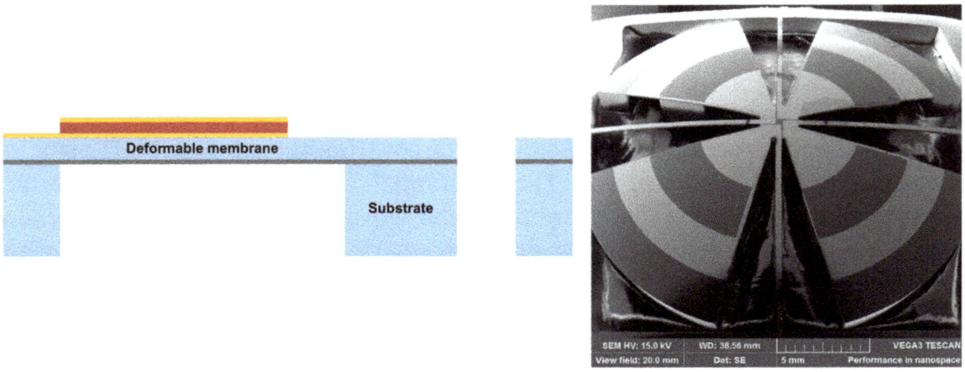

(a) (b)

Figure 1.3. (a) Schematic cross section of a typical bulk micromachining MEMS process. (b) SEM image of physical MEMS for Energy Harvesting (EH-MEMS) devices realised by means of a bulk micromachining process based on silicon [29].

If the latter solution is, for simple MEMS structures, an extension of the above-CMOS stacking of layers, in the former it is silicon (and its mechanical properties) that is to be exploited as a structural material. It is straightforward that silicon mechanical structures of MEMS devices fabricated through bulk micromachining can be completed by selective deposition of conductive/insulating layers, in order to deploy proper electrodes and feeding lines to enable electromechanical transduction.

A typical schematic cross section of a MEMS bulk micromachining process is reported in figure 1.3(a), while a Scanning Electron Microscopy (SEM) image of a physical silicon-based MEMS for Energy Harvesting (EH-MEMS) realised with such a technology [29] is shown in figure 1.3(b).

In the reported example, the initial silicon substrate is 390 μm thick, while, after the etching step, the silicon deformable membranes are 20 μm thick. The diameter of the suspended mechanical resonator in figure 1.3(b) is 8 mm.

1.1.3 Applications of MEMS sensors and actuators

To complete the overview of microsystem technologies on a general perspective, and before embarking on an in-depth discussion around RF-MEMS, a few references and examples of MEMS sensors' and actuators' market applications are going to be briefly provided.

In the first place, it should be highlighted that, if microsystem technologies started to develop in the 1970s (as discussed earlier), the first MEMS-based commercial product, namely, a surface micromachined accelerometer by Analog Devices [30], only showed up in the early 1990s, nearly two decades later. The rationale beneath the lengthy time-to-market characterising the early exploitation of MEMS in the field of sensors and actuators, is a sizable mosaic composed of numerous tiles, featuring crucial pivots, like reliability [31], packaging, integration, as well as costs and, last but not least, market readiness. These aspects will be treated with deeper specificity later, when discussing RF-MEMS.

Starting from the first commercial accelerometer mentioned in [30], MEMS-based sensors and actuators started to spread into diverse systems and devices, securing their presence in significant market segments. Accelerometers themselves, for instance, became, through the 1990s, a de facto standard in the automotive sector, being employed as deceleration sensors activating the inflation of airbags in case of car crashes [32]. Concerning microsystem-based actuators, another successful exploitation of MEMS technology is related to micro-mirrors and, in particular, to their arrangement in high-density matrices of individual micro-devices. MEMS micro-mirrors have been commercially employed since the second half of the 1990s to form optical images on projectors lenses, as well as on movie theatre screens. In this regard, Digital Micromirror Devices (DMDs) are well-known, and named in such a fashion because they are not controlled in an analogue way, but rather with two-state (ON/OFF) driving signals [33, 34].

In more recent years, MEMS accelerometers and gyroscopes, i.e. inertial devices sensitive to rotations around axes and to gravity, experienced a boost in terms of market volumes, significantly larger than traditional applications in the automotive sector. In the last decade, with the emergence of home video game consoles interacting with human motion, and later with the massive spread of smartphones and tablets, MEMS-based Inertial Measurement Units (IMUs) became standard components provided by a wide variety of Original Equipment Manufacturers (OEMs) [35].

Just to mention a few other classes of devices, MEMS are spreading in the switching/multiplexing of optical signals (actuators), miniaturised microphones (sensors) and, despite not being fully mature yet, loudspeakers (actuators), Energy Harvesters (EH) for environmental sources (sensors), pressure/gas/temperature sensors, strain gauges/deformation sensors, and so on. The targeted fields of applications are quite diverse, as they range from automotive to consumer electronics, as well as from space/defence to the medical/health sector.

Eventually, it can be stated with a certain confidence that trends in the exploitation of MEMS will keep ramping up in the years to come. The crucial paradigms of the Internet of Things (IoT) [36] and of the Internet of Everything (IoE) will demand for the availability of smaller, cheaper, less power hungry, multi-functional and more specialised sensors and actuators, to be integrated in increasing numbers within Smart-Cities, -Buildings, -Devices, -Factories, -Cars, -Objects, as well as the human body, e.g. through Body Area Networks (BANs).

1.2 Introduction to RF-MEMS

Bearing in mind the scenario previously described, the investigation of microsystem technologies for the realisation of RF passive components is more recent. The first scientific contributions relating to the exploitation of MEMS-like technology steps for RF passives started to appear in the landscape in the early 1990s, i.e. while MEMS accelerometers were establishing themselves as valuable commercial products. However, early examples of actual RF-MEMS devices only started to populate the scientific literature in the second half of the 1990s.

At an early stage, miniaturisation of microwave and millimetre-wave transmission lines, and their implementation in micromachining technologies based on silicon, emerged as a quite promising research field [37], thanks to the outstanding performance figures in terms of their low loss and compactness compared to traditional solutions [38]. The possibility of integrating fixed RF signal manipulation functions, e.g. through the realisation of stubs [39], appeared as an additional strength of silicon-based waveguides. Among the various families of transmission line configurations available, and well-known for decades [40], micro-fabrication technologies are particularly suited for planar devices. Therefore, most of the attention and interest concerning their miniaturisation was around the Coplanar Waveguide (CPW) and microstrip implementations of transmission lines.

Given these premises, the fundamentals of the aforementioned waveguide configurations are going to be synthetically recalled here. A 3D schematic view of a CPW and microstrip configuration is shown in figure 1.4(a) and figure 1.4(b), respectively. In the former, a central metallisation acts as the RF signal line, while two wider metallised patches are meant to be reference ground planes for the travelling RF signal. The central line and ground planes are separated by a gap, and all the metal layers lie on the same side of the substrate [41]. As the RF signal

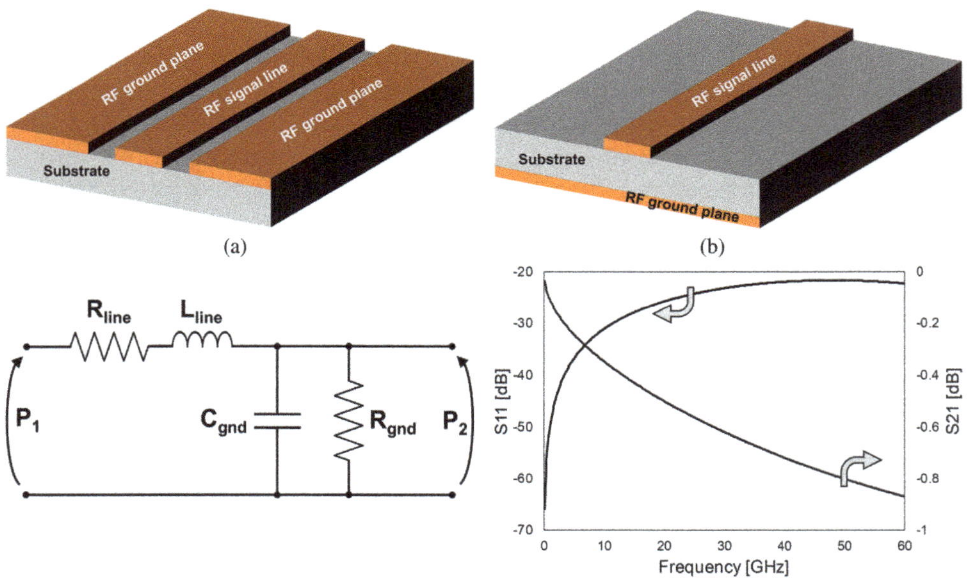

Figure 1.4. (a) Schematic 3D view of a transmission line in Coplanar Waveguide (CPW) configuration. (b) Schematic 3D view of a transmission line in a microstrip configuration. (c) Equivalent lumped element network of a CPW/microstrip line. P_1 and P_2 are the input/output terminations, respectively. (d) Typical S-parameters versus frequency characteristic of a CPW/microstrip line concerning reflection (S11 at P_1) and transmission through the line (S21 at P_2).

propagates along the waveguide, the electromagnetic field is confined between the central line and the ground planes, partially through the dielectric material underneath the metal layers, and partially through the air above them. Very often, a thin insulating layer is deposited on the substrate prior to the electroplating/evaporation of the CPW itself. This helps to reduce losses to the substrate. In the microstrip configuration, instead, the RF signal line is placed on top of the substrate, while a unique reference ground plane is metallised on the opposite face of the wafer. In this case, as the RF signal propagates along the waveguide, the electromagnetic field is mainly confined within the substrate, between the two metal layers (signal line and ground plane).

A basic lumped element network description useful to model the RF behaviour of CPWs and microstrip lines is proposed in [42] and depicted in figure 1.4(c). Across the waveguide input and output ports P_1 and P_2, a resistance and inductance are inserted, representing the series resistive (R_{line}) and inductive (L_{line}) contributions of the metal RF line, respectively. On the other hand, the shunt-to-ground capacitance and resistance model the capacitive coupling (C_{gnd}) and the resistive losses (R_{gnd}) between the RF line and the ground plane/s, respectively, through the substrate material and across air. The values of the resistive and reactive components in figure 1.4(c) are correlated to the physical properties of the transmission line, and can be parameterised in quite a straightforward fashion, in order to account for the most relevant geometrical features of the CPW or microstrip waveguide, like the length, gap, substrate thickness, and so on [43].

The RF behaviour of a typical CPW/microstrip line in terms of scattering parameters (S-parameters) versus frequency [42] is shown in figure 1.4(d). The curves result from the simulation of a CPW by means of a Finite Element Method (FEM) software tool, in the frequency range from DC to 60 GHz. The S11 parameter indicates the fraction of RF signal reflected at the input port of the CPW. Given that it is small on the whole frequency range (better than −22 dB), most of the RF signal flows into the waveguide. The S11 and S22 (reflection at the input and output ports, respectively) are particularly suited to provide an indication of matching between the characteristic impedance of the RF source and of the transmission line. Low values of S11/S22 mean good impedance matching, indeed. The S21 parameter indicates the amount of RF power reaching the output port of the CPW (transmission). Since its worst value (around −0.9 dB) is quite close to 0 dB (i.e. ideal zero losses), the attenuation of the RF signal introduced across the waveguide is limited overall across the analysed frequency span.

Beside the exploitation of a typical surface micromachining step, like selective deposition of thin metal films, additional techniques have begun to be explored with the aim of improving the RF characteristics of miniaturised CPWs and microstrip lines. For instance, shallow tranches were etched in the gap between the RF signal line and the reference ground planes, in order to reduce the losses due to penetration of the electromagnetic field through the substrate, as reported in [44]. In other examples, bulk micromachining was used to remove most parts of the silicon substrate, yielding CPWs suspended above a thin membrane, resulting in a significant reduction of losses and parasitic coupling effects, as discussed in [45, 46].

Shortly after, MEMS technology began to be demonstrated for the realisation of micro-switches [47] and variable capacitors (varactors) [48], as well as tunable filters [49], resonators [49] and programmable phase shifters [50], thus starting to address the crucial feature of reconfigurability. All these aspects are going to be discussed in more detail through the following subsections.

1.2.1 Switches and simple passives in RF-MEMS technology

Based upon the discussion developed in the above sections, it is straightforward that the reconfigurability of a certain microsystem device can be enabled by inclusion of fabrication steps purposely conceived for such a target. Recalling what was mentioned at the beginning, in a case where the surface micromachining process is used, such a specific step is the exploitation of a sacrificial layer, meant to define and then release suspended structures. In contrast, when dealing with bulk micro-machining, the MEMS structure has to be properly etched in order to be released and made free to move.

A few fundamental notions concerning actuation mechanisms are going to be recalled in the following pages. Subsequently, examples of physical RF-MEMS basic passive components will be discussed.

A brief review of actuation mechanisms

From a functional point of view, multi-physical coupling through which mechanical behaviour of movable RF-MEMS parts is controlled (and their characteristics reconfigured) can take place basically according to four different actuation principles: electrostatic, electromagnetic, piezoelectric, and thermoelectric [51]. These different mechanisms are going to be briefly explained.

- ***Electrostatic actuation.*** Two electrodes, one fixed and one movable, are necessary, and they must face each other, as in a typical parallel plate capacitance configuration. When a voltage drop is applied across the two faces, the electrostatic attraction force makes the movable electrode approach the fixed one. Above a certain biasing threshold, called 'pull-in voltage', the movable part collapses onto the underlying fixed one. Figure 1.5 shows a

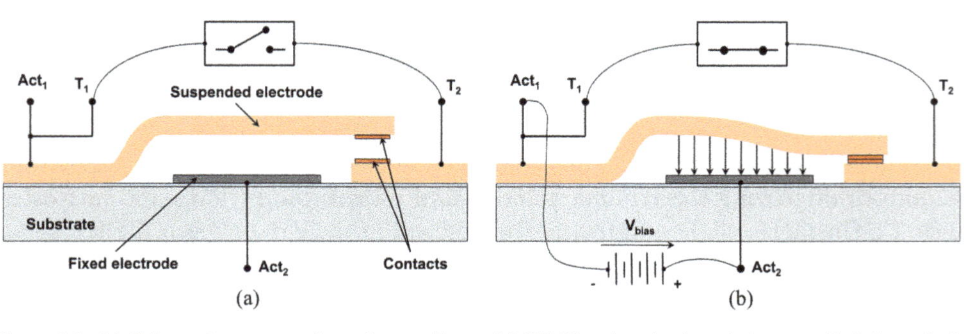

Figure 1.5. (a) Schematic cross section of a cantilevered MEMS series ohmic switch controlled through the electrostatic principle, in the rest position (OFF state). (b) Schematic cross section of the actuated or pulled-in position (ON state), when a bias voltage is imposed between the fixed and the floating electrode.

schematic cross section of an electrostatically controlled cantilever MEMS series ohmic switch [52].

In more detail, figure 1.5(a) reports the cantilever switch in its rest position (no bias imposed). In the latter condition, the input/output terminals (T_1 and T_2) are disconnected, and the micro-relay is in the OFF state, as indicated by the switch symbol above the schematic. However, when a voltage V_{bias} larger than the pull-in threshold is imposed between the movable (Act_1) and fixed (Act_2) electrodes, the contact between T_1 and T_2 is closed, and the switch commutes to the ON state, as indicated in figure 1.5(b).

- ***Electromagnetic actuation***. The suspended MEMS membrane has either to be made of or coated with a ferromagnetic material in order to be sensitive to magnetic field variations. In addition, a magnetic field must be generated by driving a current across a coil, and the former has to surround the deformable membrane. In such a way, when a bias current is imposed, the MEMS part deforms due to the interaction between the magnetic-sensitive material and the external induced magnetic field [53]. A schematic cross section of a cantilever series ohmic MEMS switch driven through electromagnetic actuation is reported in figure 1.6.

In particular, figure 1.6(a) shows a schematic of a MEMS switch in the rest position, i.e. when no bias current is driven across the terminations Act_1 and Act_2. In this case, high impedance is detected between the switch input and output ports, named T_1 and T_2, and the switch results in being OPEN (OFF state). On the other hand, when a current is driven through the coil, a magnetic field builds around the MEMS and the latter deforms until reaching pull-in, as shown in figure 1.6(b). In such a circumstance, the impedance between T_1 and T_2 commutes to a very low value, due to the physical contact between the two metal patches under the cantilever free end, and the switch is CLOSED (ON state).

- ***Piezoelectric actuation***. The suspended MEMS membrane must be covered/coated by/with a thin-film of material holding piezoelectric properties. As

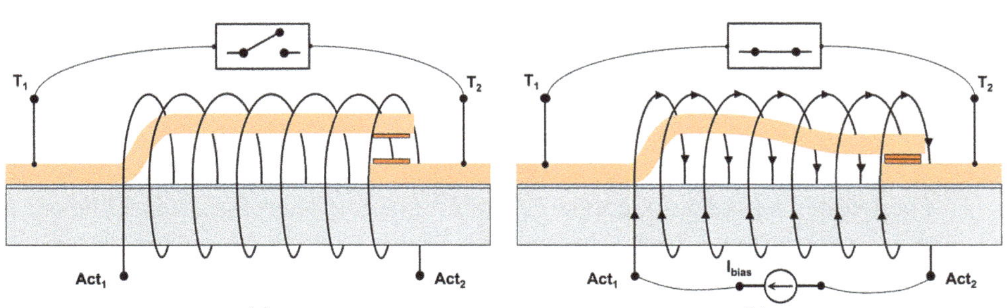

(a) (b)

Figure 1.6. (a) Schematic cross section of a cantilevered MEMS series ohmic switch controlled through the electromagnetic principle, in the rest position (OFF state). (b) Schematic cross section of the actuated or pulled-in position (ON state) when a bias current is driven through the coil that induces a magnetic field around the movable MEMS membrane (the latter coated with a ferroelectric material).

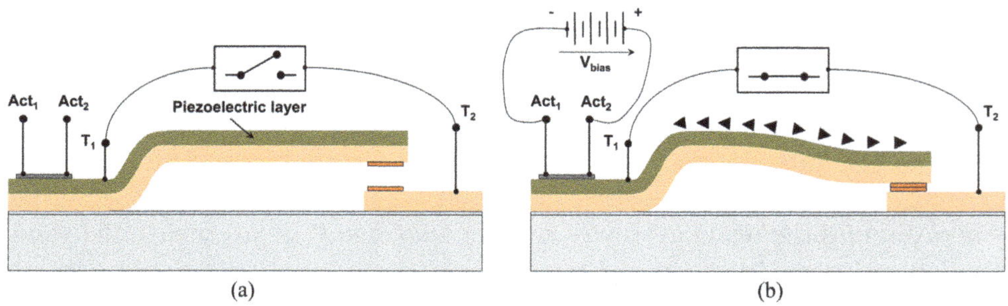

Figure 1.7. (a) Schematic cross section of a cantilevered MEMS series ohmic switch controlled through the piezoelectric principle, in the rest position (OFF state). (b) Schematic cross section of the actuated or pulled-in position (ON state) when the piezoelectric film is subjected to a bias voltage.

known, piezoelectric materials, which fundamentally behave electrically as insulators, exhibit the property of deforming/expanding when subjected to a voltage drop across their opposite faces [54]. As the piezoelectric thin-film is typically patterned above the MEMS structural part (made of bulk silicon of other supporting material), its expansion due to the piezoelectric effect results in a downward (momentum-induced) displacement [55]. A schematic cross section of a cantilever series ohmic MEMS switch driven through piezoelectric actuation is shown in figure 1.7.

In the MEMS rest position, depicted in figure 1.7(a), the switch is OPEN (OFF state). By contrast, when a bias voltage is imposed between Act_1 and Act_2, the piezoelectric material expands and induces commutation of the micro-relay to the CLOSED condition (ON state) between T_1 and T_2, as shown schematically in figure 1.7(b).

- ***Thermoelectric actuation***. In this case, the thermal expansion property of materials is exploited to drive the MEMS movable part(s). An electric current is driven across the suspended membrane that heats up due to its resistance and, therefore, expands because of the temperature increase [56]. Alternatively, thermal expansion of the suspended membrane can also be induced generating the heat not directly into the MEMS part itself, but, for instance, embedding micro-heaters underneath the device. This latter solution makes it possible to use materials with much higher resistivity (like polycrystalline silicon) than metals typically employed for the structural parts of microsystems (like gold, copper, aluminium, etc). Therefore, it is sufficient to drive a fairly low current through the heater in order to obtain the desired increase of temperature [57, 58]. A schematic cross section of a cantilever series ohmic MEMS switch driven through thermoelectric actuation is shown in figure 1.8.

In more detail, figure 1.8(a) reports the switch in the rest, i.e. OPEN, position (OFF state). In contrast, when a bias voltage is imposed across Act_1 and Act_2, a

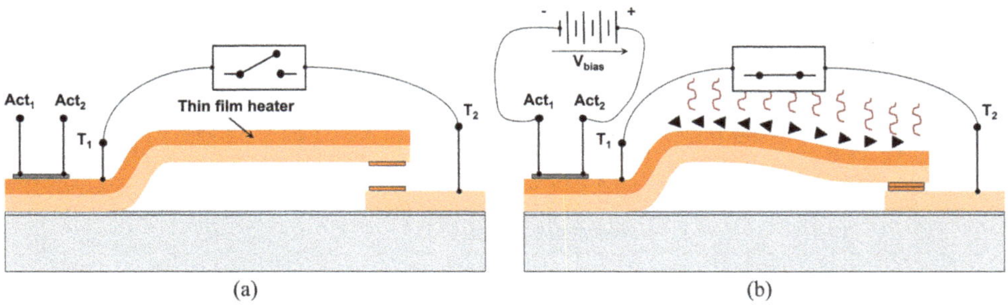

Figure 1.8. (a) Schematic cross section of a cantilevered MEMS series ohmic switch controlled through the thermoelectric principle, in the rest position (OFF state). (b) Schematic cross section of the actuated or pulled-in position (ON state) when a bias current is driven across the suspended membrane, causing the heating and the subsequent thermal expansion of the MEMS structure.

current is driven through the MEMS, and the generated heat brings commutation of the micro-relay in the CLOSED position (ON state), shown schematically in figure 1.8(b).

Among these reviewed mechanisms, electrostatic actuation is certainly the most commonly used for controlling RF-MEMS devices. There are multiple motivations for this choice. One reason being that, at the technology level, electrostatic actuation does not require the deposition of exotic materials, e.g. with piezoelectric or ferromagnetic properties, therefore easing the manufacturing process and also limiting the costs. Furthermore, at the operation level, this kind of driving method does not induce irreversible changes in the mechanical properties of the MEMS, which may happen, for instance, with thermoelectric actuation. Additionally, the effort in terms of energy employed to control MEMS devices is lower if compared with other methods, among which thermoelectric and electromagnetic actuations are definitely the most power hungry. As in electrostatically controlled MEMS, the physical contact between the movable and fixed electrodes must be avoided to prevent a short-circuit, so that no current flows through the device, leading to virtually zero power consumption of the micro-relay, in both ON/OFF configurations. In fact, small current leakages are always present along the DC biasing lines. Nonetheless, they lead in any case to very limited amounts of power necessary to drive the MEMS. In addition, the electrostatic actuation of MEMS will be often referenced in the practical examples shown later in this work. In light of all these considerations, further technical discussion around electrostatics is briefly reported below.

An electrostatically controlled MEMS device can be effectively represented as a parallel plate capacitor, with one fixed and one movable plate, as discussed in [59] and reported in figure 1.9.

A one Degree Of Freedom (1 DoF) parallel plate schematic is shown in figure 1.9(a). The lower plate is mechanically constrained, while the upper one is joined to a mechanical spring whose elastic constant is k. The spring allows displacement along the vertical x axis. The area of both plates is A, while their

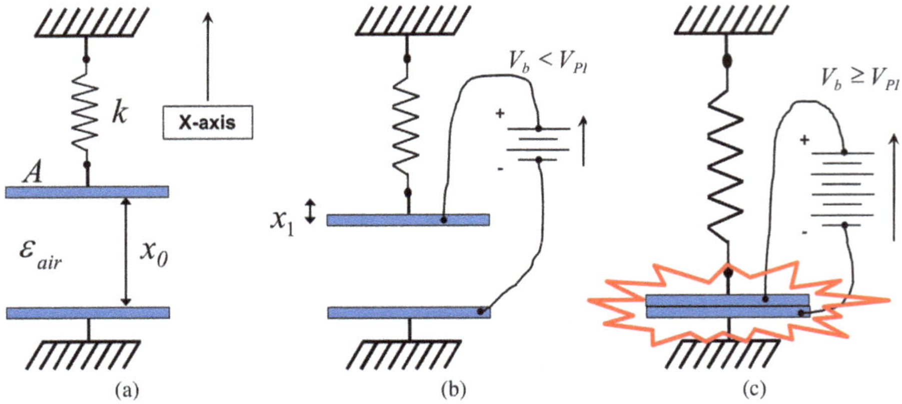

Figure 1.9. (a) Schematisation of an electrostatically controlled MEMS device as a parallel plate capacitor, with one plate fixed and the other movable, when no bias voltage is applied ($V_b = 0$). (b) Displaced movable plate when a bias voltage lower than the pull-in threshold ($V_b < V_{PI}$) is imposed across the two plates. (c) Collapsed (pulled-in) movable plate when a bias voltage equal or larger than the pull-in threshold ($V_b \geqslant V_{PI}$) is imposed across the two plates.

initial distance is x_0. They are immersed in air (dielectric constant ε_{air}) and the effects of gravity on the upper plate and spring can be neglected.

When a biasing voltage V_b is imposed across the two plates, and when it is below the pull-in threshold ($V_b < V_{PI}$), the situation is reported in figure 1.9(b). Due to electrostatic interaction, the upper plate moves downward of x_1. The electrostatic attraction force F_{el} is expressed by equation (1.1):

$$F_{el} = \frac{1}{2}\frac{\varepsilon_{air}A}{(x_0 - x_1)^2}V_b^2 = \frac{1}{2}\frac{C(x)V_b^2}{(x_0 - x_1)} \tag{1.1}$$

At the same time, still referring to figure 1.9(b), the spring elongates of x_1, giving rise to the mechanical restoring force F_{mech}, opposed to F_{el}, expressed by Hooke's law as follows:

$$F_{mech} = -kx_1 \tag{1.2}$$

Such a situation refers to an equilibrium condition since, given a certain $V_b < V_{PI}$, F_{mech} counteracts F_{el}, and the movable plate remains steadily at a distance equal to x_0–x_1 from the underlying fixed electrode. Nonetheless, it should be noted that, while F_{mech} linearly depends on the distance between the plates, commonly referred to as the air gap, F_{el} depends on its square value. By solving the system of equations (1.1) and (1.2) with respect to the voltage, it is possible to derive the pull-in voltage (V_{PI}), expressed as follows:

$$V_{PI} = \sqrt{\frac{8}{27}\frac{kx_0^3}{\varepsilon_{air}A}} \tag{1.3}$$

Such a threshold bias level makes the movable plate collapse abruptly above the underlying fixed plate, as reported in figure 1.9(c). In other words, the vertical displacement of the upper plate towards the lower one can be controlled in an analogue fashion, as long as $V_b < V_{PI}$, corresponding to one-third of the initial air gap (x_0). From a physical point of view, V_{PI} is the limiting level for which F_{el} becomes too large to be counteracted by F_{mech}, and the system becomes unstable. As figure 1.9(c) is a simplified schematic, no insulating layer is indicated between the two collapsed plates. In fact, physical contact must be avoided when pull-in occurs, as it would short-circuit the two plates. For this reason, in real MEMS, a thin insulating layer (of thickness t_{ins}) is always deposited above the underlying fixed electrode. Alternatively, or in conjunction with such a layer, electrically floating stoppers (i.e. elevated posts) can also be deployed, in order to allow clearance between the pulled-in electrodes. Once pull-in takes place, if the bias voltage is progressively decreased, there exists another threshold value corresponding to the release (detaching) of the collapsed plate, named pull-out voltage (V_{PO}), which is expressed as follows:

$$V_{PO} = t_{ins}\sqrt{\frac{2kx_0}{\varepsilon_{ins}A}} \tag{1.4}$$

As mentioned before, since F_{el} depends on the square of the imposed voltage, V_{PO} is typically much smaller than V_{PI}. This means that the pull-in/pull-out (actuation/ release) characteristic of an electrostatically controlled MEMS device exhibits a certain hysteresis, as clearly emerges from the plot in figure 1.10.

When decreasing the bias voltage, right after pull-out occurs, the vertical displacement characteristic reconnects with the one previously obtained for increasing V_b levels (of course lower than V_{PI}). Eventually, since F_{el} depends on the square of the imposed voltage, as indicated by equation (1.1), the pull-in/pull-out

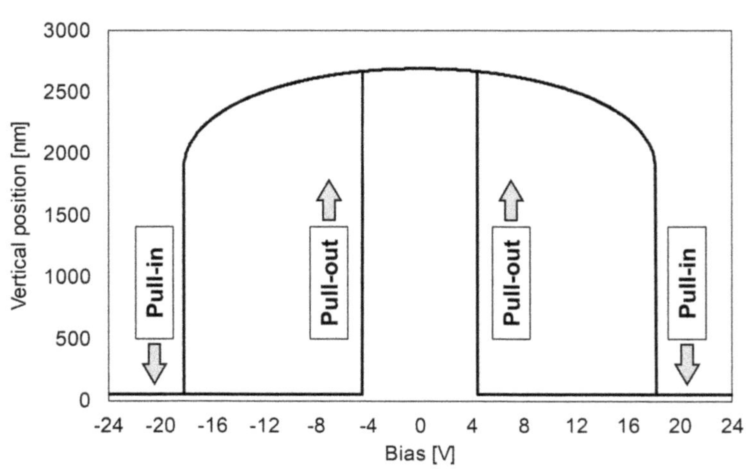

Figure 1.10. Typical pull-in/pull-out characteristic, i.e. vertical displacement versus imposed bias, of an electrostatically controlled MEMS device.

characteristic does not change if positive or negative bias levels are applied. This characteristic is straightforward when looking at the plot in figure 1.10, which is symmetric with respect to the 0 V vertical axis.

Next, examples of physical RF-MEMS simple components discussed in the literature are going to be reported. For the sake of homogeneity in explanation, all the photographs of the physical samples shown in the next pages refer to electrostatically controlled RF-MEMS devices realised in the same surface micromachining technology platform, detailed in [60, 61]. In all of the following microphotographs, the MEMS devices and CPWs/microstrip lines are made of electrodeposited gold, with a thickness ranging between 2 and 5 μm. The thickness of the sacrificial layer and, therefore, of the air gaps, is always around 3 μm. A buried polycrystalline silicon layer (high resistivity) is exploited to realise fixed DC biasing electrodes, underneath the suspended MEMS membranes. Moreover, an additional aluminium buried layer (low resistivity) is employed to implement RF signal underpasses and contact areas. A schematic cross section of this technology was previously reported in figure 1.2(a). Nonetheless, examples referring to other technologies discussed in the literature will also be reported.

RF-MEMS ohmic and capacitive switches

Briefly summarising what was discussed earlier, micro-fabrication technologies enable the manufacturing of miniaturised waveguides, mainly in CPW and microstrip configuration. By adding MEMS-specific fabrication steps, it is possible to realise suspended thin membranes, which can be driven/controlled according to different transduction mechanisms. The focus of this work is on electrostatically controlled RF-MEMS. Therefore, most of the examples that are going to be discussed in the following will refer to such a type of multiphysics coupling.

Having said that, the fundamental building block that enables the reconfigurability of RF-MEMS is represented by the micro-relay (or switch). The switching function, despite being based on a two-state (OPEN/CLOSED) configuration, can be implemented according to different fashions, as it can be ohmic or capacitive, as well as series or shunt [62]. Of course, these listed features can be paired together according to all of the possible combinations. As representative examples, series ohmic and shunt capacitive RF-MEMS switches are going to be discussed in detail.

Figure 1.11 shows a microphotograph of a cantilevered RF-MEMS series ohmic switch, electrostatically controlled in CPW configuration.

The movable membrane is placed in-line on the RF signal path. It is anchored on one side and free to move on the other, as shown in the close-up in figure 1.11(b). As a contact area is placed underneath the suspended tip, metal-to-metal contact is established when the MEMS is pulled-in. Therefore, the switch is OPEN when the MEMS is OFF (rest position), while it is CLOSED when the MEMS is ON (pulled-in), as reported in the simplified schematics in figure 1.12(a) and figure 1.12(b), respectively. The cantilever exhibits measured pull-in and pull-out voltages of 65 and 50 V, respectively. The experimental RF behaviour in terms of the S-parameters is also reported in figure 1.12, and refers to a frequency range from DC up to 40 GHz. In particular, figure 1.12(c) shows the reflection (S11) and isolation (S21) when the

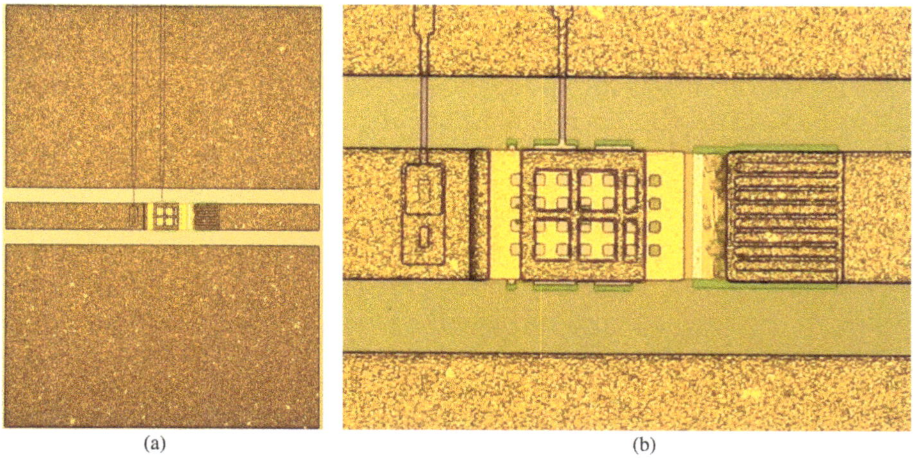

Figure 1.11. (a) Microphotograph of an RF-MEMS cantilevered series ohmic switch in CPW configuration. (b) Close-up of the micro-relay. The suspended cantilever is 180 μm long and 100 μm wide.

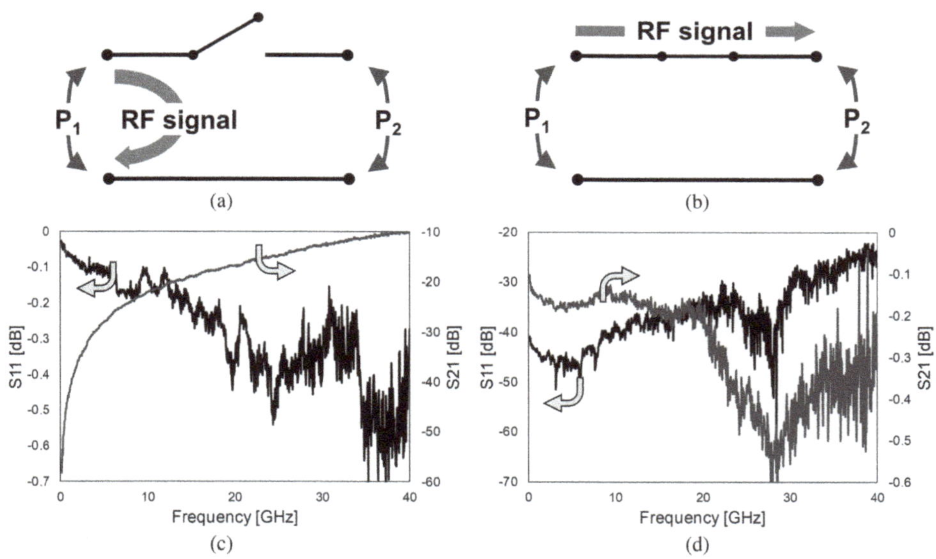

Figure 1.12. (a) Simplified schematic of the series ohmic micro-relay in figure 1.11 in the OPEN state (MEMS switch OFF). (b) Simplified schematic of the micro-relay in the CLOSED state (MEMS switch ON). (c) Measured reflection (S11) and isolation (S21) of the micro-relay in the OPEN state (MEMS switch OFF) from DC up to 40 GHz. (d) Measured reflection (S11) and transmission (S21) of the micro-relay in the CLOSED state (MEMS switch ON) from DC up to 40 GHz.

MEMS micro-relay is OFF, while figure 1.12(d) reports the reflection (S11) and transmission (S21) when the MEMS micro-relay is ON.

When OPEN, most of the power is reflected, as indicated by S11 in figure 1.12(c), which ranges between ~0 and −0.6 dB. Isolation (S21) is better than −25 dB up to 10 GHz, and better than −10 dB up to 40 GHz. The S21 worsens as the frequency increases, because parasitic series capacitance is present between the suspended MEMS cantilever tip and the underlying contact area. When in the CLOSED state, the S11 is better than −25 dB up to 40 GHz, indicating a quite good impedance matching between the RF source and the MEMS device (see figure 1.12(d)). On the other hand, loss in transmission (S21) is very limited, being better than −0.5 dB up to 40 GHz. This result indicates the good quality of the ohmic contact between the pulled-in MEMS and the underlying metal area.

Figure 1.13 shows a microphotograph of an RF-MEMS shunt capacitive switch in a CPW configuration. In this case, the MEMS micro-relay is a membrane hinged at both ends (clamped–clamped configuration) to the ground planes, placed transversally across the RF line.

The behaviour of the shunt capacitive switch is dual with respect to the series ohmic one. First, no ohmic contact is established between the MEMS movable membrane and the RF underpass in any of the ON/OFF configurations, but, instead, a two-state capacitor. The capacitance realises a variable impedance path to the RF ground, rather than between the input and output terminations [62]. More details are reported in figure 1.14. When the MEMS is OFF, the distance between the floating capacitance plate and the underlying one is maximum, therefore the shunt capacitance is minimum (C_{min}), as shown in figure 1.14(a). Such a small capacitance realises a high-impedance path to ground that lets most parts of the RF signal travel between the input and output of the device (CLOSE switch). On the other hand, when the MEMS is ON (pulled-in), the shunt capacitance reaches the

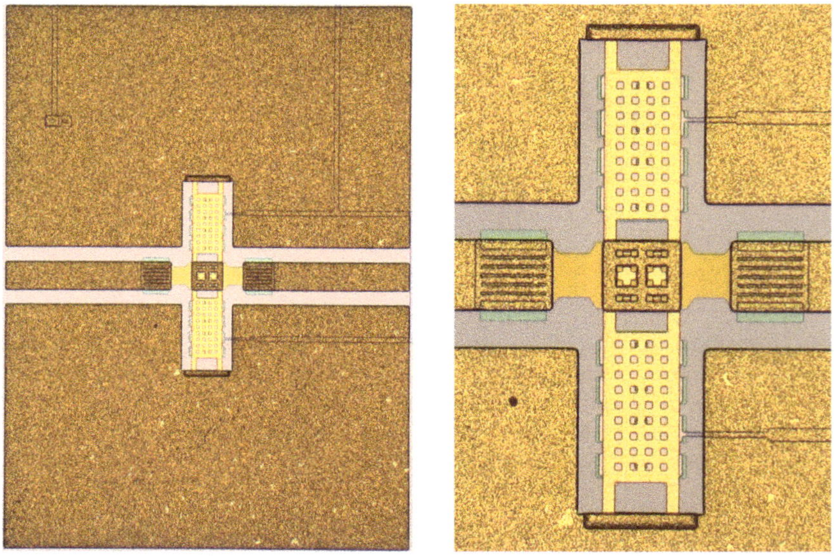

(a) (b)

Figure 1.13. (a) Microphotograph of an RF-MEMS clamped–clamped shunt capacitive switch in a CPW configuration. (b) Close-up of the micro-relay. The suspended double-hinged membrane is 180 μm long and 100 μm wide.

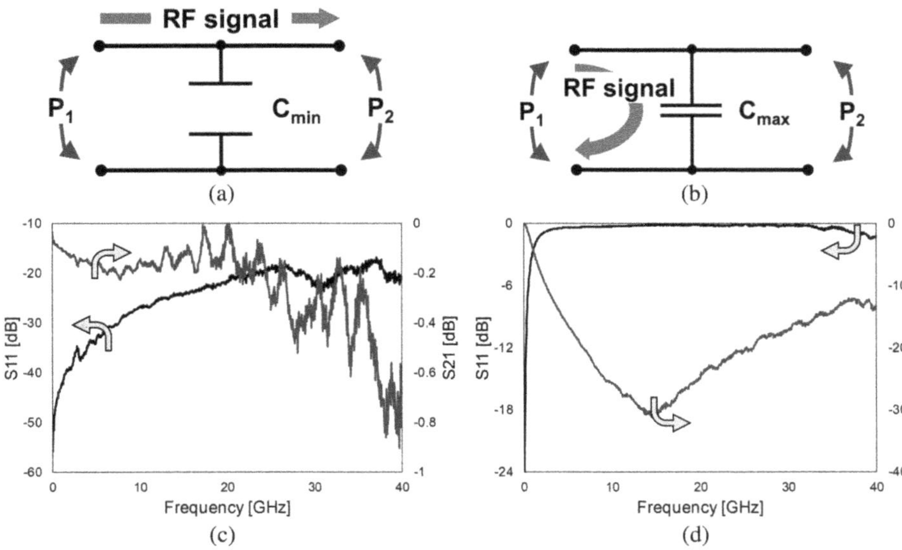

Figure 1.14. (a) Simplified schematic of the shunt capacitive micro-relay in figure 1.13 in the CLOSED state (MEMS switch OFF). (b) Simplified schematic of the micro-relay in the OPEN state (MEMS switch ON). (c) Measured reflection (S11) and transmission (S21) of the micro-relay in the CLOSED state (MEMS switch OFF) from DC up to 40 GHz. (d) Measured reflection (S11) and isolation (S21) of the micro-relay in the OPEN state (MEMS switch ON) from DC up to 40 GHz.

maximum value (C_{max}), as shown in figure 1.14(b). In this case, the large capacitance establishes a low-impedance path to ground that diverts most part of the RF signal travelling across the device, shorting it to ground (OPEN switch). The measured S-parameters of the shunt capacitive switch in figure 1.13 are reported in figure 1.14, referring to the range from DC up to 40 GHz.

In particular, the S-parameters for the MEMS in the OFF state (CLOSE switch) are reported in figure 1.14(c). Reflection (S11) is better than −20 dB up to 40 GHz, proving good impedance matching of the device. Moreover, loss in transmission (S21) is better than −0.8 dB up to 40 GHz. In contrast to the ohmic switch previously discussed, the S21 for the CLOSE switch exhibits increasing loss in the higher portion of the range. This is due to the shunt capacitance that, despite low (C_{min}), causes shorting to ground of a small part of the signal. In any case, having loss better than −1 dB up to 40 GHz is a very good performance result. Figure 1.14(d) shows the S-parameters for the MEMS in the ON state (OPEN switch). Starting from a few GHz, most part of the RF signal is reflected, as indicated by the S11 curve. On the other hand, isolation (S21) exhibits its best value, i.e. −30 dB, around 15 GHz. It is straightforward that capacitive switches, in contrast to ohmic micro-relays, are significantly influenced by the resonant behaviour of the reactive elements [61, 62]. Therefore, they cannot exhibit remarkable performance on a very wide band. Nonetheless, by exploiting their resonant characteristic, they can be optimised to outperform ohmic devices in very well defined frequency ranges.

As mentioned above, micro-relays are fundamental bricks enabling the pronounced reconfigurability of RF-MEMS technology. In light of this consideration, micro-switches have always been widely discussed in the literature, since the early days of RF-MEMS, with respect to various aspects of their design, topology, RF and electromechanical characteristics, reliability, power handling, and so on. For instance, the work reported in [63] discusses an RF-MEMS switching unit with pronounced long-term operability and cycling up to 1 billion in the K-band (18–27 GHz). Moreover, the switch concept discussed in [64] refers to an in-package commercial RF-MEMS device, suitable for 4G mobile applications. On the other hand, [65, 66] discuss novel realisations of RF-MEMS switches with improved performance. Concerning pull-in voltage reduction, designs exhibiting actuation levels as low as 5–7 V were demonstrated in the literature [67].

RF-MEMS variable capacitors (varactors)

From the conceptual point of view, an RF-MEMS variable capacitor (commonly referred to as varactor) is not significantly different from a capacitive switch. Therefore, when the variable capacitance is inserted in shunt-to-ground configuration, its behaviour is well described by the simplified circuit schematics previously shown in figure 1.14(a) and figure 1.14(b). Bearing in mind the discussion previously developed around the pull-in effect in electrostatically controlled MEMS, the varactor can be tuned in an analogue way (i.e. continuously) just from the rest position (zero bias) to the pull-in threshold, i.e. ranging across one-third of the overall air gap. After pull-in, the capacitance will abruptly commute to the maximum value.

The microphotograph in figure 1.15(a) shows an RF-MEMS varactor based on an electrostatically controlled floating gold electrode, kept suspended by four meander-shaped flexible beams.

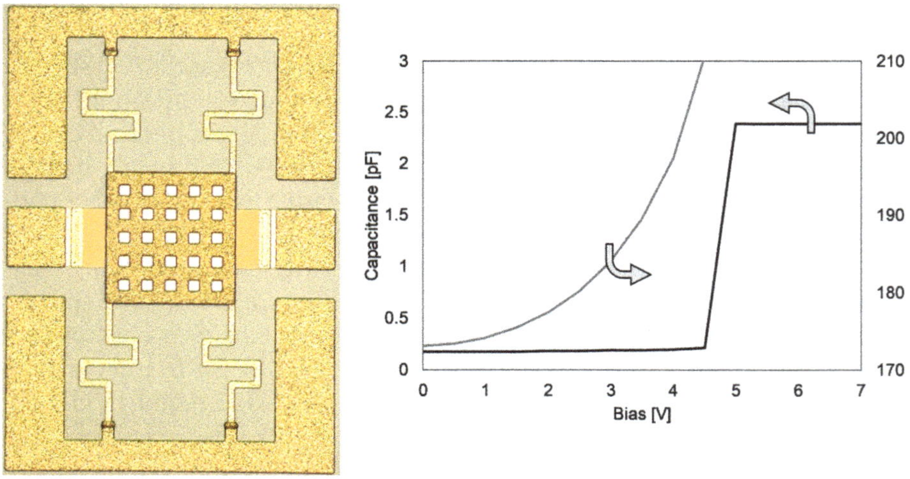

(a) (b)

Figure 1.15. (a) Microphotograph of an RF-MEMS clamped–clamped shunt variable capacitor (varactor) in a CPW configuration. (b) Measured capacitance versus bias voltage (C–V) characteristic. The capacitance can be controlled in an analogue fashion until the pull-in threshold is reached.

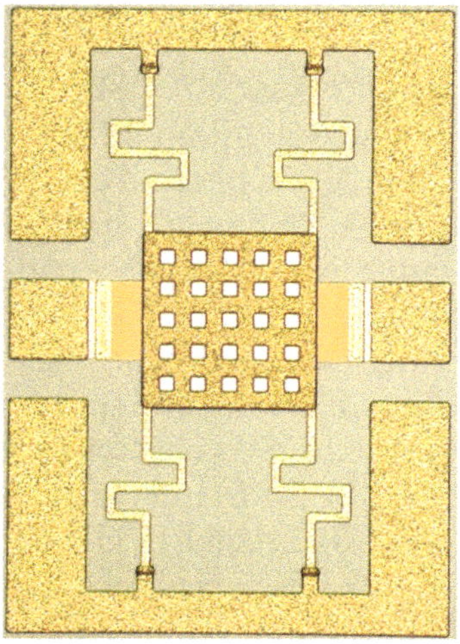

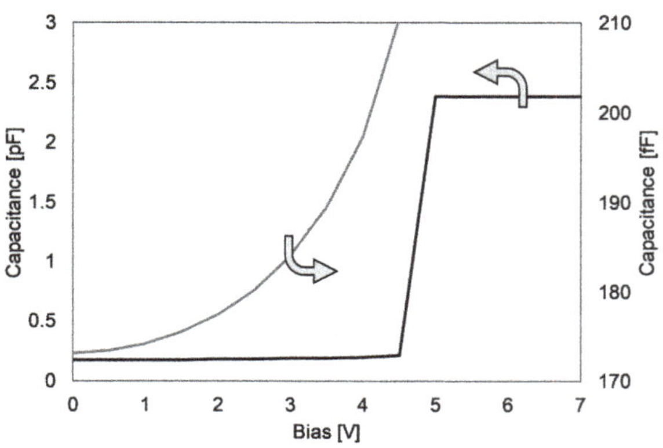

A fixed counter-electrode, lying beneath the movable MEMS, realises the second capacitor plate [67]. The plot reported in figure 1.15(b) shows the measured capacitance versus bias voltage (C–V) characteristic. Before pull-in, the capacitance ranges between ~170 fF and 210 fF (right vertical axis), and can be controlled continuously. When pulled-in, instead, the capacitance steps abruptly to ~2.4 pF (left vertical axis). Therefore, the ratio between the ON and OFF state capacitance (C_{on}/C_{off}) of the varactor in figure 1.15 is around 14, i.e. larger than one order of magnitude.

As for RF-MEMS switches, the scientific literature has always been populated by several contributions regarding varactors. To this purpose, an alternative design enabling a low-voltage controlled varactor is discussed in [68]. On the other hand, focusing on the extension of the varactor tuning range, the solution reported in [69] exploits the built-in intrinsic stress of the MEMS constitutive material, while [70] proposes a solution at the design level, featuring a double actuation mechanism. Again, concerning the linearity improvement in RF-MEMS varactors response, the work discussed in [71] exploits a double varactor with anti-bias control in order to improve such a characteristic. On the side of medium-/long-term reliability, a double DC biasing pulse is reported in [72] with the aim of decreasing the amount of charge entrapped in the insulating layer and, in turn, the so-called voltage screening (i.e. drifting of the pull-in/pull-out characteristic due to charge injection and dipoles orientation within the insulators).

RF-MEMS (variable) inductors

Given the discussion developed up to now, it is straightforward that RF-MEMS technology is also suitable for the realisation of high-performance inductors. The possibility to obtain suspended coils, as through a surface micromachining process featuring a sacrificial layer, as well as to manufacture metal lines above a thin substrate, as can be performed through etching the substrate from the back side (bulk micromachining), leads to a significant reduction of parasitic effects and, therefore, to an increase of the quality factor (Q-factor). The microphotograph in figure 1.16(a) depicts a suspended coil inductor in a CPW configuration. The plot in figure 1.16(b) shows the characteristic impedance of the inductor (S11) on a Smith chart, with reference to the frequency range from DC to 30 GHz.

The characteristic is always predominantly inductive, as confirmed by the trace mainly rotating in the upper half of the Smith chart. Another example of an air-suspended inductor, based on a different coil design, is reported in [73]. Despite the fact that the characteristic of tunability is not as critical for inductors as it is for capacitors, different approaches to tune inductance, triggered by the flexibility of RF-MEMS technology, have been reported in the literature. A fairly popular approach to enable the tunability of RF-MEMS inductors is the exploitation of suspended coils' non-planarity induced by the presence of residual stress within the patterned material. This leads to out-of-plane (i.e. vertical) displacement of adjacent coil turns, causing an inductance decrease if compared to the case of planarity. However, by driving a DC current through the coil, the heating induces a release of the intrinsic stress that temporarily improves the planarity and, therefore, increases the inductance value. Similarly, it is possible starting from a planar device in the rest

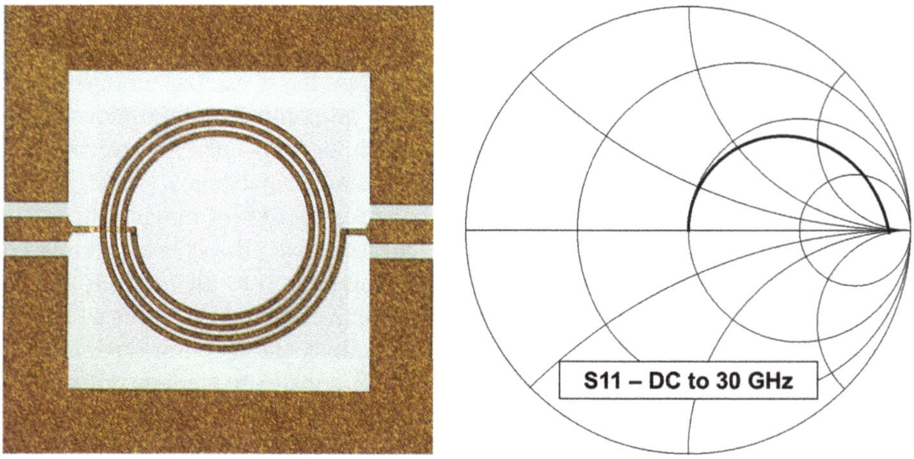

(a) (b)

Figure 1.16. (a) Microphotograph of an RF-MEMS suspended coil inductor in a CPW configuration. (b) Smith chart of the input characteristic impedance (S11) from DC to 30 GHz.

position to induce out-of-plane deformation by means of a bias current that reduces the inductance. Both these solutions are discussed in [74] and [75], respectively. A more exotic method to achieve inductor tunability is reported in [76]. In this case, a liquid is injected in the core, thus modifying its permeability and, therefore, the overall inductance value. More information and examples concerning high-performance (tunable) RF-MEMS inductors are developed in [77].

1.2.2 Complex reconfigurable passives in RF-MEMS technology

The discussion developed in the previous subsection outlined the most diffused classes of basic components realised in RF-MEMS technology. The common denominator of such groups is the implementation of a basic ON/OFF switching function (achievable in various ways), and/or of an actuation function that yields continuous (analogue) tunability within a certain range. Capitalising on the aforementioned basic components, and duplicating or combining them according to certain criteria within a unique physical device, it is possible to realise more complex RF-MEMS passive devices and networks, able to implement manipulation/treatment functions of RF/microwave/millimetre-wave signals, with highly pronounced tunability/reconfigurability. In the following sections, the most common classes of complex RF-MEMS networks will be reported.

RF-MEMS switching units and matrices

Starting from the most elemental RF-MEMS component, i.e. the micro-relay, its duplication and arrangement within a unique device enables the expansion of the switching function from an ON/OFF configuration between one input and output termination to multiple input and output branches. The microphotograph reported

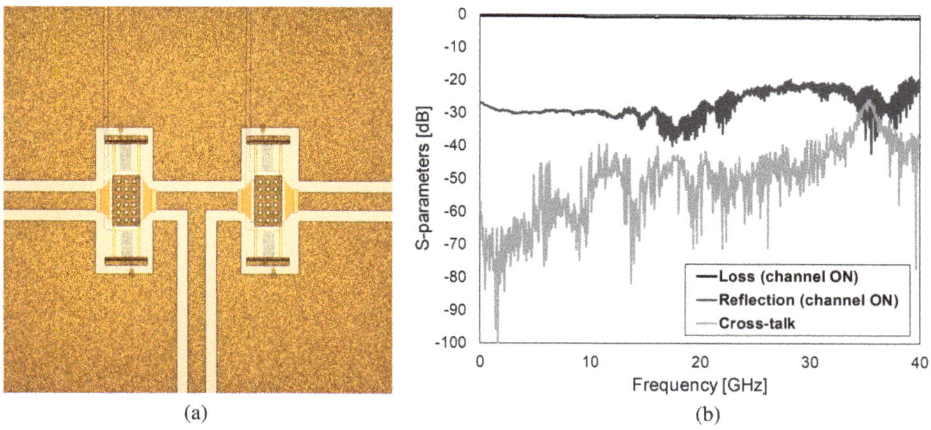

Figure 1.17. (a) Microphotograph of an RF-MEMS Single Pole Double Throw (SPDT) in a CPW configuration. (b) Measured reflection, loss and isolation between adjacent channels (i.e. cross-talk), from DC to 40 GHz.

in figure 1.17(a) shows an RF-MEMS Single Pole Double Throw (SPDT) in a CPW configuration. In summary, it is a T-type switch with one input and two output terminations, each of the latter being controlled by an electrostatically driven RF-MEMS series ohmic switch. Depending on the ON/OFF configuration of the two (independently controllable) micro-relays, the input signal can be driven to each of the two outputs, to both or none of them.

The plot in figure 1.17(b) shows the measured S-parameter behaviour of the SPDT in the frequency range from DC to 40 GHz, and refers to a configuration in which one output channel is ON (CLOSE switch; pulled-in) while the other is OFF (OPEN switch; rest position). The transmission loss (S21) between the input and the conducting output branch is better than −1.1 dB and the reflection (S11) is better than −22 dB up to 40 GHz. Moreover, the isolation between the two output channels (S23), commonly addressed as cross-talk, ranges between −75 and −27 dB, and is always better than −45 dB up to 35 GHz.

The complexity of the switching unit can be increased, thus extending the order of the realised function. Starting from the configuration in figure 1.17, the number of output branches can be enlarged, leading, for instance, to Single Pole Four Throw (SP4T) switching units, in which there are four output terminations, as well as, more in general, to Single Pole Multiple Throw (SPMT) configurations. In addition, commutation can be performed across multiple inputs and outputs, leading to actual switching matrices (e.g. 2×2, 4×4, or $N \times N$) that can be effectively designed and manufactured in RF-MEMS technology. Several examples are reported in the literature, proving the achievement of remarkable characteristics, both for simpler switching units like SP4T [78–80], as well as for various order switching matrices [81–84].

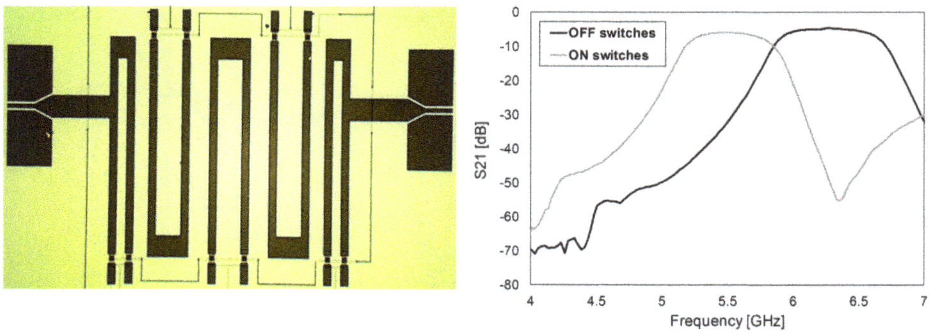

(a) (b)

Figure 1.18. (a) Microphotograph of an RF-MEMS hairpin filter in a microstrip configuration. (b) Measured bandpass characteristic (S21) in two different network configurations.

RF-MEMS tunable filters

Following the same approach, i.e. adding switching functionalities, it is possible to enable tunability for other classes of passive devices, as in the case of RF filters. For instance, the hairpin filter configuration is well-known in the RF engineering community, and commonly exploited to shape a certain bandpass characteristic between the input and output terminations [85]. Such a class of filters was made tunable by implementing it in RF-MEMS technology, as discussed in [86] and depicted by the microphotograph in figure 1.18(a).

The hairpin filter exploits the inductive contribution of U-shaped bends and their capacitive coupling to shape the bandpass response. In the RF-MEMS hairpin filter (microstrip configuration) in figure 1.18, the length of each U-shaped element can be increased by adding a patch selected by cantilevered series ohmic switches, similar to the device previously reported in figure 1.11(a). Such a variation of geometry modifies the impedance of each U-shaped element, leading to a shift of the passed band. The measured transmission (S21) characteristic of the filter is shown in figure 1.18(b). When the MEMS switches are not actuated (short U-shaped elements), the passed band is centred around 6.3 GHz. On the other hand, when the MEMS micro-switches are pulled-in (long U-shaped elements), the passed band is centred around 5.5 GHz, i.e. about 1 GHz below.

The scientific literature is full of several significant contributions exploiting RF-MEMS devices to realise high-performance tunable/switchable RF filters. Differing from other classes of RF-MEMS complex networks, hybridisation/integration of MEMS with other technologies has been explored quite broadly. To this regard, several examples are reported concerning reconfigurable filters entirely realised in RF-MEMS technology [87–89]. Besides this option, different ways to obtain integrated RF-MEMS (commercially available) micro-switches with filters realised, for instance, in Printed Circuit Board (PCB) technology, were also investigated, in order to enable the desired tunability [90–93]. Furthermore, another technological solution of particular interest concerns the realisation of 3D (evanescent-mode) resonant cavities with a high Q-factor, which exploits RF-MEMS tunable elements (mainly varactors) to modify the filter bandpass characteristic [94–97].

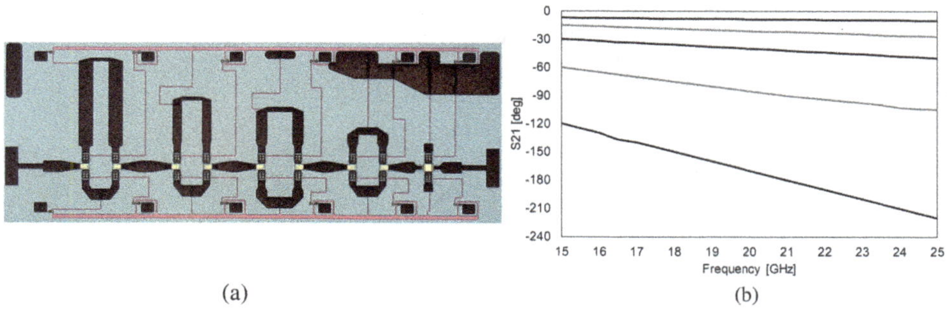

Figure 1.19. (a) Microphotograph of an RF-MEMS 5-bit reconfigurable phase shifter in a microstrip configuration. (b) Input/output phase shift (S21) for different network configurations from 15 GHz to 25 GHz.

RF-MEMS reconfigurable phase shifters

Another class of passive devices that significantly benefits from the exploitation of RF-MEMS technology is that of reconfigurable phase shifters, particularly suited in the driving chain of electronically steerable antennas. An example of an RF-MEMS 5-bit reconfigurable phase shifter in a microstrip configuration is discussed in [98], and its microphotograph is shown in figure 1.19(a). The device features five switchable stages (i.e. 5-bit), in which two paths of different lengths can be selected (per each module) by means of RF-MEMS series ohmic switches. The longer denotes the path the RF signal has to travel across, and the larger denotes the phase shift of the output signal with respect to the input, observable through the S21 parameter. The reconfigurable phase shift of each stage (bit) is added to the others, as the five blocks are cascaded, thus leading to 32 possible configurations.

The plot reported in figure 1.19(b) shows the measured phase shift (S21) of the RF-MEMS network for a few different configurations, in the frequency range from 15 GHz to 25 GHz. Besides the previously discussed example, the scientific literature reports a wide variety of multi-state RF-MEMS phase shifters. In particular, relevant efforts were devoted to the development of multi-state digital devices [99–101], as well as of continually tunable networks [102, 103]. Also, interestingly, monolithic solutions in which the RF-MEMS phase shifter is designed and manufactured together with miniaturised reconfigurable antennas within the same technology platform have been reported [104–106].

RF-MEMS impedance-matching tuners

The availability of high-performance tunable reactive components, like varactors and inductors, as well as the ease of selection/deselection of fixed high Q-factor capacitances/inductances by means of low-loss ohmic switches, has also stimulated the exploitation of RF-MEMS technology for the realisation of impedance-matching tuners. An example of an impedance-matching network entirely realised in RF-MEMS technology is discussed in [107], and a microphotograph of a physical sample is reported in figure 1.20(a). The device, designed in a CPW configuration, features eight switching stages, based on cantilever-type RF-MEMS ohmic switches

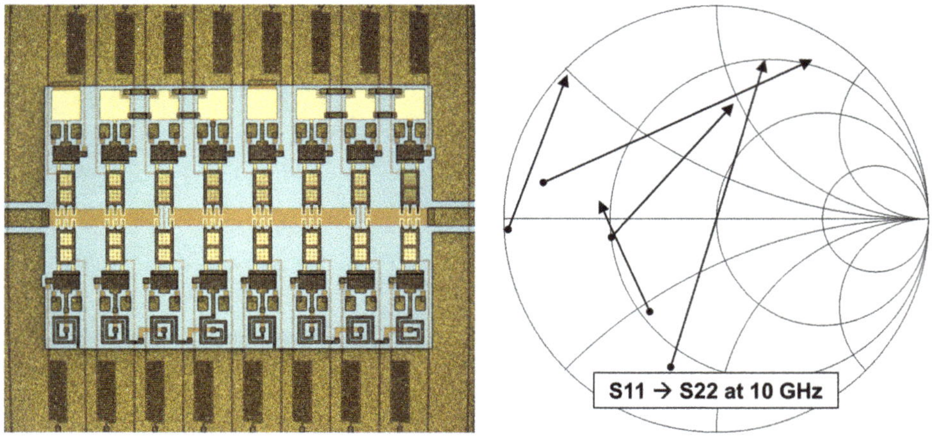

(a) (b)

Figure 1.20. (a) Microphotograph of an RF-MEMS 8-bit reconfigurable impedance-matching tuner in CPW configuration. (b) Smith chart showing the transformation operated by the device between the input (S11) and output (S22) characteristic impedance in a few configurations, at 10 GHz.

that, depending on their ON/OFF configuration, select different reactive components that load the RF line.

The RF-MEMS network is provided with two banks of reactive components, i.e. Metal-Insulator-Metal (MIM) capacitors, as visible in the upper part of figure 1.20(a), and in air-suspended inductors, as visible in the lower part of figure 1.20(a). These reactive elements are designed to be inserted both in series or a shunt configuration on the central RF line, depending on which switches are pulled-in and which are left in their rest position. In summary, the RF-MEMS network realises a double cascaded LC-ladder scheme [85], in which any series or shunt reactive element can be capacitive, inductive, both in parallel or neither selected, thus enabling 256 different impedance transformations. The Smith chart shown in figure 1.20(b) reports just a few among all functions performed by the impedance tuner. In the plot, each arrow shows how, per each configuration, the input characteristic impedance (S11) is transformed at the output (S22).

Despite not being densely populated, as is the case for other classes of devices, the scientific literature discusses several relevant examples of RF-MEMS impedance-matching tuners, proving quite extensive coverage of the Smith chart [108–111].

RF-MEMS programmable power attenuators

A category of complex networks that, despite being sporadically studied in the early years of RF-MEMS and recently attracting more attention in the research and industrial scientific community, is that of programmable (step) power attenuators for RF, microwave and millimetre-wave signals. One of the first examples discussed in the literature is reported in [112], while a microphotograph of the fabricated RF-MEMS network is shown in figure 1.21(a).

The RF-MEMS network features polycrystalline silicon buried resistors of different values, inserted along the RF line (series configuration). Moreover, cantilever-type

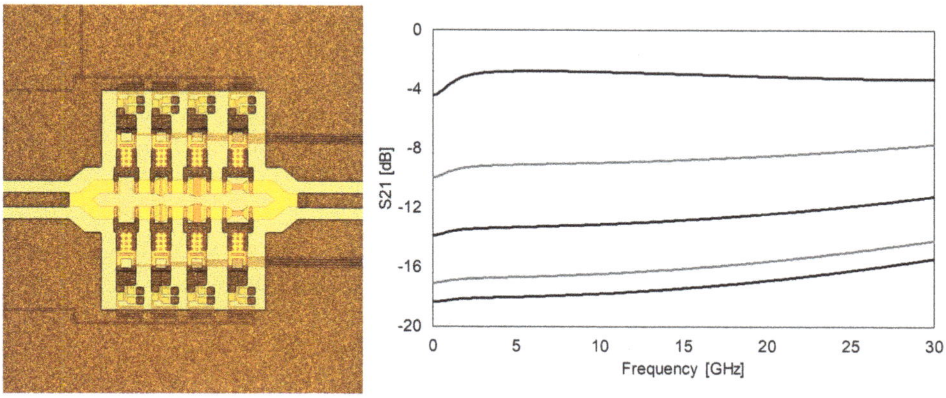

(a) (b)

Figure 1.21. (a) Microphotograph of an RF-MEMS 4-bit programmable step attenuator in a CPW configuration. (b) Measured attenuation (S21) realised by the device in a few configurations, from DC to 30 GHz.

MEMS ohmic switches, with contact fingers before and after each loading resistor, are placed transversally along the device. When a MEMS switch is actuated (pulled-in), the fingers establish ohmic contact with the underneath metal pads, and a very-low-resistance path makes the RF signal flow through the gold switch membrane itself, rather than through the polycrystalline silicon buried resistor. In other words, the loading resistor is shorted, and therefore the attenuation implemented by the whole network is decreased. With particular reference to the device in figure 1.21(a), the whole attenuation can be stepped according to 4-bit, i.e. 16 different attenuation levels. The plot in figure 1.21(b) shows the measured attenuation (S21) in a few network configurations, ranging from DC to 30 GHz. Despite the fact that the characterised frequency range is rather wide, all the traces exhibit a quite flat characteristic and attenuation levels that can be set from around −3 to −20 dB.

Other implementations in RF-MEMS technology of multi-state step attenuators have been recently demonstrated in the literature, in some cases for measured frequency ranges as high as 110 GHz [113–115].

Miscellaneous RF-MEMS

In order to conclude this introductory chapter on RF-MEMS technology, other examples of basic components and complex networks, not covered by the previously reported classes, are going to be briefly listed.

The first category is of mechanical resonators based on MEMS technology. The reasons why they are not always classified as RF-MEMS are multiple. For one thing, often the frequency of operation is not in the RF/microwave/millimetre-wave range, but is lower. In addition, MEMS mechanical resonators are rarely tunable, and they are typically not framed in a waveguide configuration and do not feature switches or varactors. Beside these considerations purely related to a matter of definition, MEMS mechanical resonators are high-performance transducers exploiting a double conversion between physical domains, in order to operate a frequency selection in a very narrow band. In more detail, a signal with a certain spectrum feeds the resonator, operating electrical to mechanical transduction. The MEMS device resonates at a certain mechanical frequency that is transduced back from a mechanical to electrical domain. The output electrical signal has a very narrow band with respect to the input one, and the frequency selection is operated by the mechanical characteristics of the MEMS resonator. Such a very selective filtering function is critical to devices like oscillators, which are meant to generate a reference (RF) frequency with pronounced stability versus time, and a bandwidth as narrow as possible around the carrier. MEMS mechanical resonators with a high Q-factor and remarkable stability against time and temperature have been widely reported in the literature [116–120]. The principle of the forth-and-back transduction between the electrical and mechanical domain is exploited also in other classes of frequency-filtering devices, like the so-called Surface Acoustic Wave (SAW) and Bulk Acoustic Wave (BAW) filters and (thin) Film Bulk Acoustic Resonators (FBARs). In this case, the input electrical signal (with a certain spectrum) is transduced into an acoustic wave that travels across a certain material, and then is transduced back into a narrowband electrical signal [121–124].

Another category of RF-MEMS devices is that of electromagnetic resonators and LC-tanks, necessary, for instance, to operate a precise selection of the RF frequency generated by Voltage Controlled Oscillators (VCOs). Differing from mechanical resonators, in this case there is no transduction from the electrical/electromagnetic to mechanical domain to operate selection of functionalities. The reactive (capacitive and inductive) characteristics of the device determine its specific resonance that, therefore, shapes the output signal. Electromagnetic resonators and LC-tanks significantly benefit from the intrinsic reconfigurability of RF-MEMS, extensively discussed above. For example, it is sufficient to realise part of the capacitive contribution of LC-tanks by means of an RF-MEMS varactor, in order to enable wide tunability of the filtering function operated by the whole device. The scientific literature describes various implementations of electromagnetic resonators and LC-tanks in RF-MEMS technology, both concerning design concepts in planar technology [125, 126], as well as exploiting 3D (evanescent-mode) resonant cavities featuring tunable elements [127, 128].

Other RF-MEMS devices of interest in modern RF components and systems are directional couplers/splitters. These passives are meant to couple different signals in order to perform certain mixing functionalities, e.g. of critical importance in RF transmitters/receivers (transceivers), as well as in Automated Test Equipment (ATE) systems, like Vector Network Analysers (VNAs). As it is easy to envisage, the intrinsic tunability of RF-MEMS enables pronounced reconfigurability of couplers/splitters, both concerning selection of different RF signals to be mixed together, as well as with reference to the extent (in terms of dB) according to which such signals must be coupled or split. As for other categories of RF-MEMS, the scientific literature on directional couplers/splitters is populated by several significant contributions, discussing various design concepts and diverse solutions in terms of micro-fabrication technology platforms [129–135].

Returning to programmable phase shifters, a sub-class of RF-MEMS devices is the so-called True-Time Delay Lines (TTDLs). These reconfigurable networks are, in fact, phase shifters with a quite pronounced linearity of the phase shift over frequency. This characteristic ensures True-Time Delay (TTD), which is a constant (i.e. invariant) delay with respect to frequency. Relevant papers discuss TTDLs in RF-MEMS technology, as well as the achieved remarkable characteristics in terms of reconfigurability [136–140].

Finally, to conclude this chapter, the flexibility of RF-MEMS technology in implementing treatment functionalities of RF/microwave/millimetre-wave signals has unequivocally emerged, in enabling high-performance, pronounced reconfigurability/tunability and wideband operation over frequency. In the following approaches, similar to those discussed in the previous sections, RF designers and engineers now have the possibility to conceive innovative design concepts dedicated to the implementation of RF-MEMS passives with characteristics that transcend those reported up to now. Additionally, blending more functionalities, such as, for example, reconfigurable phase shifting with multi-level step attenuation, within the same RF-MEMS physical device, is also a viable option holding a not-so-hidden potential, especially bearing in mind emerging 5G as well as beyond-5G applications.

1.3 Conclusion

This chapter developed a general discussion on Micro-ElectroMechanical Systems (MEMS) and MEMS for Radio Frequency passives (RF-MEMS) technologies. First, the inception of the concept of microsystems was analysed with reference to the evolution of semiconductor technologies, highlighting common features as well as how they can be differentiated. To this regard, a brief discussion was developed around the evolution of (standard) semiconductor technologies, also including the driving trend of Moore's law, which basically described, and still continues to describe, decades of transistors' technology evolution. It was also remarked that, in parallel to the maturation and consolidation of standard semiconductor technologies, experimentation of micro-fabrication steps devoted to obtaining microsystems, i.e. micro-devices with mechanical properties, began in the late 1960s. Then, the first examples of actual MEMS devices, reported in the literature in the second half of the 1970s, were also discussed.

The most widely exploited technology process flows for the realisation of MEMS, namely surface and bulk micromachining, were also described, focusing on their fundamental characteristics and on the typical features distinguishing one from the other. Still referring to the intricate relationship between MEMS and standard semiconductor technologies, the concepts of '*More Moore*' and of '*More than Moore*' were debated, putting under the spotlight the opposite trends that microsystems and semiconductors follow concerning miniaturisation and customisation.

The concept of RF-MEMS was then introduced, stressing the relative novelty of this kind of microsystem exploitation as compared to sensors and actuators, such as, for example, inertial sensors (accelerometers and gyroscopes) and micro-mirrors.

The four main driving mechanisms to control RF-MEMS devices (electrostatic, electromagnetic, piezoelectric, thermoelectric) were reviewed. Fundamental physical considerations were also included regarding electrostatic actuation of RF-MEMS, it being one of the most diffused, as well as that exploited in the practical examples that will be listed throughout this book.

Subsequently, the most common categories of RF-MEMS devices were introduced, focusing both on simple components as well as on complex high-order reconfigurable networks. Concerning the former, micro-relays (or switches), variable capacitors (varactors) and inductors were reported. With reference to the latter, instead, several classes of complex RF-MEMS networks were discussed, among which are complex switching units, programmable step attenuators, impedance-matching tuners, and so on. For all of the aforementioned categories of RF-MEMS, typical design implementations and experimentally observed performance/characteristics have been shown.

References

[1] Brock D C and Moore G E 2006 *Understanding Moore's Law: Four Decades of Innovation* 1st edn (Philadelphia, PA: Chemical Heritage Foundation) p 122

[2] *Converter Passion ADC Performance Evolution: Low-Voltage Operation—Part 2* https://converterpassion.wordpress.com/tag/cmos/)

[3] Zhang G Q and van Roosmalen A 2009 *More than Moore—Creating High Value Micro/Nanoelectronics Systems* 1st edn (Berlin: Springer) p 332

[4] Hobstetter J H 1960 Mechanical properties of semiconductors *Properties of Crystalline Solids* 1st edn (West Conshohocken, PA: ASTM) p 40

[5] Beams J W, Freazeale J B and Bart W L 1955 Mechanical strength of thin films of metals *Phys. Rev. Lett.* **100** 1657–61

[6] Neugebauer C A 1960 Tensile properties of thin, evaporated gold films *J. Appl. Phys.* **31** 1096–101

[7] Blakely J M 1964 Mechanical properties of vacuum-deposited gold *J. Appl. Phys.* **35** 1756–9

[8] Bassous E 1978 Fabrication of novel three-dimensional microstructures by the anisotropic etching of (100) and (110) silicon *IEEE Trans. Electron. Dev.* **25** 1178–85

[9] Wen H K, Hynecek J and Boettcher S F 1979 Development of a miniature pressure transducer for biomedical applications *IEEE Trans. Electron. Dev.* **26** 1896–905

[10] Roylance L M and Angell B J 1979 A batch-fabricated silicon accelerometer *IEEE Trans. Electron. Dev.* **26** 1911–17

[11] Chen P, Muller R, Shiosaki T and White R 1979 WP-B6 silicon cantilever beam accelerometer utilizing a PI-FET capacitive transducer *IEEE Trans. Electron. Dev.* **26** 1857

[12] Petersen K E 1979 Micromechanical membrane switches on silicon *IBM J. Res. Dev.* **23** 376–85

[13] Petersen K E 1977 Micromechanical light modulator array fabricated on silicon *Appl. Phys. Lett.* **31** 521

[14] Petersen K E 1982 Silicon as a mechanical material *Proc. IEEE* **70** 420–57

[15] Bustillo J, Howe R and Muller R 1998 Surface micromachining for microelectromechanical systems *Proc. IEEE* **86** 1552–74

[16] Howe R T and Muller R S 1983 Polycrystalline silicon micromechanical beams *J. Electrochem. Soc.* **130** 1420–23

[17] Howe R and Muller R 1986 Resonant-microbridge vapor sensor *IEEE Trans. Electron. Dev.* **33** 499–506

[18] Fan L S, Tai Y C and Muller R S 1987 *Pin Joints, Gears, Springs, Cranks, and Other Novel Micromechanical Structures* (Oak Ridge, TN: OSTI) https://osti.gov/scitech/biblio/5974687

[19] Fan L S, Tai Y C and Muller R S 1989 IC-processed electrostatic micromotors *Sensors Actuators* **20** 41–7

[20] Jaeger R C 1988 *Introduction to Microelectronic Fabrication* 1st edn (Boston, MA: Addison-Wesley) p 232

[21] Gualtieri D Interdisciplinarity in Physics *Tikalon Blog* http://tikalon.com/blog/blog.php?article=2012/interdiscipline)

[22] Li H, Li D, Hu Y, Pan P, Li T and Feng J 2016 UV-LIGA microfabrication for high frequency structures of a 220GHz TWT amplifier *Proc. IEEE Int. Vacuum Electron. Conf. (IVEC) (Monterey, CA, April)* pp 131–3

[23] Rusch C, Börner M, Mohr J, Zwick T, Chen Y and De Los Santos H J 2013 Electrical tuning of dielectric resonators with LIGA-MEMS *Proc. Europ. Microw. Integrated Circ. Conf. (Nuremberg, Oct. 2013)* pp 316–9

[24] Johnstone R W and Parmaswaran A 2004 *An Introduction to Surface-Micromachining* 1st edn (Berlin Springer) p 189

[25] Franssila S 2010 *Introduction to Microfabrication* 2nd edn (New York: Wiley) p 534

[26] Iannacci J, Tschoban C, Reyes J, Maaß U, Huhn M, Ndip I and Pötter H 2016 RF-MEMS for 5G mobile communications: a basic attenuator module demonstrated up to 50 GHz *Proc. IEEE SENSORS (Orlando, FL, Oct.–Nov. 2016)* pp 131–3

[27] Pal P and Sato K 2017 *Silicon Wet Bulk Micromachining for MEMS* 1st edn (Boca Raton, FL: CRC Press) p 424

[28] Tilli M, Motooka T, Airaksinen V-M, Franssila S, Paulasto-Krockel M and Lindroos V 2015 *Handbook of Silicon Based MEMS Materials and Technologies* 2nd edn (Norwich, NY: William Andrew) p 826

[29] Iannacci J, Sordo G, Serra E and Schmid U 2016 The MEMS four-leaf clover wideband vibration energy harvesting device: design concept and experimental verification *Microsyst. Technol.* **22** 1865–81

[30] Sherman S J, Tsang W K, Core T A and Quinn D E 1992 *A low cost monolithic accelerometer Proc. of VLSI Circ. Symp. (Seattle, WA, June 1992)* pp 34–5

[31] Iannacci J 2015 Reliability of MEMS: a perspective on failure mechanisms, improvement solutions and best practices at development level *Displays* **37** 62–71

[32] Sensors Online *The Growing Presence of MEMS and MST in Automotive Applications* https://dialnet.unirioja.es/servlet/articulo?codigo=486778

[33] National Inventors Hall of Fame *Larry Hornbeck, Digital Micromirror Device* https://invent.org/inductees/larry-hornbeck

[34] Douglass M R 1998 Lifetime estimates and unique failure mechanisms of the Digital Micromirror Device (DMD) *Proc. of IEEE Int. Reliab. Phys. Symp. (Reno, NV, April 1998)* p 9–16

[35] *Growth, Trends, COVID-19 Impact, and Forecasts* https://mordorintelligence.com/industry-reports/global-mems-based-inertial-measurement-unit-market

[36] Econocom *How the Internet of Things Is Revolutionising Industry* https://blog.econocom.com/en/blog/how-the-internet-of-things-is-revolutionising-industry/

[37] McGrath W R, Walker C, Yap M and Tai Y C 1993 Silicon micromachined waveguides for millimeter-wave and submillimeter-wave frequencies *IEEE Microw. Guid. Wave Lett* **3** 61–3

[38] Katehi L P B, Rebeiz G M, Weller T M, Drayton R F, Cheng H J and Whitaker J F 1993 Micromachined circuits for millimeter- and sub-millimeter-wave applications *IEEE Antennas Propag. Mag.* **35** 9–17
[39] Weller T M and Katehi L P B 1995 Compact stubs for micromachined coplanar waveguide *Proc. 25th Europ. Microw. Conf. (Bologna, Sept. 1995)* pp 589–93
[40] Mahmoud S F 1991 *Electromagnetic Waveguides: Theory and Applications* 1st edn (London: Peter Peregrinus) p 240
[41] Iannacci J 2013 RF passive components for wireless applications *Handbook of MEMS for Wireless and Mobile Applications* 1st edn (Cambridge: Woodhead Publishing) pp 100–35
[42] Pozar D M 2005 *Microwave Engineering* 2nd edn (New York: Wiley) p 716
[43] Wadell B 1991 *Transmission Line Design Handbook* 1st edn (Norwood, MA: Artech House)
[44] Yang S, Hu Z, Buchanan N B, Fusco V F, Stewart J A C, Wu Y, Armstrong B M, Armstrong G A and Gamble H S 1998 Characteristics of trenched coplanar waveguide for high-resistivity Si MMIC applications *IEEE Trans. Microw. Theory Tech* **46** 623–31
[45] Farrington N E S and Iezekiel S 2011 Design and simulation of membrane supported transmission lines for interconnects in a MM-wave multichip module *Prog. Electromag. Res. B* **27** 165–86
[46] Shi Y, Lai Z, Xin P, Shao L and Zhu Z 2001 Design and fabrication of micromachined microwave transmission lines *Proc. SPIE* **4557** 477–84
[47] Goldsmith C L, Yao Zhimin Eshelman S and Denniston D 1998 Performance of low-loss RF MEMS capacitive switches *IEEE Microw. Guid. Wave Lett.* **8** 269–71
[48] Feng Z, Zhang W, Su B, Harsh K F, Gupta K C, Bright V and Lee Y C 1999 Design and modeling of RF MEMS tunable capacitors using electro-thermal actuators *Proc. IEEE MTT-S Int. Microw. Symp. (Anaheim, CA, June 1999)* pp 1507–10
[49] Katehi L P B, Rebeiz G M and Nguyen C T- 1998 MEMS and Si-micromachined components for low-power, high-frequency communications systems *Proc. IEEE MTT-S Int. Microw. Symp. (Baltimore, MD, June 1998)* pp 331–3
[50] Malczewski A, Eshelman S, Pillans B, Ehmke J and Goldsmith C L 1999 X-band RF MEMS phase shifters for phased array applications *IEEE Microw. Guid. Wave Lett.* **9** 517–9
[51] Liu C 2011 *Foundations of MEMS* 2nd edn (London: Pearson Education) p 560
[52] Lee H S, Leung C H, Shi J and Chan S C 2004 Micro-electro-mechanical relays-design concepts and process demonstrations *Proc. 50th IEEE Holm Conf. on Electrical Contacts and the 22nd Int. Conf. on Electrical Contacts (Seattle, WA, Sept. 2004)* pp 242–7
[53] Cho I-J, Song T, Baek S-H and Yoon E 2005 A low-voltage and low-power RF MEMS series and shunt switches actuated by combination of electromagnetic and electrostatic forces *IEEE Trans. Microw. Theory Tech* **53** 2450–7
[54] Safari A and Akdoğan E K 2008 *Piezoelectric and Acoustic Materials for Transducer Applications* 1st edn (New York: Springer) p 482
[55] Kawakubo T, Nagano T, Nishigaki M, Abe K and Itaya K 2005 Piezoelectric RF MEMS tunable capacitor with 3V operation using CMOS compatible materials and process *Proc. IEEE Int. Electron Dev. Meet. IEDM (Washington, DC, Dec. 2005)* pp 294–7
[56] Daneshmand M, Fouladi S, Mansour R R, Lisi M and Stajcer T 2009 Thermally-actuated latching RF MEMS switch *Proc. IEEE Int. Microw. Symp. MTT-S Digest (Boston, MA, June 2009)* pp 1217–20

[57] Iannacci J, Faes A, Repchankova A, Tazzoli A and Meneghesso G 2011 An active heat-based restoring mechanism for improving the reliability of RF-MEMS switches *Microelectron. Reliab.* **51** 1869–73
[58] Iannacci J, Repchankova A, Faes A, Tazzoli A, Meneghesso G and Dalla Betta G F 2010 Enhancement of RF-MEMS switch reliability through an active anti-stiction heat-based mechanism *Microelectron. Reliab.* **50** 1599–603
[59] Senturia S D 2001 *Microsystem Design* 1st edn (New York: Springer) p 689
[60] Giacomozzi F, Mulloni V, Colpo S, Iannacci J and Margesin B 2011 Faes A flexible fabri-cation process for RF MEMS devices Rom. *J. Inf. Sci. Technol. (ROMJIST)* **14** 259–68
[61] Iannacci J, Resta G, Farinelli P and Sorrentino R 2012 RF-MEMS components and networks for high-performance reconfigurable telecommunication and wireless systems *Trans. Tech. Pub. Adv. Sci. Tech* **81** 65–74
[62] Iannacci J 2013 *Practical* Guide *to RF-MEMS* 1st edn (Weinheim: Wiley-VCH) p 372
[63] Carty E, Fitzgerald P, McDaid P, Stenson B and Goggin R 2016 Development of a DC to K-band ultra long on-life RF MEMS switch with integrated driver circuitry *Proc. 11th Europ. Microw, Integrated Circ. Conf. EuMIC (London, Oct. 2016)* pp 444–7
[64] Seki T, Yamamoto J, Murakami A, Yoshitake N, Hinuma K-i Fujiwara T, Sano K, Matsushita T, Sato F and Oba M 2013 An RF MEMS switch for 4G Front-Ends *Proc. IEEE MTT-S Int. Microw. Symp. IMS (Seattle, WA, June 2013)* pp 131–3
[65] Chu C, Liao X and Yan H 2017 Ka-band RF MEMS capacitive switch with low loss, high isolation, long-term reliability and high power handling based on GaAs MMIC technology *IET Microw. Antennas Propag.* **11** 942–8
[66] Ma L Y, Soin N and Nordin A N 2016 *A novel design of low-voltage low-loss K-band RF-MEMS capacitive switch Proc. Symp. on Design, Test, Integr. and Packaging of MEMS/MOEMS DTIP (Budapest, May–June 2016)* pp 151–5
[67] Iannacci J, Gaddi R and Gnudi A 2007 *Non-linear electromechanical RF model of a MEMS varactor based on veriloga© and lumped-element parasitic network Proc. Europ. Microw. Integ. Circ. Conf. (Munich, Oct. 2007)* pp 544–7
[68] Elshurafa A M, Ho P H and Salama K N 2012 Low voltage RF MEMS variable capacitor with linear C–V response *IET Electron. Lett.* **48** 392–3
[69] Nishiyama M, Konishi H, Suzuki J, Tezuka Y, Suzuki Y and Suzuki K 2007 *Extremely high capacitance ratio (C/R) RF MEMS variable capacitor with chameleon actuators Proc. Int. Solid-State Sensors, Actuators and Microsyst. Conf. TRANSDUCERS (Lyon, June 2007)* pp 631–4
[70] Cazzorla A, Sorrentino R and Farinelli P 2015 *Double-actuation extended tuning range RF MEMS Varactor Proc. Europ. Microw. Conf. EuMC (Paris, Oct. 2015)* pp 937–40
[71] Chen K, Kovacs A and Peroulis D 2010 *Anti-biased RF MEMS varactor topology for 20–25 dB linearity enhancement Proc. IEEE MTT-S Int. Microw. Symp. (Anaheim, CA, May 2010)* pp 1142–5
[72] Ikehashi T *et al* 2008 *An RF MEMS variable capacitor with intelligent bipolar actuation Proc. IEEE Int. Solid-State Circ. Conf. (San Francisco, CA, Feb. 2008)* pp 582–637
[73] Mizuochi Y, Amakawa S, Ishihara N and Masu K 2009 *Study of air-suspended RF MEMS inductor configurations for realizing large inductance variations Proc. Argentine School of Micro-Nanoelectronics, Tech. and Applications (San Carlos de Bariloche, Oct. 2009)* pp 550–5

[74] Chang S and Sivoththaman S 2006 A tunable RF MEMS inductor on silicon incorporating an amorphous silicon bimorph in a low-temperature process *IEEE Electron Dev. Lett.* **27** 905–7

[75] Zine-El-Abidine I, Okoniewski M and McRory J G 2004 *A tunable RF MEMS inductor Proc. Int. Conf. on MEMS, NANO and Smart Syst. ICMENS (Banff, Aug 2004)* pp 636–8

[76] Banitorfian F, Eshghabadi F, Manaf A A, Pons P, Noh N M, Mustaffa M T and Sidek O 2013 *A novel tunable water-based RF MEMS solenoid inductor Proc. IEEE Regional Symp. on Micro and Nanoelectron. RSM (Langkawi, Sept. 2013)* pp 58–61

[77] Hikmat O F and Mohamed Ali M S 2017 RF MEMS inductors and their applications—a review *IEEE J. Microelectromech. Syst.* **26** 17–44

[78] Casini F, Farinelli P, Mannocchi G, DiNardo S, Margesin B, De Angelis G, Marcelli R, Vendier O and Vietzorreck L 2010 *High performance RF-MEMS SP4T switches in CPW technology for space applications Proc. Europ. Microw. Conf. (Paris, Sept. 2010)* pp 89–92

[79] Yang H-H, Yahiaoui A, Zareie H, Blondy P and Rebeiz G M 2014 *A compact high-isolation DC-50 GHz SP4T RF MEMS switch Proc. IEEE MTT-S Int. Microw. Symp. IMS (Tampa, FL, June 2014)* pp 141–4

[80] Zareie H and Rebeiz G M 2014 Compact high-power SPST and SP4T RF MEMS metal-contact switches *IEEE Trans. Microw. Theory Tech.* **62** 297–305

[81] Chan K Y, Ramer R and Mansour R R 2012 Novel miniaturized RF MEMS staircase switch matrix *IEEE Microw. Wirel. Comp. Lett.* **22** 117–19

[82] Daneshmand M and Mansour R R 2011 RF MEMS satellite switch matrices *IEEE Microw. Mag.* **12** 92–109

[83] Fomani A A and Mansour R R 2009 Monolithically integrated multiport RF MEMS switch matrices *IEEE Trans. Microw. Theory Tech.* **57** 3434–41

[84] Fomani A A and Mansour R R 2009 *Miniature RF MEMS switch matrices Proc. IEEE MTT-S Int. Microw. Symp. IMS (Boston, MA, June 2009)* pp 1221–4

[85] Besser L and Gilmore R 2003 *Practical RF Circuit Design for Modern Wireless Systems: Passive circuits and systems* 1st edn (Norwood, MA: Artech House) p 576

[86] Ocera A, Farinelli P, Mezzanotte P, Sorrentino R, Margesin B and Giacomozzi F 2006 *A novel MEMS-tunable hairpin line filter on silicon substrate Proc. Europ. Microw. Conf. (Manchester, Sept. 2006)* pp 803–6

[87] Reines I, Park S J and Rebeiz G M 2010 Compact low-loss tunable X-band bandstop filter with miniature RF-MEMS switches *IEEE Trans. Microw. Theory Tech.* **58** 1887–95

[88] Hsu H H, Margomenos A D and Peroulis D 2011 *A monolithic RF-MEMS filter with continuously-tunable center-frequency and bandwidth Proc. IEEE Top. Meet. on Silicon Monolithic Integr. Circ. in RF Syst. (Phoenix, AZ, Jan. 2011)* pp 169–72

[89] Shah U, Sterner M and Oberhammer J 2011 *Basic concepts of moving-sidewall tuneable capacitors for RF MEMS reconfigurable filters Proc. Europ. Microw. Integ. Circ. Conf. (Manchester, Oct. 2011)* pp 526–9

[90] Shojaei-Asanjan D and Mansour R R 2017 The sky's the limit: A switchable RF-MEMS filter design for wireless avionics intracommunication *IEEE Microw. Mag.* **18** 100–6

[91] Wang H, Anand A and Liu X 2017 *A miniature 800–1100-MHz tunable filter with high-Q ceramic coaxial resonators and commercial RF-MEMS tunable digital capacitors Proc.*

IEEE Wireless and Microw. Tech. Conf. WAMICON (Cocoa Beach, FL, April 2017) pp 131–3

[92] Kumar N and Singh Y K 2017 RF-MEMS-based bandpass-to-bandstop switchable single- and dual-band filters with variable FBW and reconfigurable selectivity *IEEE Trans. Microw. Theory Tech.* pp 1–14

[93] Hickle M D, Li J, Psychogiou D and Peroulis D 2016 A high-performance pathway: a 0.95/ 2.45-GHZ switched-frequency bandpass filter using commercially available RF MEMS tuning elements *IEEE Microw. Mag.* **17** 34–41

[94] Stefanini R, Chatras M, Pothier A, Guines C and Blondy P 2013 *High-Q 3D tunable RF MEMS filter with a constant fractional bandwidth Proc. Europ. Microw. Integr. Circ. Conf. (Nuremberg, Oct. 2013)* pp 312–5

[95] Chan K Y, Ramer R and Mansour R R 2017 A switchable iris bandpass filter using RF MEMS switchable planar resonators *IEEE Microw. Wirel. Comp. Lett.* **27** 34–6

[96] Park S J, Reines I, Patel C and Rebeiz G M 2010 High-Q RF-MEMS 4–6-GHz tunable evanescent-mode cavity filter *IEEE Trans. Microw. Theory Tech.* **58** 381–9

[97] Schulte B, Ziegler V, Schoenlinner B, Prechtel U and Schumacher H 2011 *RF-MEMS tunable evanescent mode cavity filter in LTCC technology at Ku-band Proc. of Europ. Microw. Integr. Circ. Conf. (Manchester, Oct. 2011)* pp 514–7

[98] Bastioli S, Di Maggio F, Farinelli P, Giacomozzi F, Margesin B, Ocera A, Pomona I, Russo M and Sorrentino R 2008 *Design manufacturing and packaging of a 5-bit K-band MEMS phase shifter Proc. Europ. Microw. Integr. Circ. Conf. (Amsterdam, Oct. 2008)* pp 338–41

[99] Ramli N A and Arslan T 2017 *Design and simulation of a 2-bit distributed S-band MEMS phase shifter Proc. Int. Conf. Thermal, Mech. and Multi-Physics Simul. and Exp. in Microelectron.and Microsyst. EuroSimE (Dresden, April 2017)* pp 1–5

[100] Dey S and Koul S K, *2014 10 –35-GHz frequency reconfigurable RF MEMS 5-bit DMTL phase shifter uses push-pull actuation based toggle mechanism Proc. IEEE Int. Microw. and RF Conf. IMaRC (Bangalore, Dec. 2014)* pp 21–4

[101] Bakri-Kassem M, Mansour R R and Safavi-Naeini S 2014 *A novel latching RF MEMS phase shifter Proc. Europ. Microw. Integr. Circ. Conf. (Rome, Oct. 2014)* pp 668–71

[102] McFeetors G and Okoniewski M 2004 *Analog tunable microwave phase shifters using RF MEMS Proc. Int. Symp. on Antenna Tech. and Appl. Electromagnetics and URSI Conf. (Ottawa, July 2004)* pp 141–4

[103] Unlu M, Demir S and Akin T 2013 A 15–40-GHz frequency reconfigurable RF MEMS phase shifter IEEE Trans. Microw. *Theory Tech.* **61** 2865–77

[104] Abumunshar A J, Nahar N K, Hyman D and Sertel K 2017 18 *–40GHz low-profile phased array with integrated MEMS phase shifters Proc. Europ. Conf. on Antennas and Propagation EUCAP (Paris, March 2017)* pp 2800–1

[105] Yeo W G, Nahar N K and Sertel K 2013 *Phased array antenna with integrated MEMS phase shifters for Ka-band SATCOM Proc. IEEE Antennas and Propagation Soc. Int. Symp. APSURSI (Orlando, FL, July 2013)* pp 105–6

[106] Das A, Puri M and Sengar J S 2013 *A novel monolithic integrated phased array antenna using 4-bit distributed MEMS phase shifter and triangular patch antenna Proc. Int. Conf. on Adv. in Comput., Commun. and Informatics ICACCI (Mysore, Aug. 2013)* pp 913–8

[107] Iannacci J, Masotti D, Kuenzig T and Niessner M 2011 *A reconfigurable impedance matching network entirely manufactured in RF-MEMS technology* Proc. (SPIE) **8066** 280–91

[108] Festo A E, Folgero K, Ullaland K and Gjertsen K M 2009 A six bit, 6–18 GHz *RF-MEMS impedance tuner for 50 Ω systems Proc. Europ. Microw. Conf. EuMC (Rome, Sept.–Oct. 2009)* pp 1132–5

[109] Vaha-Heikkila T, Varis J, Tuovinen J and Rebeiz G M 2004 *A V-band single-stub RF MEMS impedance tuner Proc. Europ. Microw. Conf. (Amsterdam, Oct. 2004)* pp 1301–4

[110] Fouladi S, Domingue F and Mansour R 2012 *CMOS-MEMS tuning and impedance matching circuits for reconfigurable RF front-ends Proc. IEEE/MTT-S Int. Microw. Symp. (Montreal, June 2012)* pp 131–3

[111] Larcher L, Brama R, Ganzerli G, Iannacci J, Margesin B, Bedani B and Gnudi A 2009 *A MEMS reconfigurable quad-band class-E power amplifier for GSM standard Proc. IEEE Int. Conf. on Micro Electro Mech. Syst. (Sorrento, Jan. 2009)* pp 864–7

[112] Iannacci J, Faes A, Mastri F, Masotti D and Rizzoli V 2010 *A MEMS-based wide-band multi-state power attenuator for radio frequency and microwave applications Proc. TechConnect World, NSTI Nanotech (Anaheim, CA, June 2010)* pp 328–31

[113] Sun J and Li Z 2016 *A broadband 3-bit MEMS digital attenuator Proc. Int. Conf. on Electron. Design ICED (Phuket, Aug 2016)* pp 442–5

[114] Iannacci J, Huhn M, Tschoban C and Pötter H 2016 RF-MEMS technology for future mobile and high-frequency applications: Reconfigurable 8-bit power attenuator tested up to 110 GHz *IEEE Electron Dev. Lett.* **37** 1646–9

[115] Iannacci J, Huhn M, Tschoban C and Pötter H 2016 RF-MEMS technology for 5G: Series and shunt attenuator modules demonstrated up to 110 GHz *IEEE Electron Dev. Lett.* **37** 1336–9

[116] Courcimault C G and Allen M G 2005 *High-Q mechanical tuning of MEMS resonators using a metal deposition-annealing technique Proc. Int. Conf. on Solid-State Sensors, Actuators and Microsyst. TRANSDUCERS (Seoul, June 2005)* pp 875–8

[117] Pillai G, Tan W S, Chen C C and Li S S 2016 *Modeling of zero TCF and maximum bandwidth orientation for lithium tantalate RF MEMS resonators Proc. IEEE Ann. Int. Conf. on Nano/Micro Engineered and Molecular Syst. NEMS (Sendai, April 2016)* pp 450–4

[118] Pacheco S, Zurcher P, Young S, Weston D and Dauksher W 2004 *RF MEMS resonator for CMOS back-end-of-line integration Proc. Top. Meet. on Silicon Monolithic Integ. Circ. in RF Syst. (Atlanta, GA, Sept. 2004)* pp 203–6

[119] Cassella C, Chen G, Qian Z, Hummel G and Rinaldi M 2017 RF passive components based on aluminum nitride cross-sectional Lamé-mode MEMS resonators *IEEE Trans. Electron Dev.* **64** 237–43

[120] Yuan Q, Luo W, Zhao H, Peng B, Yang J and Yang F 2015 Frequency stability of RF-MEMS disk resonators *IEEE Trans. Electron Dev.* **62** 1603–8

[121] Hashimoto K-Y 2009 *RF Bulk Acoustic Wave Filters for Communications* 1st edn (Norwood: Artech House) p 275

[122] Enz C C and Kaiser A 2013 *MEMS-based Circuits and Systems for Wireless Communication* 1st edn (New York: Springer) p 332

[123] Aigner R 2008 *SAW and BAW technologies for RF filter applications: A review of the relative strengths and weaknesses Proc. IEEE Ultrasonics Symp. (Beijing, Nov. 2008)* pp 582–9

[124] Koohi M Z, Lee S and Mortazawi A 2016 *Design of BST-on-Si composite FBARs for switchable BAW filter application Proc. Europ. Microw. Conf. EuMC (London, Oct. 2016)* pp 1003–6

[125] Gaddi R, Gnudi A, Franchi E, Guermandi D, Tortori P, Margesin B and Giacomozzi F 2005 Reconfigurable MEMS-enabled LC-tank for multi-band CMOS oscillator *Proc. IEEE MTT-S Int. Microw. Symp. IMS (Long Beach, CA, June 2005)* pp 141–4

[126] Cazzorla A, Sorrentino R and Farinelli P 2015 *MEMS based LC tank with extended tuning range for multi-band applications Proc. IEEE Mediterranean Microw. Symp. MMS (Lecce, Nov.–Dec. 2015)* pp 141–4

[127] Small J, Arif M S, Fruehling A and Peroulis D 2013 A tunable miniaturized RF MEMS resonator with simultaneous high Q (500–735) and fast response speed (<10–60 μs) *IEEE J. Microelectromech. Syst.* **22** 395–405

[128] Irshad W and Peroulis D 2011 *A 12–18 GHz electrostatically tunable liquid metal RF MEMS resonator with quality factor of 1400–1840 Proc. IEEE MTT-S Int. Microw. Symp. IMS (Baltimore, MD, June 2011)* pp 141–4

[129] Psychogiou D, Yang Z and Peroulis D 2012 *RF-MEMS enabled power divider with arbitrary power division ratio Proc. Europ. Microw. Conf. (Amsterdam, Oct. 2012)* pp 53–6

[130] Ocera A, Farinelli P, Cherubini F, Mezzanotte P, Sorrentino R, Margesin B and Giacomozzi F 2007 *A MEMS-reconfigurable power divider on high resistivity silicon substrate Proc. IEEE/MTT-S Int. Microw. Symp. IMS (Honolulu, HI, June 2007)* pp 501–4

[131] Ocera A, Gatti R V, Mezzanotte P, Farinelli P and Sorrentino R 2005 *A MEMS programmable power divider/combiner for reconfigurable antenna systems Proc. Europ. Microw. Conf. (Paris, Oct. 2005)* pp 141–4

[132] Marcaccioli L, Farinelli P, Tentzeris M M, Papapolymerou J and Sorrentino R 2008 *Design of a broadband MEMS-based reconfigurable coupler in Ku-band Proc. Europ. Microw. Conf. (Amsterdam, Oct. 2008)* pp 595–8

[133] Gurbuz O D and Rebeiz G M 2015 A 1.6–2.3-GHz RF MEMS reconfigurable quadrature coupler and its application to a 360° reflective-type phase shifter *IEEE Trans. Microw. Theory Tech.* **63** 414–21

[134] Ocera A, Farinelli P, Mezzanotte P, Sorrentino R, Margesin B and Giacomozzi F 2007 *Nov el RF-MEMS widely-reconfigurable directional coupler Proc. Europ. Microw. Conf. (Munich, Oct. 2007)* pp 122–5

[135] Shah U, Sterner M and Oberhammer J 2013 High-directivity MEMS-tunable directional couplers for 10–18-GHz broadband applications *IEEE Trans. Microw. Theory Tech.* **61** 3236–46

[136] De Angelis G *et al* 2008 *Packaged single pole double thruSPDT) and true time delay linesTTDL) based on RF MEMS switches Proc. Int. Semicond. Conf. CAS (Sinaia, Oct. 2008)* pp 227–30

[137] Fu D, Bey Y A, Domier C, Luhmann N C and Liu X 2015 *A Q-band RF-MEMS tapered true time delay line for fusion plasma diagnostics systems Proc. IEEE MTT-S Int. Microw. Symp. IMS (Phoenix, AZ, May 2015)* pp 131–3

[138] Van Caekenberghe K and Vaha-Heikkila T 2008 An analog RF MEMS slotline true-time-delay phase shifter *IEEE Trans. Microw. Theory Tech.* **56** 2151–9

[139] Hacker J B, Mihailovich R E, Kim M and DeNatale J F 2003 A Ka-band 3-bit RF MEMS true-time-delay network *IEEE Trans. Microw. Theory Tech.* **51** 305–8

[140] Kim M, Hacker J B, Mihailovich R E and DeNatale J F 2001 A DC-to-40 GHz four-bit RF MEMS true-time delay network *IEEE Microw. Wirel. Comp. Lett.* **11** 56–8

Chapter 2

Expectations versus actual market of RF-MEMS: analysis and explanation of a repeatedly fluctuating scenario

2.1 General considerations of the market in the technology sector

This chapter will focus on the analysis of the Radio Frequency Micro-ElectroMechanical Systems (RF-MEMS) potentialities in terms of market penetration, by reviewing the expectations and forecasts taking turns in the last 20 years, as well as by interpreting the reasons for them not being fully achieved. Nonetheless, in order for such a detailed study to be effective, it is necessary to first introduce a couple of basic concepts belonging to marketing and management science.

The first notion is the one of the hype curve, which is a graphic tool elaborated by Gartner Inc., an information technology research and advisory company. Such a curve depicts the typical cycle of acceptance related to a new technology, from its first development to consolidated employment in market applications [1]. The usual trend of the hype curve is shown in figure 2.1.

When a novel technology starts to be intensively investigated at the research level and its remarkable performance/characteristics are demonstrated in lab-based scenarios (a, 'Technology trigger'), its visibility increases and, in parallel to it, expectations of market placement and outcomes become higher (b, 'Peak of inflated expectations'). Subsequently, when the efforts of research start approaching real application cases and scenarios, aspects of the new technology that were not critical for demonstration purposes, but that, on the other hand, are crucial from the market perspective, begin to emerge. Among them, the most important ones are typically limited in number, and very often the same, like integration/compatibility with other (already existing) technologies, reliability and manufacturing/production costs, as well as yield. The aforementioned context causes the initially inflated expectations to

doi:10.1088/978-0-7503-4199-8ch2

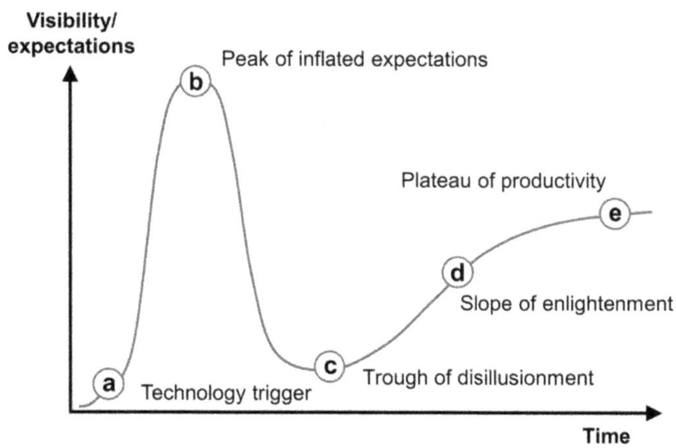

Figure 2.1. Typical hype curve behaviour, describing the evolution of a novel technology against time, since its first investigation (in the basic research scenario), to the maturation and employment in market and commercial applications. Reproduced from [17], with permission from Springer.

drop (c, 'Trough of disillusionment'). At the same time, such a scenario provides the scientific and industrial community with proper motivation to address and overcome the limitations of the new technology against real application scenarios (d, 'Slope of enlightenment'). Finally, when the technology is mature enough, it starts to be steadily employed and available on the market, leading to the final stage of the hype curve (e, 'Plateau of productivity'). The whole hype cycle exhibits a different duration depending on several factors, such as, for instance, technology complexity, market acceptance, market needs as well as readiness, and so on. In any case, the time unit of the horizontal axis in figure 2.1 is typically in the order of a magnitude of years.

Still referring to the ways in which new technologies penetrate the market, if such paths can be multiple and quite diverse, in essence they fall within two main scenarios, labelled *technology push* and *market driven* [2].

In the technology push scenario, a novel solution is introduced into the market neither because of a factual need to be addressed nor of a certain issue to be overcome. In such a context, the added value related to the employment of a certain technology or component can be the extension of system functionality or, more trivially, the perception of distinctiveness it provides to the final user, as happens with the concept of brand. An example of the technology push approach can be found when cameras were integrated in the previous generations of mobile phones. Around 15–20 years ago, probably no one, among the millions of handset users, was expecting to take pictures and make videos with their cellular phones. More generally, it is unlikely that anybody was even thinking to use handsets to do something other than placing calls and typing short (purely textual) messages. Despite not responding to an actual market need, cameras started to be integrated in mobile handsets, and people started to become accustomed to the possibility of accessing and visualising multimedia content through their portable phones.

In other words, still referring to the aforementioned example, the technology push approach does not respond to a market need, but, conversely, aims at its generation.

On the other bank of the river stands the market pull/driven scenario. In this case, the introduction of an innovative technology, or the improvement of an existing one, is driven by specific market needs, such as issues to be overcome or more stringent specifications to be met. Despite looking odd at first glance, a valid example of the market pull scenario is still the one of integrated cameras in mobile handsets. As a matter of fact, after stepping across the user-acceptance phase, i.e. the time interval from the introduction of a novel technology to its recognition as being valuable and necessary by most users, it is probable that nobody is currently thinking of using a mobile handset just for calling and text messaging. In other words, pushing cameras into mobile phones contributed, in a matter of a few years, to create a strong market need driving innovation in smartphone technologies for vision, thus aiming at a better quality of images and videos, smaller and more integrated devices, reduced power consumption, and, last but not least, lower prices.

Now that the fundamental concepts concerning the market acceptance of novel technologies and solutions have been introduced, the evolution of the expected versus real market of RF-MEMS is going to be analysed.

2.2 RF-MEMS on the market: vision of the early days

Since early discussions in the scientific literature, comprehensively reported before in this work, RF-MEMS passive components and networks exhibited remarkable performance and characteristics, definitely boosted if compared against those typical of their counterparts realised in standard (semiconductor) technologies. This aspect provided an enormous stimulus to scientists and engineers in proliferating visions around future (at that time) exploitations of RF-MEMS technology in radio and wireless applications and systems. Several valuable works and ideas were published by distinguished researchers, both in the academic and private sector. Among the former, Clark T-C Nguyen provided a quite extensive contribution that is going to be referenced here as an explanatory example.

In the initial phase, the work developed by Nguyen pioneered the field of Micro-ElectroMechanical Systems (MEMS) mechanical resonators with a very high quality factor (Q-factor), to be employed in highly selective Radio Frequency (RF) oscillators [3, 4]. In parallel to such a development at the device level, possible ways RF-MEMS technology might have exerted an impact on radio transmitters/receivers (transceivers) began to be analysed. To this purpose, one of the most diffused transceiver architectures, namely the super-heterodyne transmitter/receiver [5], was taken as a reference. Such an architecture operates frequency up conversion (transmitter end) and down conversion (receiver end) of the baseband signal, to an Intermediate Frequency (IF) band lower than the RF band exploited for signal broadcasting. The IF eases the treatment of the signal to be transmitted or received (e.g. filtering, amplification, modulation/demodulation, and so on). According to the vision of Nguyen, the impact of RF-MEMS technology was expected to be twofold,

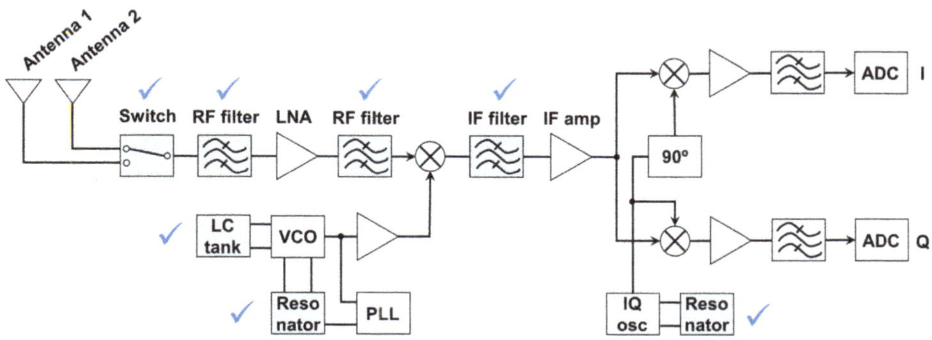

Figure 2.2. Super-heterodyne radio receiver block diagram [3]. Check-marked elements could be replaced with passive components based on RF-MEMS technology, thus improving performance and characteristics of the whole system. Reproduced from [17], with permission from Springer.

with a limited impact on transceiver architectures at first, and with a partial redesign of them in a more advanced phase.

In the first stage, RF-MEMS basic passive components would have replaced traditional devices in several parts of the super-heterodyne transceiver, indeed improving its performance and characteristics [4–6]. The block diagram in figure 2.2 shows a typical configuration of a super-heterodyne receiver, as reported and discussed by Nguyen.

All the coloured (and check-marked) diagram subblocks could have been replaced by implementations of passive components in RF-MEMS technology. In particular, looking at the receiver from the antenna to the output of the demodulated received signal—that is, both In-phase (I) and Quadrature (Q)—a MEMS switching unit could be used to select the proper antenna. Moreover, MEMS varactors and inductors are suitable to realise high Q-factor filters (both RF and IF) and reconfigurable LC tanks to tune the oscillation frequency of the Voltage Controlled Oscillator (VCO). Finally, MEMS resonators could replace the typical quartz-based devices within oscillators.

As widely discussed prior to this point, besides the realisation of RF-MEMS basic passives, MEMS technology is suitable for implementing more complex elements, as well. In this regard, high-order switching matrices, reconfigurable phase shifters, impedance matching tuners, programmable step attenuators, and so on, have already been discussed.

In the second stage forecasted by Nguyen, the availability of such networks would have led to rethinking the architecture of a transceiver, rather than just replacing some of its elemental components with RF-MEMS counterparts. Sticking to the same example as before, the super-heterodyne receiver architecture could have been rearranged and simplified on the basis of RF-MEMS complex networks, as discussed in [6–9] and shown in figure 2.3.

In this altered scheme, a multi-channel selector (i.e. a bank of MEMS switches) with several filtering functions would simplify the hardware complexity of the whole platform. The typical low loss of MEMS-based devices and networks compared to

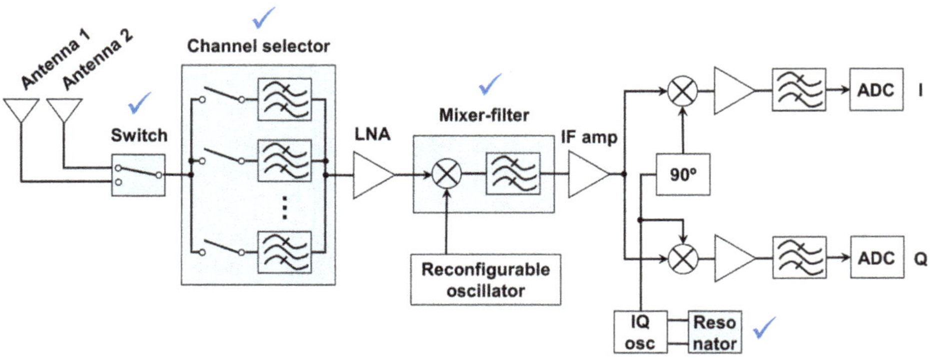

Figure 2.3. Super-heterodyne radio receiver block diagram, based on a modified architecture featuring complex blocks (check-marked) entirely implemented in RF-MEMS technology (i.e. multi-channel selector, reconfigurable oscillator, mixer-filter) as discussed in [3]. Reproduced from [17], with permission from Springer.

standard technologies would reduce the number of Power Amplifiers (PAs) needed to regenerate the signal. Moreover, the high reconfigurability of MEMS could be exploited to realise a widely tunable oscillator, thus extending the range of possible received signals that can be mixed and demodulated by the receiver, through the integration of a mixer-filter IF block also based on RF-MEMS technology.

As is straightforward to figure out, transceivers are just one of the numerous RF applications upon which RF-MEMS were expected to exert significant impact. For example, in modern radar systems, electronic antenna beam steering has replaced the old rotary mechanical antennas, leading to a significant reduction of complexity and space occupation, as well as to improved robustness of the whole system [10]. In spite of these advantages, phase shifters in standard technology (necessary to enable dynamic orientation of the antenna beam) are typically quite lossy. Therefore, redundant PAs, duplicated per each of the antenna array delay lines, must be inserted. A block diagram of a radar system, close to the transmitting antennas array and featuring standard phase shifters, is depicted in figure 2.4(a) and discussed in [11].

Evidently, a PA is necessary before and after the phase-shifter element on each branch, in order to regenerate the attenuated RF signal. For mid-power radar systems, traditional phase shifters could be replaced by high-performance, low-loss realisations in RF-MEMS technology. In this case, the power consumption and hardware complexity of the system could be considerably reduced, since a unique amplifying stage on each branch would be sufficient, as reported in figure 2.4(b).

Besides the few examples discussed here in detail, RF-MEMS were expected to secure successful applications in many other contexts, with reference to the main fundamental market segments of telecommunications (both mobile handsets and infrastructure), industrial appliances, as well as aerospace and defence applications.

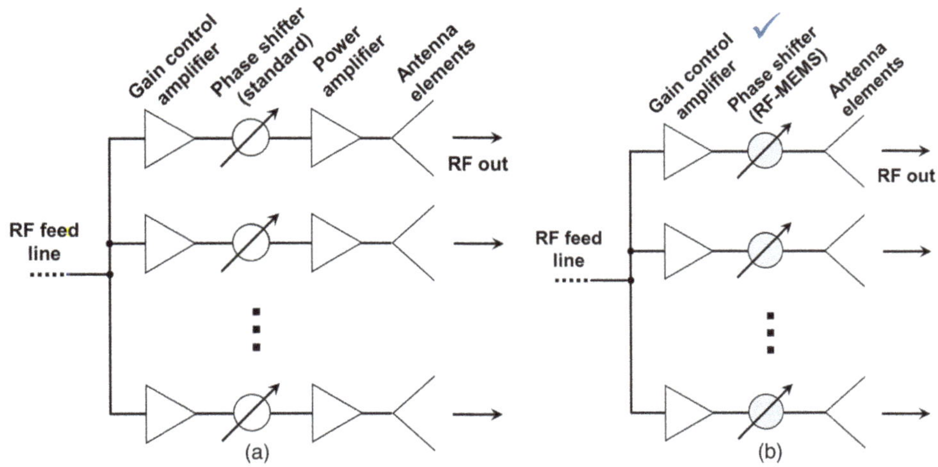

Figure 2.4. Block diagram scheme of a radar system employing electronically steerable antenna beams [11]. (a) Scheme employing phase shifters in standard technology. Power Amplifiers (PAs) are necessary on each branch in order to regenerate the signal attenuated by lossy phase-shifting elements. (b) Scheme employing phase shifters in RF-MEMS technology. Due to the very low loss of such components, the power consumption and hardware complexity of the system can be significantly reduced. Reproduced from [17], with permission from Springer.

2.3 Fluctuating RF-MEMS market forecasts

In light of what has just been deployed in the previous section, RF-MEMS technology, especially in the first years of its experimental demonstration, was definitely not lacking ideas, perspectives and visions around its successful employment within commercial applications and, therefore, concerning market penetration. This way of interpreting the possibilities offered by MEMS for RF passives is confirmed and corroborated by the first market forecast releases, signed by influential analysts who specialised in market observations. However, stimulating hints to deepen such an analysis emerge if one looks at the progression of the mentioned forecasts, keeping as a timeframe the first decade after RF-MEMS started to appear in the scientific literature, i.e. from the first years of the 2000s.

In early analyses, released in 2004–2005, the market volume was expected to ripple between 700 $M (millions of US Dollars, USD) and 1000 $M in 2009 [12] (see figure 2.5(a)). Stepping forward in time, in 2006 WTC Consulting predicted a downsized market volume of RF-MEMS for mobile applications fluctuating around 10 $M in 2009 and 70 $M in 2011 [13] (see figure 2.5(b)). Later, in August 2010, IHS Inc. consolidated the RF-MEMS market figure to a few $M in 2009, and predicted a volume of 225 $M in 2014 [14] (see figure 2.5(c)). Nonetheless, the same IHS Inc. in 2012 shrunk the forecast for the 2014 market to less than 100 $M [15] (see figure 2.5(d)). In addition, Yole Développement in 2013 estimated a market volume for RF-MEMS of around 50 $M in 2014 and of less than 350 $M in 2018 [16] (see figure 2.5(e)).

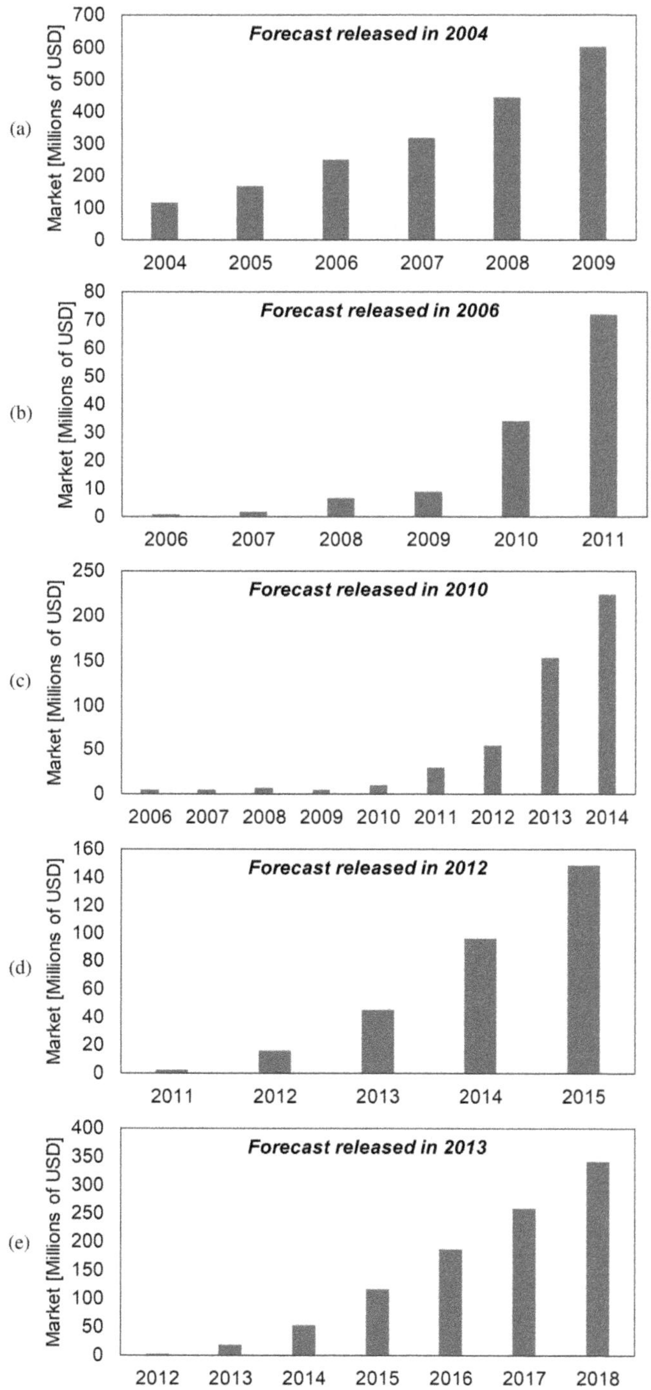

Figure 2.5. Evolution of RF-MEMS market forecasts released in (a) 2004 [12], (b) 2006 [13], (c) 2010 [14], (d) 2012 [15], (e) 2013 [16].

Abstracting from the numbers and summarising the data listed in a qualitative fashion, RF-MEMS market forecasts were (somehow) systematically overestimated, and expectations were disappointed for about a decade. Moreover, recalling the concept of the hype curve previously reported in figure 2.1, and keeping in mind the plots in figure 2.5, it is as if RF-MEMS were characterised by a double peak of inflated expectations, the first one taking place around 2004–2005, and the second displaying around 2010, as discussed in [17]. This seems quite peculiar if compared to how hype curves typically develop, and is intimately related to the factors that are going to be discussed in the following section.

2.4 Analysis of RF-MEMS market expectations (disappointment)

Providing a unique and straightforward interpretation to frame the multiple fluctuations of RF-MEMS market forecasts is not a simple task. Nonetheless, in-depth analysis of the scientific literature and of its alternating trends over the years enables one to form sound explanations. That being said, the reasons for the repeatedly missed market achievements of RF-MEMS can be classified as belonging to two main categories, namely, *intrinsic* and *extrinsic*. The former factors are closely related to the technology and functional aspects of RF passives based on microsystems. On the other hand, the latter are linked to the employment of RF-MEMS components within more complex subsystems and systems. Referring again to the hype curve of RF-MEMS, it can be inferred that intrinsic factors were the main cause for the drop of expectations after the first peak, taking place around 2006. Instead, extrinsic factors could be regarded as the primary reason for the second drop of expectations, right after 2010. Nonetheless, as explained in depth in the following pages, extrinsic factors were present from the beginning, despite not being taken into proper consideration.

2.4.1 Intrinsic factors

Prior to enumerating the most relevant intrinsic factors causing a drop of expectations around market absorption of RF-MEMS, it is appropriate to aggregate the proper background in which they have to be framed. To this purpose, the concept of a development chain, tailored to the topic analysed in this work, is going to be discussed briefly. The latter is a schematic flow, not necessarily unique in its deployment, embracing all the phases and levels of intervention, bringing a certain design concept from inception to physical demonstration and functional validation. A plausible implementation of a development chain for RF-MEMS is shown in figure 2.6.

A novel device idea has to be extensively modelled and simulated, in order to figure out the most suitable technology solutions to realise it; and, after the process flow is chosen, one needs to optimise the RF-MEMS characteristics, bearing in mind the most relevant existing trade-offs. Subsequently, all of the knowledge gathered through modelling and simulation is distilled into the layout of a physical device by means of the design phase, followed by the fabrication step. Once RF-MEMS specimens are available, they undergo lab-based experimental testing, in order to

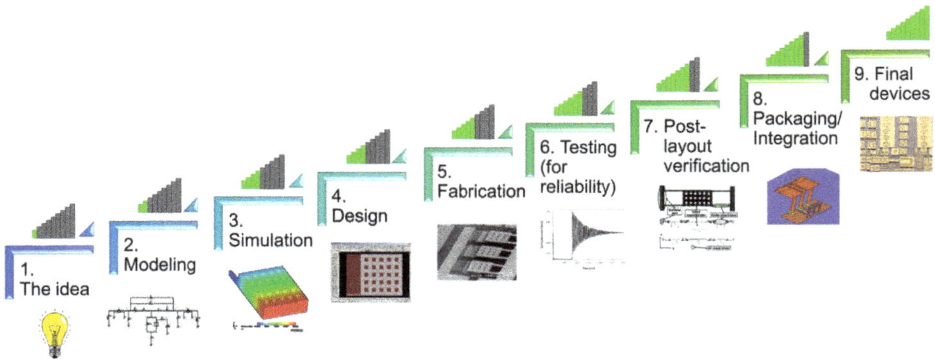

Figure 2.6. Typical arrangement of the RF-MEMS development chain that, through various steps, leads from inception of a novel design concept to the physical realisation and experimental validation.

verify their operation and characteristics. Such a testing phase can be simply devoted to the functional verification of the fabricated samples, or it can be aimed at the reliability assessment of the Devices Under Test (DUTs). In the latter circumstance, the specifications imposed by the final application scenario of the RF-MEMS device are already known/defined, and DUTs are tested in such a way to verify if they are able to meet them or not. When lab-based experiments aim to verify RF-MEMS against final application requirements, this phase is typically addressed as testing for reliability. In general terms, when a novel design concept is manufactured and experimentally verified for the first time, the testing phase is more focused on the demonstration of the device functionality, rather than on the assessment of reliability, because of a twofold motivation. On one hand, a few characteristics of the DUTs will differ to a certain extent from the nominal ones, targeted during the previous simulation step. Therefore, a redesign of improved layouts and additional fabrication batches will be necessary. On the other hand, when dealing with novel design concepts, it is plausible that neither the final application is clear in all its aspects, nor, in turn, the set of specifications to be targeted.

The testing phase is of particular relevance for the subsequent post-layout verification. During that, the set of predictive models, previously exploited for the initial modelling and simulation, are tuned based on the collected experimental datasets. This enables improved accuracy of the behavioural models that are then used to converge toward an optimised design, to be fabricated and tested again. It is straightforward that the more consolidated the exploitation of a certain technology is to realise RF-MEMS devices, the less extensive the effort has to be in the post-layout verification phase.

Progressing along the development chain, the subsequent phase concerns the packaging and integration of RF-MEMS. Because of their complexity, such steps could be two individual phases of the whole chain. Moreover, both packaging and integration require additional validation and optimisation cycles. Concerning the former, a proper technology flow must be defined at first, and has to be consolidated through subsequent fabrication runs, as happens for the RF-MEMS technology itself. In addition, issues related to the package-to-device interfacing should also be evaluated, both at the technology and functional levels. When dealing with integration, instead, the most suitable solution, among those available, has to be identified, carefully evaluating all of the aspects that might exert a negative influence on the performance and reliability of the RF-MEMS devices.

As mentioned before, emerging in the previous discussion, the development chain is not characterised by a unique unrolling setting, even when referring to the same device. For instance, the level of technology maturity, as well as the expertise in designing certain classes of devices, exert relevant impact on the effort to be invested in each single phase of the chain, as well as on the number of cycles and loops across a few stages, leading to optimised performance and characteristics. However, single phases of the flow in figure 2.6, like the packaging, might require a standalone and ad hoc development chain, if solid expertise is not already available.

Eventually, coming back to the initial discussion, the intrinsic factors that are going to be analysed in more detail are the following: *reliability*, *packaging* and *integration*. As a matter of fact, because they were not given proper consideration in the early days of MEMS for RF passives, they became one of the fundamental reasons triggering the first drop of expectations around the market breakout of RF-MEMS, taking place in 2004 and 2005.

2.4.1.1 Reliability

Right after demonstration of the remarkable RF passives' characteristics achievable through exploitation of microsystem technology, the reliability of RF-MEMS started to emerge as a major issue. In actuality, given their multi-physical behaviour, MEMS are exposed to a wide range of malfunctions and failure mechanisms (both reversible and irreversible) which, while quite common in material and mechanical engineering, are substantially unknown in the community of electronic and RF engineers [18]. As discussed in detail in [19], the most critical reliability issues at the functional level can be grouped as thermomechanical, electrical and environmental failures. They are briefly described as follows:

(a) Contact wear (thermomechanical) of surfaces repeatedly in contact under conditions of large stresses (e.g. temperature and current densities) [20].
(b) Fatigue (thermomechanical) in brittle and ductile materials resulting in progressive load bearing decrease, eventually bringing catastrophic failures [21].
(c) Hardening (thermomechanical) of ductile materials (e.g. metals and alloys) taking place when overstress above the yield limit is reached [19].
(d) Delamination (thermomechanical) of deposited layers occurring both because of processing defects or interfacial high stress levels [22].
(e) Creep (thermomechanical) of metal layers undergoing time-dependent loading at elevated temperatures [23].
(f) Dielectric charging (electrical) due to the presence of large electric fields across thin dielectric films leading to accumulation of charges [24] that induce voltage screening [25] and/or stiction [26] (i.e. missed release of the switch to the rest position when the DC bias is removed).

(g) Electromigration (electrical) leading to the formation of voids or hillocks due to high current densities in thin-film conductors [27].
(h) Electrostatic discharge (electrical) [28], dielectric breakdown [29], corona effects, and other electrically driven irreversible failures due to the very small gaps typical of MEMS.
(i) Micro-welding (electrical) due to accumulation of high current densities in very small areas, leading to formation of heat-induced welds of metals and joints [30].
(j) Self-heating (electrical) caused by large currents and/or RF power flows through the MEMS device [31]. Self-heating favours the occurrence of thermomechanical failure mechanisms previously mentioned, like contact wear and creep.
(k) Self-actuation (electrical) due to large RF power delivered to MEMS devices, able to generate an effective (spurious) DC voltage larger than the pull-in threshold [32]. Failure modes labelled as (i), (j) and (k) fall under the denomination of power handling [33].
(l) Corrosion (environmental), depending on the MEMS operation environmental characteristics (e.g. moisture, microorganisms, contaminants), and leading to a variety of failures [19].
(m) Humidity (environmental), leading to various degradation effects and failures, among which is stiction induced by capillary forces acting on the surfaces in contact [34].
(n) Radiation (environmental) in harsh environments like space, inducing modifications of device properties, spanning from reversible/irreversible drifts to catastrophic failures [35, 36].

Undoubtedly, in the beginning the interest of the research community was mainly captured by demonstration of the remarkable characteristics of RF-MEMS versus standard semiconductor components. However, limited effort was dedicated to the reliability aspects listed above. Nevertheless, as soon as compliance of RF-MEMS devices started to be verified against sets of specifications originating from real applications, the demand for such a technology for further development aimed towards improving reliability emerged in a forceful fashion. Adopting this need as a fundamental driver, significant work was carried out at the research and development level in subsequent years, which aimed to address one or more of the aforementioned failure mechanisms. As a result, the lifetime, operability and medium-/long-term performance stability of RF-MEMS components were significantly improved. To this purpose, table 2.1 summarises a few relevant solutions proposed and discussed in the literature to improve robustness against specific reliability issues.

As mentioned before, reliability was not the only aspect to be underestimated in the early stages of investigation of RF-MEMS technology. Packaging and integration are now going to be discussed.

Table 2.1. Summary of a few relevant solutions proposed in the literature to improve the reliability of various RF-MEMS devices versus specific failure mechanisms and performance drifts.

Target device	Addressed failure mode/s	Solution	References
Resonator	Long-term stability of resonant frequency	Low-temperature processing of silicon thin-films	[37]
Lateral comb-drive resonator	Performance drifts driven by aging, device defects, harsh conditions, etc.	Custom active adaptive controller	[38]
MEMS with cyclic contact	Contact wear and stiction	Atomic Layer Deposition (ALD) of thin-film coatings	[39]
Resonator in a vacuum	Q-factor decrease due to degradation of in-package vacuum conditions	Use of getter films within the sealed cavity to absorb gases	[40]
Capacitive switch	Charge accumulation in the dielectric film, voltage screening and failure for stiction	Deposition of SiO_2/Si_3N_4 double-layer dielectric stacks instead of single layer	[41]
Ohmic switch	Stiction failure due to micro-welding induced by large current densities	Active restoring mechanism (embedded micro-heater) activated in case of stiction	[42]
Capacitive switch	Stiction due to charge accumulation	Toggle design enabling push/pull active electrostatic control	[43]
Single Pole Single Throw (SPST)	Contact wear and other mechanical failures induced by cycled abrupt pull-in contact	Shaping of the biasing waveform to enable soft-landing/-contact of surfaces at pull-in	[44]
Capacitive switch	Contact wear and other mechanical failures induced by cycled abrupt pull-in contact	Resistive/capacitive braking schemes via design of embedded loading resistor and capacitor plate geometry	[45]
Lateral Bulk Acoustic Resonator (LBAR)	Resonant frequency drift due to temperature variations	Temperature compensation active control circuitry	[46]
Resonator	Resonant frequency drift due to temperature variations	Temperature Coefficient of Frequency (TCF) reduction by heavy n-type and p-type doping	[47]

2.4.1.2 Packaging

Also cross-linked in relation to reliability, the issues of packaging and encapsulation caused the initial enthusiasm triggered by RF-MEMS to be tempered. As MEMS devices are composed by movable micro-membranes, they are very fragile and exposed to harmful environmental factors, such as mechanical shocks, moisture, dust particles and various contaminants. Therefore, RF passives in microsystem technology need to be properly isolated from the surrounding environment by being housed within a protective (possibly hermetic or semi-hermetic) capping [48–50]. In the case of RF-MEMS, the application of a package increases, on one hand, the complexity of technology and the manufacturing cost; just to help build a more circumstanced idea around the impact of the package, in one of the first analyses it was estimated to be as high as 80% of the final product price [51]. On the other hand, the presence of a protective cap puts the outstanding RF performance of MEMS-based passives in jeopardy, due to the longer paths that the signal has to travel across, increased parasitic effects, discontinuities, and so on. Therefore, the package should be carefully designed and accounted for as an actual part of the device from the beginning of development, thus making the design and modelling phases more challenging [52–55].

To complete the discussion on the packaging of RF-MEMS, the widely employed technique of Wafer-Level Packaging (WLP) is shown schematically in figure 2.7, and discussed below as a relevant example for explanatory purposes.

The WLP solution employs an entire wafer for the capping part, of the same dimension as the RF-MEMS device wafer to be packaged. The cap needs a proper scheme of signal redistribution paths that must correspond to the electrical pads of each MEMS device. The cap and device wafers must be carefully aligned to each other, as reported in figure 2.7(a). A cross-sectional schematic view of what happens at the single device level is shown in figure 2.7(b). Through Wafer Vias (TWVs) for signal redistribution are visible. In factual terms, TWVs are holes etched across the whole wafer thickness, then filled with metal, intended to contact RF-MEMS signal pads. Optionally, a shallow recess can be etched in the cap wafer backside, in order to yield housing for the RF-MEMS movable elevated parts. After bonding, the capping and device wafers come into contact with each other, as depicted in figure 2.7(c). Physical adhesion between the two parts is ensured by a reflow of solder bumps (see figure 2.7(d)), which realise at the same time electrical interconnection between the RF-MEMS in-package devices and the external world. Once wafer-to-wafer adhesion is established, the packaged RF-MEMS wafer is ready to be cut into pieces (singulation), thus making available chips (containing one or a limited number of RF-MEMS devices) ready to be mounted, i.e. integrated, in a more complex system or subsystem.

Unquestionably, WLP, albeit commonly employed, is not the only solution to package RF-MEMS. Other techniques and technologies exist, with a wide deck of advantages and disadvantages with respect to WLP. Such solutions are not reported here, as their discussion falls out of the scope of this work. Nonetheless, they are extensively covered by other worthy contributions easily traceable in the scientific literature [56–58].

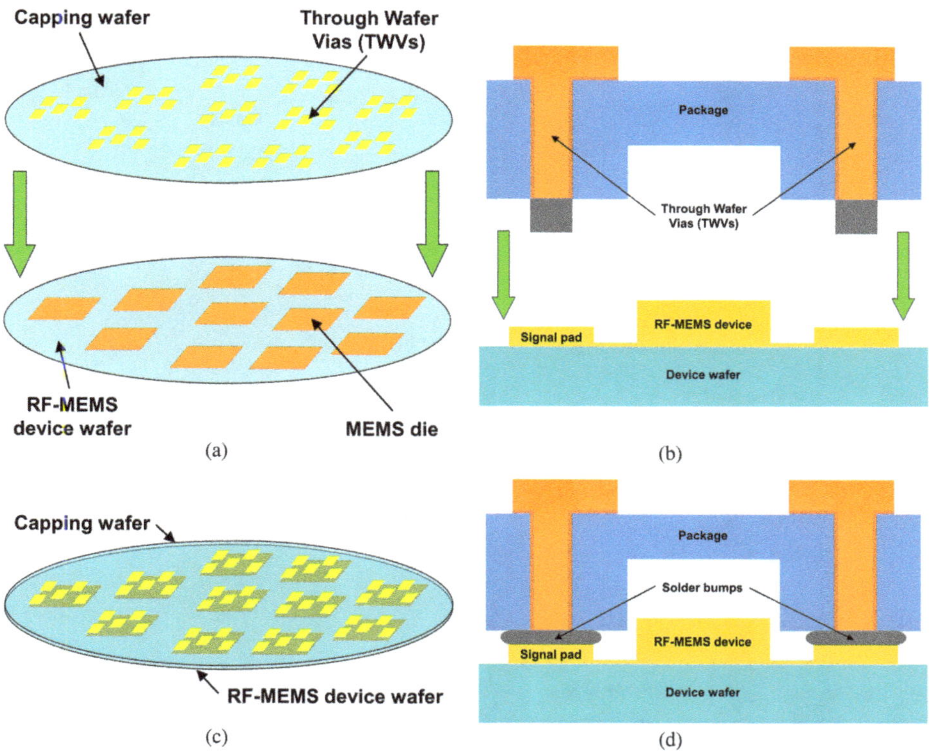

Figure 2.7. Schematic views of the Wafer-Level Packaging (WLP) technique. (a) Wafer-level schematic of the RF-MEMS device wafer and capping wafer alignment with Through Wafer Vias (TWVs) for signals redistribution. (b) Device-level cross-sectional schematic of the alignment between the RF-MEMS device and the package (TWVs are visible). (c) Wafer-level schematic of the aligned and bonded RF-MEMS device and capping wafers. (d) Device-level cross-sectional schematic of an in-package RF-MEMS device. Reflow of solder bumps is performed to ensure both electrical interconnections and physical adhesion between the device wafer and the package.

2.4.1.3 Integration

Finally, yet importantly, MEMS technology is typically incompatible with standard semiconductor platforms, such as Complementary Metal Oxide Semiconductor (CMOS) technology. Consequently, in-package RF-MEMS passive components need to be integrated with active electronics at subsystems or system level, i.e. on a board. This step is typically performed through Surface Mount Technologies (SMTs), and ad hoc circuitry must be developed and deployed as well, in order to operate/drive RF-MEMS, raising, also in this circumstance, complexity and costs [59–64].

As for the previous discussion on packaging, a couple of examples of RF-MEMS integration are briefly reported next, in order to help one to understand the multitude of issues and hurdles related to such a delicate phase of the development chain. The case studies mentioned are shown schematically in figure 2.8. In both circumstances, integration of the in-package RF-MEMS device into a more complex

(a)

(b)

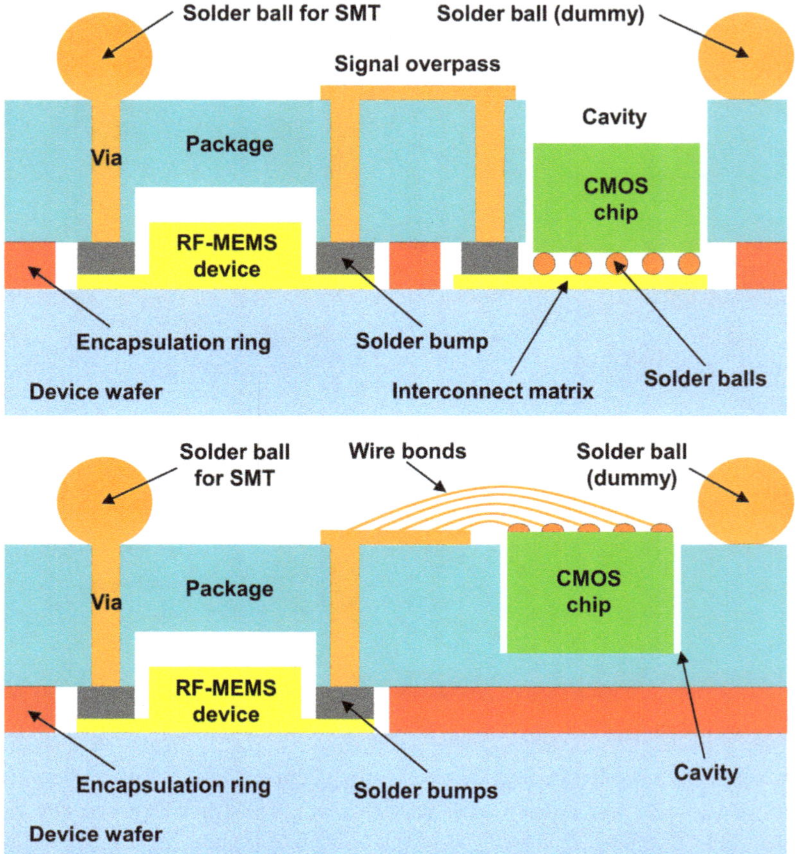

Figure 2.8. Schematic cross-sections of two possible CMOS/RF-MEMS hybrid in-package integration solutions, for Surface Mounting Technology (SMT). (a) The electrical interconnection matrix is deployed on the RF-MEMS device wafer. A through-wafer recess in the package allows housing of the CMOS chip. (b) The CMOS chip is housed within a shallow recess in the package. Electrical interconnections are ensured by wire bonding.

system is meant to be performed through SMT. To this regard, solder balls deployed on the package's front side are meant to undergo reflow to ensure adhesion and electrical interconnection between the in-package RF-MEMS and the system. From a conceptual point of view, the latter step is equivalent to the reflow of solder bumps previously reported in figure 2.7(d) to establish adhesion and electrical continuity between the RF-MEMS device wafer and the package.

Nonetheless, apart from ensuring in-package RF-MEMS integration, the two examples reported in figure 2.8 realise an additional functionality, albeit through different strategies. As a matter of fact, in both cases the CMOS active circuitry, necessary for controlling and driving the RF-MEMS device, is integrated and interfaced through the package, thus carrying out the so-called System in Package (SiP) strategy, despite in this case the system order of complexity being quite reduced. Moving into a little more detail, in the example reported in figure 2.8(a) a matrix of electrical interconnections is designed and realised on the RF-MEMS device wafer. The CMOS chip is mounted upside-down upon such contacts thanks to a through-wafer recess etched in the package, meant to provide proper housing. The adhesion and electrical continuity between the CMOS chip and the underlying pads is ensured by a reflow of solder balls. However, in the example reported in figure 2.8(b), the recess etched in the cap is shallow and does not develop across the whole wafer thickness. In this case, the CMOS chip is placed into the housing with the contacts upward. The electrical interconnection between the CMOS driving circuitry and the RF-MEMS is ensured by wire bonds.

In conclusion, bearing in mind the previous sections, it is straightforward that the readiness of RF-MEMS technology for market absorption was still far from maturity in the first years of its investigation. In any event, once the scientific community became aware of the aforementioned issues, plenty of effort was spent on them, thus significantly reducing the gap of RF-MEMS in terms of reliability, encapsulation and integration/compatibility with standard semiconductor active technologies, and boosting, in turn, expectations and optimism at the market level for years to come.

Unfortunately, RF-MEMS technology still had impediments to overcome, as will be discussed in the following section.

2.4.2 Extrinsic factors

The frame of reference depicted in the previous pages appears to be significantly complex. The just-reviewed critical intrinsic factors, namely, reliability, packaging and integration, were clearly underestimated in the early days of RF-MEMS, in which their remarkable characteristics were inflating expectations around massive market breakout, which was meant to be imminent, at least according to the initial market forecasts. However, the massive effort brought by the scientific community in the following years drove paramount improvement of RF-MEMS technology against all the weaknesses mentioned in the previous sections; making RF passives in MEMS technology, in factual terms, ready—from a technical point of view—to be employed in commercial applications. Nonetheless, what has been discussed up to here is just part of the picture.

If the discussion of intrinsic factors impairing the success of RF-MEMS was complex and rich in ramifications, the one around extrinsic factors is much simpler. However, this does not mean that this class of factors exerted less influence on the missed penetration of MEMS-based RF passives into the market. In fact, most probably, extrinsic factors represent the substantial reason for the thread of disappointments collected by RF-MEMS across more than a decade, since they were present from the beginning, but fundamentally underestimated, as were the intrinsic factors.

Looking at the evolution of mobile phones' communication standards [65], starting from the early 1980s with 1G (i.e. First Generation) to the latest 3.5G and 4G across the first and second decade of the 2000s, the scenario appears to be quite extended, as is reported in figure 2.9.

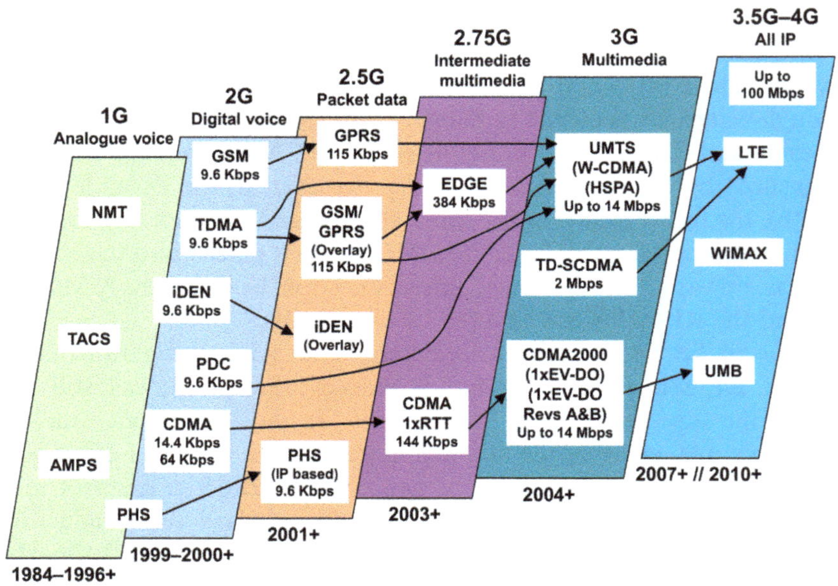

Figure 2.9. Evolution of mobile communication standards from 1G, in the early 1980s, to 3.5G and 4G, across the first and second decade of the 2000s [65].

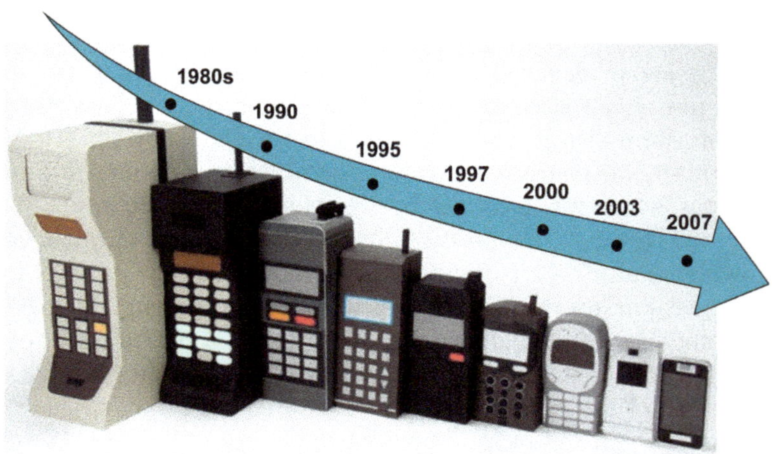

Figure 2.10. Schematic view of the seamless trend in size reduction of mobile handsets, from the 1980s to the first decade of the 2000s.

Across the decades, crucial steps forward were taken, such as the transition from analogue to digital signals, the seamless increase of transmitted data volumes, as well as the increase of other-than-voice services, venturing into the field of multimedia exchange. In parallel to the evolution of communication standards and to the increase of services, mobile handsets followed a seamless path in size reduction (as shown in figure 2.10), weight, power consumption and costs.

The crystal-clear fact of this scenario is that the evolution just discussed (for certain aspects similar to a revolution) and reported in figure 2.9 and figure 2.10, took place without RF-MEMS. In other words, in the huge consumer market segment of mobile phones, there has never been a strong enough need for RF passives with boosted performance, prompting the investigation of technologies other than standard ones, opening up the floor to issues such as reliability, packaging and integration, as well as to higher costs for the final single component.

In conclusion, the combination of intrinsic and extrinsic factors that caused multiple drops of inflated expectations in the hype curve, led the scientific community (circa 2010) to the unspoken opinion that RF-MEMS technology was mainly meant for niche and small-volume applications, such as in the space and defence sector. In the latter, indeed, the driver of remarkable performance rules over increased price and technical hurdles at the integration level.

2.5 The RF-MEMS market

The discussion that is going to be developed is divided in two subsections. The first one frames how the RF-MEMS market evolved in recent years, finally closing the gap between research and industry. The second one briefly reports on the market situation today and on its perspectives for the future.

2.5.1 Evolution of the market in recent years

Quite unexpectedly, the market segment of smartphones, which started to expand massively in a few years, brought in some elements partially altering the landscape described in the previous pages. With the widespread diffusion of mobile phones in the mid-1990s, the quality of the voice signal experienced a degrading trend, hopping from one generation to the next (i.e. GSM, 2.5G, and so on), as stated in [66] and shown in figure 2.11. The increasing use of cellular communications, the introduction of full-screen devices (with touch technology) and the integration of the antenna inside the handset together determined this decrease in performance.

In addition, as smartphones became smaller and thinner, it turned out to be difficult to integrate proper circuitry to boost performance, to counteract dropped

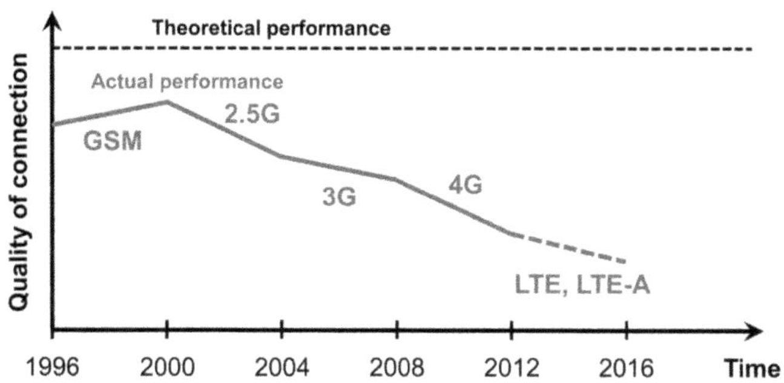

Figure 2.11. Trend in the decrease of connection quality, stepping from the first generations of mobile handsets in the mid-1990s to modern 4G-LTE smartphones [66].

calls and improve voice quality. In [66] it is estimated that the ratio of theoretical-versus-actual RF signal quality was decreasing at a pace of about 1 dB per year for over a decade. In light of all these factors, the antennas of recent smartphones were not working efficiently anymore, leading to slower download speeds, reduced voice quality, lower energy efficiency and more dropped calls [67]. As a matter of fact, the demand for compact and device-integrated antennas decreased the efficiency and made them more sensitive to interaction with external items, such as the head of the speaker or the hand holding the device. In light of this scenario, fixed impedance matching between the antenna and the RF Front End (RFFE) was not an optimal solution anymore, as the characteristic impedance of the transmitting/receiving aerial underwent slight changes.

The availability of high-performance reconfigurable impedance matching tuners started to be desirable, and RF-MEMS were shown to be a proper technology solution to implement such complex devices. In 2012, information about the employment of RF-MEMS adaptable impedance tuners manufactured by WiSpry within the Samsung Focus Flash Windows smartphone was announced [68]. Later, in autumn 2014, Cavendish Kinetics (CK) announced the commercial adoption of its RF-MEMS-based SmarTune™ antenna tuning solution in the Nubia Z7 smartphone, manufactured by the Chinese ZTE Corporation [69]. A few months later, CK again went public with the information that SmarTune products were adopted in five different Chinese smartphone models, promising signal enhancement of 2–3 dB, resulting in boosted data rates (up to 2×) and improved battery life (up to 40%) [70]. To demonstrate that this was not just a flash in the pan, Qorvo has more than quadrupled its RF-MEMS sales in the next three years (2015–2017) from 145 $M to 585 $M, as reported in [71]. Such a remarkable achievement has been reached by extensively integrating RF-MEMS switches within RFFEs, in order to extend their reconfigurability and, in turn, their operability. This brought the same hardware to communicate simultaneously and in an effective fashion, over Wi-Fi, Bluetooth, and cellular bands belonging to 3G and 4G networks.

Bearing in mind the overall discussion developed throughout this chapter, the most sceptical observers might call into question that the just-listed examples are simply another peak of inflated expectations preceding the next (and probably last) drop of enthusiasm in the hype curve of RF-MEMS. Nonetheless, a fundamental element of discontinuity with respect to the past must be taken into careful consideration. Starting from the exacerbation of the negative trend discussed in figure 2.11, the surge in market need for RF passives with improved characteristics and boosted reconfigurability was unmistakable. In other words, recalling what was previously discussed in this chapter, the approach to exploiting RF-MEMS began to shift from a mainly technology push scenario to a market-driven philosophy.

Coming back for a moment to the vision of Nguyen [3] (previously reviewed in figure 2.2 and figure 2.3), it looks like it is somehow taking place, despite a reversal in development. The first market exploitation of RF-MEMS technology was, in fact, related to complex reconfigurable networks (impedance tuners at first), which was supposed to be the second phase of RF passives in MEMS technology (see figure 2.3). The spreading of these networks has been increasing the demand for high-performance, reliable and low-cost basic RF-MEMS components (mainly switches and tunable capacitors), thus pushing advanced research and engineering of such building blocks as well.

2.5.2 The market situation today and its future perspectives

Moving now attention to what has been going on in the RF-MEMS market scenario in more recent times, an important item certainly deserving of mention is that, in fall 2019, Qorvo acquired CK [72], proving further how RF-MEMS are turning into a Key Enabling Technology (KET) for modern RF transceivers and 5G applications. Besides RF-MEMS-based impedance tuners technology [73], already discussed in the previous pages, Qorvo-CK has been extending its products, focusing on high-performance micro-switches solutions, to be employed both for tuning antenna aperture [74] as well as for transfer switching devices, devoted to the selection among multiple antennas in modern smartphones [75].

As far as switching solutions are concerned, it has to be stressed how the trends of recent years confirm what was stated above, around a sort of reversed deployment of the early visions of Nguyen, with complex RF-MEMS networks gaining room in the market at first, and basic components (i.e. micro-relays) following them. Such a kind of capsized evolution makes perfect sense from the point of view of the market and of its *rules*. In fact, as soon as RF-MEMS technology demonstrated its appropriateness in the context of consumer applications, i.e. antenna to RFFE adaptive impedance tuning in smartphones, efforts to fill the still-existing technology gaps around the crucial components enabling large reconfigurability of RF-MEMS, i.e. switches, were suddenly reinvigorated.

To this end, Analog Devices (AD) (www.analog.com) has in recent years began to include in their catalogue a couple of RF-MEMS-based switching items. The products, identified as ADGM1304 and ADGM1004 [76], are both Single Pole Four Throw (SP4T) units, with the latter one providing Electrostatic Discharge (ESD) protection, as well. What is more relevant to note is that both solutions are provided along with an integrated driver, in order to feed electrostatic MEMS switches with the proper DC level, within a package compatible with Surface Mount Technologies (SMTs). As reported in the data sheets [77, 78], both AD switches work in the frequency range from DC up to 13–14 GHz. Typical losses are 0.26–0.45 dB at 2.5 GHz, while isolation is 24 dB at 2.5 GHz. The maximum input RF power, depending on the device, ranges between 32 dBm and 36 dBm.

Concerning reliability, both AD products report an average time to failure in continuous actuation condition of 7.2 years. Also relevant, it is interesting to observe what the data sheets report about ON/OFF switching. In cold switching conditions, 1 billion cycles are guaranteed, while the situation changes quite radically when an RF power is delivered while the MEMS switches actuate/release (hot switching condition). In the latter case, 50% of the tested population of devices failed at 5.16 billion, 3.21 million and 390 kilo cycles, when the delivered RF power was 10 dBm, 15 dBm and 20 dBm, respectively.

Still sticking to the ground of RF-MEMS micro-switches, the probably most relevant entry to be reported in the timespan of recent years is a company located in Irvine, California, named MenloMicro (https://menlomicro.com/). It was founded in 2016 as a spin-off of General Electrics (GE) Ventures, with the aim of directing the significant know-how previously developed by GE in the field of thin-film materials and coatings to market products [79]. Given these premises, MenloMicro targeted the delivery of the so-called 'ideal switch', i.e. a relay meeting all the most desirable features, for instance in terms of low loss when CLOSED, high isolation when OPEN, fast transition time, high reliability, high power handling, very low power consumption, small footprint and ease of integration [80].

Today, the company features a few MEMS-based switching products in its catalogue, including two SP4T units, i.e. the MM5130 [81] and the MM5120 [82], the latter one also including an integrated driver, the MM5600 Double Pole Double Throw (DPDT) [83], and the MM3100 Single Pole Single Throw (SPST) [84]. In particular, the latter product features six input/output lines that, combined together, can yield multiple configurations, like dual SP3T, triple SP2T, or even a 2-by-3 switching matrix. Depending on the specific product, the MenloMicro RF-MEMS switches work from DC up to 3–26 GHz, with losses in the range 0.3–1.3 dB, and ON-state resistance around 1 Ω. The switching speed is always below 20 μs, while the power handling, in Continuous Wave (CW) conditions, ranges from 2 W up to 25 W. What has to be stressed is that the specified lifetime is always larger than 3 billion cycles.

Stepping forth with an outlook of the current market placement of RF-MEMS technology, it is worth spending some focus on how the activity of the early commercial players in such products has evolved, transformed, or even shrunk and discontinued. To this end, OMRON discontinued their RF-MEMS-based switch products [85]. The last technical report is dated 2013 [86]. WiSpry (https://wispry.com/), now owned by AAC Technologies (https://www.aactechnologies.com/en/), moved focus on RF antenna tuning solutions based on MEMS technology [87]. Tronics (https://www.tronicsgroup.com/), currently owned by TDK (https://www.tdk.com/en/index.html), is now focusing on MEMS inertial sensing solutions. MEMtronics (https://memtronics.com/) narrowed down their focus around MEMS-based filters implementations. Differently, DelfMEMS (https://www.delfmems.com/) shut down activity in 2015/2016, while no recent news are available for what concerns Radant MEMS. In addition, in more recent years, NEDITEK (http://www.neditek.com/) appeared on the market landscape, offering a wide portfolio of RF-MEMS passive components that, however, are no longer included in their catalogue. Eventually, the RF-MEMS technology services provided by Teledyne DALSA [88] and AirMems (https://www.airmems.com/) must certainly be mentioned.

To conclude this section, it is worth highlighting some of the considerations included by YOLE Dévelopement in their 2021 market and technology report on the status of the MEMS industry [89]. For what concerns the application of MEMS technologies in the field of RF, YOLE confirms the centrality of RF-MEMS switches in the following sectors:

- **Telecommunications**, and in particular micro-switches for Base Stations (BSs) and small cells.
- **Industrial**, with RF-MEMS switches for Automated Test Equipment (ATE) and RF instruments/appliances.
- **Space**, with particular focus on the exploitation of micro-switches in redundant and communication matrices, as well as in radar and communication systems.
- **Defence**, concerning switches to be employed for communications and radars.
- **Smartphones**, with large exploitation of RF-MEMS switches for antenna tuning and reconfiguration purposes.

Eventually, considering the RF applications sector of MEMS technology as a whole, and starting from the consolidated market numbers referred to in 2020, YOLE predicts an expansion of revenues in the timespan from 2020 to 2026, from 2.05 B$ (Billion US Dollar) to 4.05 B$, with a Compound Annual Growth Rate (CAGR) of 12%.

2.6 Conclusion

This chapter studied, in a quite detailed fashion, the evolution of the RF-MEMS market, from the early days in which the technology first started to be discussed within the scientific community. As a first step, critical concepts of market analysis were introduced, in order to support the subsequent discussion. In detail, these concepts are the hype curve and technology push versus market pull/driven scenarios. Briefly, the hype curve sketches the behavioural up and down trend versus time of market expectations fed by a novel technology, since its first discussion in the scientific community to full maturity. However, the technology push and market pull are simply two different formulations of the so-called *value proposition* related to a product realised through a certain technique and/or technology. In the technology push approach, a strong market need for addressing a certain problem or issue, in fact, does not exist. Therefore, the value has leverage upon different drivers, such as, for instance, the added value in terms of distinctiveness related to a certain solution. In contrast, in a market pull/driven scenario, one or more market needs clearly exist, and the added value of a given technology solution resides right within its capability of addressing those very requirements.

After reviewing such fundamental concepts, they were applied to the case study focus of this book. The vision of the early days of RF-MEMS, concerning their expected penetration into the mobile handsets market, was reported. It was envisioned to act in a first phase at a single-components level. In other words, in standard transmitter/receiver (transceiver) architectures, RF-MEMS simple devices, like switches and variable capacitors (varactors), were meant to replace standard devices. Later on, more complex RF passive networks in MEMS technology, like switching matrices and tunable filters, would have been the pivot around which transceivers architecture was redesigned, aiming for better performance and improved reconfigurability.

As a matter of fact, all of the most positive market forecasts, released from the early years of the 2000s and for more than a decade after, resulted in hope being systematically disappointed. The most relevant causes leading to the tepid success of RF-MEMS technology were identified, split into two main classes, namely intrinsic and extrinsic factors. The former were intimately linked to aspects of the technology that were not properly accounted for in the beginning, identified to be reliability, packaging and integration of RF-MEMS. Such items have been analysed with an in-depth technical insight. The latter, instead, were related to the scarce readiness of market needs with respect to the boosted characteristics offered by RF-MEMS solutions, albeit counterbalanced by additional issues at integration level and higher costs.

Finally, the environmental changes taking place in the market in recent years were also mentioned and discussed. Such modifications brought the approach to exploit RF-MEMS from a less favourable technology push scenario to a more advantageous market pull reference frame. The most recent success stories related to RF-MEMS products employed in consumer applications were listed, proving the achievement and consolidation of an appropriate level of maturity. This will constitute the foundation for the future spread of RF-MEMS devices, which will be discussed in the following pages of this work.

References

[1] Gartner Inc July 2021 *Interpreting Technology Hype* https://gartner.com/en/research/methodologies/gartner-hype-cycle

[2] Martin M J C 1994 *Managing Innovation and Entrepreneurship in Technology-Based Firms* 1st edn (New York: Wiley) p 416

[3] Nguyen C T-C 2001 Transceiver front-end architectures using vibrating micromechanical signal processors *Proc. Top. Meet. on Silicon Monolithic Integrated Circ. in RF Sys. (Ann Arbor, MI, Sept. 2001)* pp 23–32

[4] Nguyen C T-C 2002 RF MEMS for wireless applications *Proc. Device Res. Conf. DRC (Santa Barbara, CA, June 2002)* pp 9–12

[5] Laskar J, Matinpour B and Sudipto C 2004 *Modern Receiver Front-Ends: Systems, Circuits, and Integration* 1st edn (New York: Wiley) p 221

[6] Nguyen C T-C 2006 Integrated micromechanical circuits for RF front ends *Proc. Europ. Solid-State Circ. Conf. ESSCIRC (Montreux, Sept 2006)* pp 7–16

[7] Nguyen C T-C 1998 Microelectromechanical devices for wireless communications *Proc. Int. Workshop on Micro Electro Mechan. Syst. MEMS (Heidelberg, Jan. 1998)* pp 1–7

[8] Nguyen C T-C 2007 MEMS technology for timing and frequency control *IEEE Trans. Ultrason. Ferroelectr. Freq. Control* **54** 251–70

[9] Nguyen C T-C 2013 MEMS-based RF channel selection for true software-defined cognitive radio and low-power sensor communications *IEEE Commun. Mag.* **51** 110–9

[10] Meikle H 2001 *Modern Radar Systems* 1st edn (Norwood: Artech House) 563

[11] Haridas N *et al* 2008 Reconfigurable MEMS antennas *Proc. NASA/ESA Conf. on Adaptive Hardware and Syst. (Noordwijk, June 2008)* pp 147–54

[12] Bouchaud J, Knoblich B and Wicht H 2006 RF MEMS market *Proc. German Microw. Conf. GeMiC (Karlsruhe, March 2006)* pp 1–2
[13] Johnson R C *RF MEMS Aim for Cell Phone on-a-Chip* https://memsjournal.com/2010/07/rf-mems-aims-for-cell-phone-onachip.html)
[14] MEMS Journal *Cell Phone Antenna Troubles? RF MEMS Come to the Rescue* https://memsjournal.com/2010/09/cell-phone-antenna-troubles-rf-mems-come-to-the-rescue.html)
[15] McGrath D *Teardown Finds RF MEMS in Samsung Handset* https://eetimes.com/teardown-finds-rf-mems-in-samsung-handset/#)
[16] DeLisle J-J *The Future of Connectivity: Mobile & Automobiles* https://mwrf.com/technologies/systems/article/21845463/the-future-of-connectivity-mobile-automobiles)
[17] Iannacci J 2015 RF-MEMS: an enabling technology for modern wireless systems bearing a market potential still not fully displayed *Microsyst. Tech.* **21** 2039–52
[18] Iannacci J 2014 Reliability of MEMS: a perspective on failure mechanisms, improvement solutions and best practices at development level *Displays* **37** 62–71
[19] Hartzell A L, da Silva M G and Shea H 2011 *MEMS Reliability* 1st edn (New York: Springer) p 291
[20] Chianrabutra C, Jiang L, Lewis A P and McBride J W 2013 Evaluating the influence of current on the wear processes of Au/Cr-Au/MWCNT switching surfaces *Proc. Holm Conf. on Electrical Contacts (Newport, RI, Sept. 2013)* pp 1–6
[21] Kawai T, Gaspar J, Paul O and Kamiya S 2009 Prediction of strength and fatigue lifetime of MEMS structures with arbitrary shapes *Proc. Solid-State Sens., Actuators and Microsystems Conf. TRANSDUCERS (Denver, CO, June 2009)* pp 1067–70
[22] Khanna V K 2011 Adhesion–delamination phenomena at the surfaces and interfaces in microelectronics and MEMS structures and packaged devices *J. Phys. D: Appl. Phys.* **44** 1–19
[23] Van Gils M, Bielen J and McDonald G 2007 Evaluation of creep in RF MEMS devices *Proc. Int. Conf. on Thermal, Mech. and Multi-Physics Simulation Experiments in Microelectron. and Micro-Syst. EuroSime (London, April 2007)* pp 1–6
[24] Tavassolian N, Koutsoureli M, Papaioannou G, Lacroix B and Papapolymerou J 2010 Dielectric charging in capacitive RF MEMS switches: the effect of electric stress *Proc. Asia-Pacific Microw. Conf. APMC (Yokohama, Dec. 2010)* pp 1833–6
[25] Iannacci J 2013 *Practical Guide to RF-MEMS* 1st edn (Weinheim: Wiley) p 372
[26] Zaghloul U, Bhushan B, Pons P, Papaioannou G, Coccetti F and Plana R 2011 Different stiction mechanisms in electrostatic MEMS devices: nanoscale characterization based on adhesion and friction measurements *Proc. Int. Solid-State Sensors, Actuators and Microsystems Conf. TRANSDUCERS (Beijing, June 2011)* pp 2478–81
[27] Wilson C J, Horsfall A B, O'Neill A G, Wright N G, Bull S J, Terry J G, Stevenson J T M and Walton A J 2007 Direct Measurement of electromigration-induced stress in interconnect structures *IEEE Trans. Dev. Mater. Reliab.* **7** 356–62
[28] Sangameswaran S, De Coster J, Linten D, Scholz M, Thijs S, Haspeslagh L, Witvrouw A, Van Hoof C, Groeseneken G and De Wolf I 2008 ESD reliability issues in micro-electromechanical systems (MEMS): a case study on micromirrors *Proc. Electrical Overstress/Electrostatic Discharge Symp. EOS/ESD (Tucson, AZ, Sept. 2008)* pp 249–57
[29] Tazzoli A, Barbato M, Ritrovato V and Meneghesso G 2010 A comprehensive study of MEMS behavior under EOS/ESD events: breakdown characterization, dielectric charging, and realistic cures *Proc. Electrical Overstress/Electrostatic Discharge Symp. EOS/ESD (Reno, NV, Oct 2010)* pp 1–10

[30] Tazzoli A, Iannacci J and Meneghesso G 2011 A positive exploitation of ESD events: micro-welding induction on ohmic MEMS contacts *Proc. Electrical Overstress/Electrostatic Discharge Symp. EOS/ESD (Anaheim, CA, Sept. 2011)* pp 1–8

[31] Ivira B, Fillit R-Y, Ndagijimana F, Benech P, Parat G and Ancey P 2008 Self-heating study of bulk acoustic wave resonators under high RF power *IEEE Trans. Ultrason. Ferroelectr. Freq. Control* **55** 139–47

[32] Malmqvist R, Jonsson R, Samuelsson C, Reyaz S, Rantakari P, Ouacha A, Vaha-Heikkila T, Varis J and Rydberg A 2012 Self-actuation tests of ohmic contact and capacitive RFMEMS switches for wideband RF power limiter circuits *Proc. Int. Semiconductor Conf. CAS (Sinaia, Oct. 2012)* pp 217–20

[33] Blondy P and Peroulis D 2013 Handling RF power: The latest advances in RF-MEMS tunable filters *IEEE Microw. Mag.* **14** 24–38

[34] Wu L, Noels L, Rochus V, Pustan M and Golinval J-C 2011 A micro–macroapproach to predict stiction due to surface contact in microelectromechanical systems *IEEE J. Microelectromechan. Syst.* **20** 976–90

[35] Lozano A, Palumbo F and Alurralde F 2009 Radiation effects on SOI electrostatic comb drive actuators of MEMS devices *Proc. Argentine School of Micro-Nanoelectronics, Technology and Applications EAMTA (San Carlos de Bariloche, Oct. 2009)* pp 46–49

[36] Goldsmith C L, Hwang J C M, Gudeman C, Auciello O, Ebel J L and Newman H S 2012 Robustness of RF MEMS capacitive switches in harsh environments *Proc. IEEE MTT-S Int. Microwave Symp. (Montreal, June 2012)* pp 1–3

[37] Sousa P M, Chu V and Conde J P 2012 Reliability and stability of thin-film amorphous silicon MEMS resonators *J. Micromech. Microeng.* **22** 181–8

[38] Izadian A and Famouri P 2008 Reliability enhancement of MEMS lateral comb resonators under fault conditions *IEEE Trans. Control Syst. Technol.* **16** 726–34

[39] Ashurst W R, Jang Y J, Magagnin L, Carraro C, Sung M M and Maboudian R 2004 Nanometer-thin titania films with SAM-level stiction and superior wear resistance for reliable MEMS performance *Proc. IEEE Int. Conf. on Micro Electro Mechanical Systems MEMS (Maastricht, Jan. 2004)* pp 153–6

[40] Longoni G, Conte A, Moraja M and Fourrier A 2006 Stable and reliable Q-factor in resonant MEMS with getter film *Proc. IEEE Int. Reliab. Phys. Symp. IRPS (San Jose, CA, March 2006)* pp 416–20

[41] Li G, Hanke U, Cheng Z, Min D, San H and Chen X 2011 Si_3N_4/SiO_2 dielectric stacks for high reliable capacitive RF MEMS switch *Proc. IEEE-NANO Conf. on Nanotech. (Portland, OR, Aug. 2011)* pp 496–9

[42] Iannacci J, Faes A, Repchankova A, Tazzoli A and Meneghesso G 2011 An active heat-based restoring mechanism for improving the reliability of RF-MEMS switches *Microelectron. Reliab.* **51** 1869–73

[43] Solazzi F, Tazzoli A, Farinelli P, Faes A, Mulloni V, Meneghesso G and Margesin B 2010 Active recovering mechanism for high performance RF MEMS redundancy switches *Proc. Europ. Microw. Conf. EuMC (Paris, Sept. 2010)* pp 93–6

[44] Czaplewski D A, Dyck C W, Sumali H, Massad J E, Kuppers J D, Reines I, Cowan W D and Tigges C P 2006 A soft-landing waveform for actuation of a single-pole single-throw ohmic RF MEMS switch *IEEE J. Microelectromechan. Syst. JMEMS* **15** 1586–94

[45] Ankit J, Nair P R and Alam M A 2011 Strategies for dynamic soft-landing in capacitive microelectromechanical switches *Appl. Phys. Lett.* **98** 131–3

[46] Lavasani H M, Pan W, Harrington B P, Abdolvand R and Ayazi F 2012 Electronic temperature compensation of lateral bulk acoustic resonator reference oscillators using enhanced series tuning technique *IEEE J. Solid-State Circuits* **47** 1381–93
[47] Pensala T, Jaakkola A, Prunnila M and Dekker J 2011 Temperature compensation of silicon MEMS resonators by heavy doping *Proc. IEEE Int. Ultrasonics Symp. IUS (Orlando, FL, Oct. 2011)* pp 1952–5
[48] Jourdain A, Ziad H, De Moor P and Tilmans H A C 2003 Wafer-scale 0-level packaging of (RF-)MEMS devices using BCB *Proc. Symp. on Design, Test, Integ. and Packaging of MEMS/MOEMS DTIP (Mandelieu-La Napoule, May 2003)* pp 239–44
[49] Park Y-K, Park H-W, Lee D-J, Park J-H, Song I-S, Kim C-W, Song C-M, Lee Y-H, Kim C-J and Ju B K 2002 A novel low-loss wafer-level packaging of the RF-MEMS devices *Proc. IEEE Int. Conf. on Micro Electro Mechanical Systems (Las Vegas, NV, Jan. 2002)* pp 681–4
[50] Park Y-K, Kim Y-K, Hoon K, Lee D-J, Kim C-J, Ju B-K and Park J-O 2003 A novel thin chip scale packaging of the RF-MEMS devices using ultra thin silicon *Proc. IEEE Int. Conf. on Micro Electro Mechanical Systems (Kyoto, Jan. 2003)* pp 618–21
[51] Cohn M B, Roehnelt R, Xu J-H, Shteinberg A and Cheung S 2002 MEMS packaging on a budget (fiscal and thermal) *Proc. Int. Conf. on Electronics, Circ. and Systems ICECS (Dubrovnik, Sept. 2002)* pp 287–90
[52] Iannacci J, Tian J, Sosin S, Gaddi R and Bartek M 2006 Hybrid wafer-level packaging for RF MEMS applications *Proc. Int. Wafer-Level Packaging Conf. IWLPC (San Jose, CA, Nov. 2006)* pp 106–13
[53] Iannacci J, Bartek M, Tian J, Gaddi R and Gnudi A 2008 Electromagnetic optimization of an RF-MEMS wafer-level package *Sensors Actuators* A **142** 434–41
[54] Margomenos A and Katehi L P B 2002 DC to 40 GHz on-wafer package for RF MEMS switches *Proc. IEEE Top. Meet. on Electrical Performance of Electronic Packaging (Monterey, CA, Oct. 2002)* pp 91–4
[55] Margomenos A and Katehi L P B 2003 High frequency parasitic effects for on-wafer packaging of RF MEMS switches *Proc. IEEE MTT-S Int. Microw. Symp. (Philadelphia, PA, June 2003)* pp 1931–4
[56] Lau J H, Lee C, Premachandran C S and Aibin Y 2009 *Advanced MEMS Packaging* 1st edn (New York: McGraw-Hill) 576
[57] Kuang K, Kim F and Cahill S S 2010 *RF and Microwave Microelectronics Packaging* 1st edn (New York: Springer) p 285
[58] Kuang K and Sturdivant R 2017 *RF and Microwave Microelectronics Packaging II* 1st edn (New York: Springer) p 172
[59] De Silva A P and Hughes H G 2003 The package integration of RF-MEMS switch and control IC for wireless applications *IEEE Trans. Adv. Packag.* **26** 255–60
[60] Lu A C W, Chua K M and Li H G 2005 Emerging manufacturing technologies for RFIC, antenna and RF-MEMS integration *Proc. IEEE Int. Workshop on Radio-Freq. Integr. Technol.: Integ. Circ. for Wideband Commun. and Wireless Sensor Networks (Singapore, Nov–Dec 2005)* pp 142–6
[61] Pacheco S, Zurcher P, Young S, Weston D and Dauksher W 2004 RF MEMS resonator for CMOS back-end-of-line integration *Proc. Top. Meet. on Silicon Monolithic Integ. Circ. in RF Systems (Atlanta, GA, Sept. 2004)* pp 203–6

[62] Th Rijks G S M, van Beek J T M, Ulenaers M J E, De Coster J, Puers R, den Dekker A and van Teeffelen L 2003 Passive integration and RF MEMS: a toolkit for adaptive LC circuits *Proc. Europ. Solid-State Circ. Conf. ESSCIRC (Estoril, Sept. 2003)* pp 269–72

[63] Zhang Q X *et al* 2006 Novel monolithic integration of RF-MEMS switch with CMOS-IC on organic substrate for compact RF system *Proc. IEEE Int. Electron. Dev. Meet. IEDM (San Francisco, CA, Dec. 2006)* pp 1–4

[64] Ziegler V, Siegel C, Schonlinner B, Prechtel U and Schumacher H 2005 RF-MEMS switches based on a low-complexity technology and related aspects of MMIC integration *Proc. Europ. Gallium Arsenide and other Semiconductor Application Symp. EGAAS (Paris, Oct. 2005)* pp 289–92

[65] Turner B and Orange M *3G Tutorial* https://slideshare.net/9876012345/3g-tutorial-38911321

[66] Allan R *RF MEMS Switches Are Primed for Mass-Market Applications* https://mwrf.com/technologies/active-components/article/21844989/rf-mems-switches-are-primed-for-mass-market-applications

[67] Jacobs School of Engineering *RF MEMS: New Possibilities for Smartphones* https://phys.org/pdf309079091.pdf

[68] IHS iSuppli *IHS iSuppli Teardown Analysis Service Identifies First Use of RF MEMS Part, Set to be Next Big Thing in Cellphone Radios* https://evertiq.com/design/25224

[69] Cavendish Kinetics *Nubia Adopts Cavendish Kinetics SmarTuneTM Antenna Tuning Solution for its new Z7 LTE Smartphone* http://cavendish-kinetics.com/news/news-releases/

[70] Cavendish Kinetics *Cavendish's SmarTune™ Antenna Tuning Solution Now Shipping in* 5 *Smartphone Models* https://cavendish-kinetics.com/release/cavendish-kinetics-adds-design-wins-and-ramps-shipments-of-rf-mems-tuners/

[71] Iannacci J 2018 'Surfing the hype curve of RF-MEMS passive components: towards the 5th generation (5G) of mobile networks,' *Microsyst. Technol.* https://doi.org/10.1007/s00542-018-3718-4

[72] EE Times *Qorvo Buys Cavendish Kinetics Outright* https://eetimes.com/qorvo-buys-cavendish-kinetics-outright/

[73] REF Qorvo-Cavendish Kinetics *RF MEMS SmarTuners* https://cavendish-kinetics.com/rf-mems-products/rf-mems-smartuners/

[74] Qorvo-Cavendish Kinetics *RF MEMS Smartswitches* https://cavendish-kinetics.com/rf-mems-products/rf-mems-smartswitches/

[75] Qorvo-Cavendish Kinetics *RF MEMS Transfer Switches* https://cavendish-kinetics.com/rf-mems-products/rf-mems-transferswitches/

[76] Analog Devices September 2021 *MEMS Switch Technology Products* https://analog.com/en/applications/technology/mems-switch.html

[77] Analog Devices *0 Hz/dc to 14 GHz, Single-Pole, Four-Throw MEMS Switch with Integrated Driver—Data Sheet ADGM1304* https://analog.com/media/en/technical-documentation/data-sheets/adgm1304.pdf

[78] Analog Devices *0 Hz/DC to 13 GHz, 2.5 kV HBM ESD, SP4T, MEMS Switch with Integrated Driver—Data Sheet ADGM1004* https://analog.com/media/en/technical-documentation/data-sheets/adgm1004.pdf

[79] EETimes *RF MEMS Switch Specialist Menlo Micro Raises $44M* https://eetimes.com/rf-mems-switch-specialist-menlo-micro-raises-44m/

[80] Zorpette G Aug. 2020 ‘RF MEMS deliver the ‘ideal switch’: after two decades of development, MEMS-based RF switches are finally finding real-world uses,’ *IEEE Spectr.* **57** 8–9
[81] MenloMicro *Datasheet MM5130—DC to 26 GHz High Power RF Switch* https://menlomicro.com/images/general/Menlo_MM5130_Datasheet_v2.2.pdf
[82] MenloMicro *Datasheet MM5120 DC—12 GHz RF SP4T with Integrated Driver* https://menlomicro.com/images/general/Menlo_MM5120_Preliminary_Datasheet_Rev1.6.pdf
[83] MenloMicro *Datasheet MM5600 – 40 Gbps DPDT Differential Switch with Integrated Driver* https://menlomicro.com/images/general/MM5600_DPDT_Preliminary_Datasheet_v1.2.pdf
[84] MenloMicro *Datasheet MM3100—6 Channel SPST RF Micro Switch* https://menlomicro.com/images/general/Menlo_MM3100_Preliminary_Datasheet_Rev1.1.pdf
[85] TTI *Omron RF MEMS Switch* https://tti.com/content/ttiinc/en/manufacturers/omron/resources/mems-white-paper.html
[86] OMRON White paper: RF MEMS Switch: What You Need to Know—Structure and Usage of OMRON MEMS Switch 2SMES-01 https://tti.com/content/dam/ttiinc/manufacturers/omron/PDF/2SMES-01_RFMEMSSwitch_Whitepaper.pdf
[87] AAC Technologies *RF Antenna Solutions* https://aactechnologies.com/en/index.php?m=content&c=index&a=lists&catid=95
[88] Teledyne DALSA https://teledynedalsa.com/en/products/foundry/mems/applications/rf-mems/
[89] YOLE Développement Status of the MEMS Industry 2021 https://s3.i-micronews.com/uploads/2021/07/YINTR21180-Status-of-the-MEMS-Industry-2021_Sample.pdf

IOP Publishing

RF-MEMS Technology for High-Performance Passives
(Second Edition)
5G applications and prospects for 6G
Jacopo Iannacci

Chapter 3

The 5th Generation (5G) of mobile networks: challenges and opportunities of an impelling scenario with an outlook on the future 6th Generation (6G)

3.1 General considerations on mobile telecommunication networks

Throughout this chapter, the focus is going to move from Radio Frequency Micro-ElectroMechanical Systems (RF-MEMS) technology to plausible ways of its exploitation in a market scenario as wide and articulated as that of mobile telecommunications. In the previous chapter, the increasing confirmation of RF-MEMS technology in the current market of 4th Generation Long Term Evolution (4G-LTE) standard smartphones was sketched. From now on, the scope of the discussion will be around the intrinsic potential of RF passives in Micro-ElectroMechanical Systems (MEMS) technology for the future 5th Generation (5G) of mobile networks. Such a frame of reference is quite complex and dense with ramifications. In light of these considerations, the playground of mobile communications is going to be debated according to an upward/downward cross-reversed approach. If, on one hand, the technical complexity of the analysed items will increase, on the other hand, the level of abstraction will decrease, gliding from a simplified conceptualisation of the mobile network as a whole to the breakdown of some relevant portions, or tiles, composing the mosaic. The envisaged benefit of such an approach is twofold. First, it will make the reader more confident in strolling up and down the field of 5G. Then, it will help to consolidate multiple links between the high-level conceptual reference plane of a mobile network and the low-level complexity frame of RF passives, which are comparable, in this circumstance, to a bolt in a skyscraper. Complying with this strategy, the working principles of network cellular coverage are going to be recalled first.

The fundamental concept of cellular networks is to divide the space covered by the service into limited areas, addressed as *cells*. Each cell is served by a base station, i.e. a transmitting and receiving antenna, which is connected (wired) to the central network managing all the other base stations and the data/information flows across them. The schematic reported in figure 3.1 summarises how the setting of a cellular network typically appears.

In the example shown in figure 3.1, the transmitting mobile phone (labelled 'S', speaker) is located in cell A. The signal is received by the cell A base station. Through interrogation of all the other cells composing the network, the receiving mobile phone (labelled 'L', listener) is located in cell B. Therefore, the cell B base station delivers data to that specific handset. If the person listening to the conversation moves stepping from cell B to cell C, his/her mobile phone is unlinked from the base station of cell B and linked to that of cell C, in a totally transparent way, both to the speaker and listener [1, 2]. In spite of the remarkable evolution that telecommunication standards underwent across the timespan of the last four decades, the concept sketched in figure 3.1 is still valid today.

Moving from the network to the mobile phone level of abstraction, it is appropriate to briefly recall the handsets' architecture. The level of detail can be freely set on any plane, however exerting significant impact on the complexity of the scheme and, therefore, on the ease of understanding. For the purposes of this chapter, a high-level block diagram of the smartphone is preferred, splitting its parts based upon the implemented functionalities [3], as reported in figure 3.2.

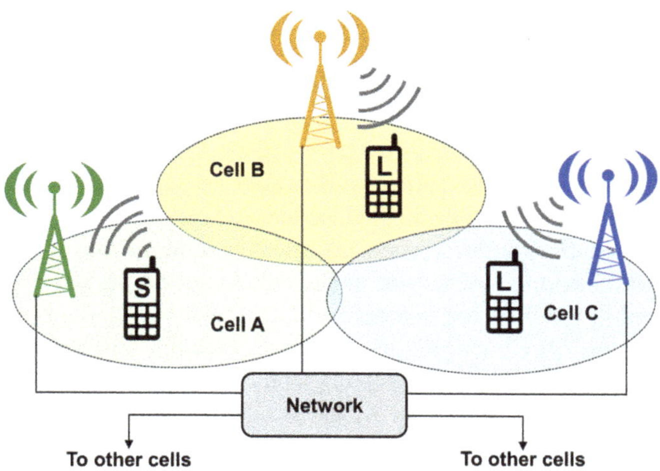

Figure 3.1. Schematic of a typical arrangement of a cellular mobile network. The area is divided into cells, each of which is covered by a base station. The label 'S' identifies the speaker (mobile phone transmitting data), while the label 'L' identifies the listener (mobile phone receiving data).

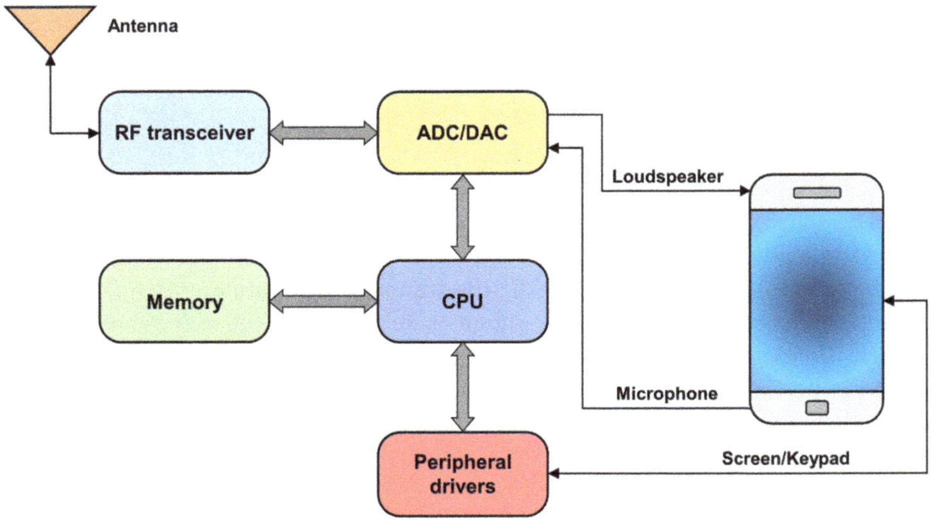

Figure 3.2. High-level block diagram of a modern smartphone. The block splitting is based on the functionalities implemented by each of them.

The RF signal is received from the antenna and treated by the RF transceiver, whose functionality and detailed architecture were previously discussed in this work. The transceiver provides the filtered and amplified baseband signal to the Analogue to Digital Converter (ADC) block. The latter, when dealing with signals to be transmitted, works as a Digital to Analogue Converter (DAC), performing the opposite function. The ADC/DAC block is controlled by the Central Processing Unit (CPU), which performs other treatment functions on the received signal in the digital domain, before driving it into the loudspeaker. Similarly, the voice collected by the microphone follows the opposite path through the ADC/DAC block (digital to analogue conversion) and the RF transceiver (filtering, amplification and frequency up conversion), before being transmitted by the antenna. The CPU also controls the memory unit and the peripheral drivers block, thus driving all the other onboard devices, like a touchscreen (input/output device) and sensors (gyroscopes, accelerometers, cameras, and so on). The schematic reported in figure 3.2 is notably simplified, as critical functions such as the encoding/decoding of signals, for instance, are not mentioned. However, it provides a sound and easy-to-parse insight into modern mobile phones' architecture.

Before concluding this introductory section on mobile applications, the level of abstraction is going to be taken for a while to the highest level. The target is to complete the overview provided up to here, with a brief insight into the overall socioeconomic environment related to mobile market applications, thus featuring all the profit and non-profit stakeholders. To this purpose, a schematic representation in figure 3.3 is provided.

The main pillars of the mobile applications ecosystem are embodied by the following entities: (a) networks and infrastructures; (b) mobile phones; (c) and end

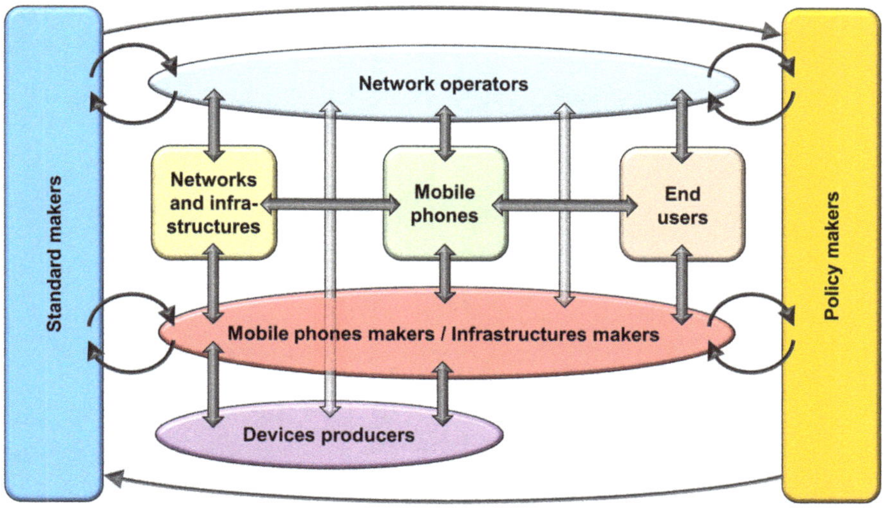

Figure 3.3. Schematisation of the socioeconomic environment featuring the most relevant categories of stakeholders, spinning around mobile applications.

users. As a matter of fact, very simply, end users buy and use mobile phones, which need proper infrastructures to work (horizontal arrows in figure 3.3). Above the pillars stand network operators, which interact directly with all the former entities (vertical arrows in figure 3.3). Network operators provide diverse services to end users, also including mobile phones. On the other hand, they manage, and very often own, mobile infrastructures. In the lower part of the scheme are mobile phone makers and infrastructure makers/providers. Mobile phone makers interact directly with end users, as they develop products complying with the needs and expectations of customers. On the other hand, infrastructure makers develop and provide network terrestrial infrastructures. At the very bottom stands device producers, often called Original Equipment Manufacturers (OEMs). Their role is to develop and sell components both to mobile phone and infrastructure makers, necessary in the assembly of their respective products. Network operators also interact with mobile phone/infrastructure makers as well as product makers (lighter vertical arrows in figure 3.3). This is because operators guide/influence the specifications and requirements mobile handsets and infrastructures have to reach.

The mobile ecosystem is populated by other two critical stakeholders, namely standard and policy makers. The former act on the definitions with which mobile communication standards have to comply, while the latter define the regulations to be applied to mobile services, in accordance to existing national and international laws, e.g. concerning frequency band allocations, electromagnetic compatibility, and so on. Just to mention a few of them, the 3rd Generation Partnership Project (3GPP) and the International Telecommunication Union (ITU) are examples of such classes of stakeholders. Standard and policy makers have mutual interaction with network operators and mobile phones/infrastructure makers, as well as between themselves.

Figure 3.4. First *portable* device for mobile radio communications (i.e. a radio signal strength testing system), dated back to 1924. Reused with permission of Nokia Corporation.

Their role is crucial both during the preliminary definition of standard specifications, as well as after the release and deployment of a certain generation of mobile services, for control/management purposes.

The latter description concludes the unwinding of the introductory elements meant to ease understanding of the next few sections. In the following, the evolution of telecommunication standards will be discussed, from the early days of mobile services, approaching in a smooth fashion the core topic of 5G.

3.2 Evolution of mobile standards and services

The focus is now going to be shifted to the progression of mobile telecommunication standards over the last four decades. The most relevant achievements, challenges and limitations of mobile services, starting from the 1st Generation (1G) to the currently in force 4G-LTE, will be listed, thus building a proper background for unfolding the scenario of targets and advancements to be reached by full deployment of 5th Generation (5G) standard. As a matter of fact, already starting from 1G, deployed across the turning point between the 1970s and the 1980s, the advancement of the technology was unambiguous with respect to the first 'mobile' device (i.e. a radio signal strength testing system), dated 1924 and shown in figure 3.4.

3.2.1 The 1st Generation—1G

The first commercial, fully automated cellular network generation (1G) for voice transmission was launched in Japan, in 1979, providing coverage of the full metropolitan area of Tokyo (more than 20 million inhabitants). The 1G network was expanded to cover the whole population of Japan within 5 years from its first deployment. The second launch of 1G networks was performed in Denmark, Finland, Norway and Sweden in 1981. Subsequent 1G networks were launched in the United Kingdom, Mexico, Canada and the United States in the early 1980s [4]. The main mobile phone cell systems standards spreading under the umbrella of 1G were the Advanced Mobile Phone System (AMPS), Nordic Mobile Telephone (NMT) and Total Access Communication System (TACS) [5].

1G established the fundamental pillars of mobile communications that, albeit undergoing significant evolution in terms of complexity, were to be exploited in the subsequent generations of cellular services. The general setting of 1G networks is the one previously depicted in figure 3.1. The transmission of voice was analogue and taking place over a licensed spectrum, i.e. a cleared spectrum exclusively allocated for mobile communications. The terrestrial infrastructure, i.e. base stations and network as reported in figure 3.1, was deployed by mobile operators [6]. 1G established seamless access to mobile handsets by means of a transparent backhaul network, as well as full integration with the Public Switched Telephone Network (PSTN). Moreover, since one channel was allocated for just a single voice signal, 1G introduced the concept of frequency reuse, as illustrated in figure 3.5.

Adjacent base stations used different frequencies in order to avoid interference. Nonetheless, in the span of a few cells, the same frequency was reused (cells with the same colour in figure 3.5). As mentioned before, 1G transmitted one voice signal per channel. Therefore, separation between different channels, i.e. among different voice conversations, was achieved by setting them apart over frequency, thus

Figure 3.5. Frequency reuse across non-adjacent cells, introduced in 1G to reduce frequency occupation without creating interferences. The same colour indicates the same frequency.

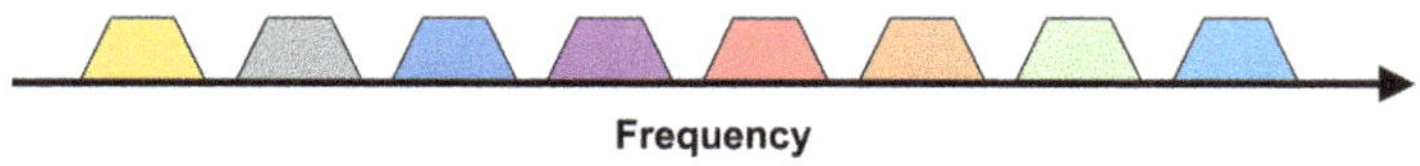

Figure 3.6. Schematic representation of Frequency Division Multiple Access (FDMA) between different voice channels adopted by 1G.

implementing so-called Frequency Division Multiple Access (FDMA), schematically shown in figure 3.6.

Each channel was allocated a 30 kHz band and needed certain clearance in terms of frequency distance from other channels, in order to avoid interference. In light of the aforementioned characteristics, 1G introduced significant innovation but was also characterised by several limitations. In the first place, the FDMA in figure 3.6 and the allocation of one signal per channel was definitely non-optimal in terms of spectrum usage efficiency. Hard handoffs in switching channel frequency, when passing from one base station to another, also caused dropped calls quite frequently. In addition, 1G was not secure at all, as voice calls were broadcasted with no encryption [7]. Further, an important point to note is that analogue handsets were bulky (with limited potential for down-scaling), power-inefficient and expensive.

3.2.2 The 2nd Generation—2G

The second generation of mobile networks (2G), released in the early 1990s, had the appropriate characteristics to trigger the spread of mobile services to large numbers of people. In contrast to the 1G standard, 2G exclusively dealt with digital signals. Stepping into the digital world enabled two significant advancements in terms of transmission efficiency. On one hand, digital voice signals could be encoded, thus significantly compressing the amount of data to be broadcasted. Just to report a few numbers, an uncompressed voice signal required 64 kb (kilobit) per second, while after encoding the necessary bitrate was reduced to 8 kb per second [6]. On the other hand, with 2G it was possible to transmit more than one voice signal over the same channel, overcoming the 1G limitation of one channel per one signal. This was possible by splitting different voice signals into packets of data and broadcasting them over a unique channel, according to so-called Time Division Multiple Access (TDMA). In particular, the Digital Advanced Mobile Phone System (D-AMPS), mainly used in North America, ensured the transmission of three different voice signals (talking one at a time) over the same 30 kHz wide channel. However, the actual wide spread of 2G mobile services took place with the introduction of the Global Systems for Mobile communications (GSM), which allowed transmission of eight different voice signals over the same 200 kHz wide channel [6]. Also, GSM allowed transmission of the first non-voice data, by means of Short Message Service (SMS) and e-mail exchange. The 2G D-AMPS and, in particular, the GSM, were both based on TDMA, so still suffered the same pitfalls as 1G due to the large frequency gaps needed between channels to avoid interference, as well as from frequently dropped calls in hard handoffs at the edge of one cell and another.

What radically improved the traffic volume capacity in the evolved protocols released under 2G, posing also as the basis of the 3rd Generation (3G), was the

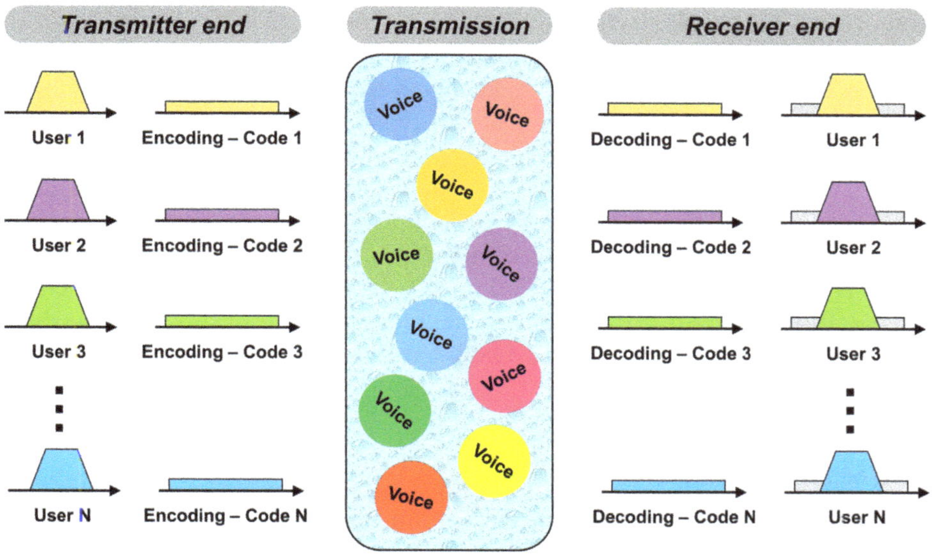

Figure 3.7. Schematisation of the CDMA transmission. Each user voice signal is encoded with a unique code and transmitted together with other voice signals. When received on the listener side, the voice signal is reconstructed (decoded) by using the same code. Signals related to other users appear as noise (below noise threshold).

development of the Code Division Multiple Access (CDMA) approach [8]. The rationale underlying CDMA is shown schematically in figure 3.7.

On the transmitter end, the voice signal of each user is encoded with a unique code, and transmitted together with another voice signal over 1.25 MHz wide channels. At the receiver end, the voice signal is reconstructed (decoded) using the same code originally exploited for encoding. This makes it possible to deliver the proper voice signal of the speaker to the proper listener, with all the other signals appearing as background noise. The CDMA exhibited several advantages in comparison to previous TDMA. First, voice capacity was significantly increased, as the whole frequency spectrum was finally exploited across its full extent. CDMA also improved the security of communications, as encoding made eavesdropping much more challenging. Furthermore, efficiency improvements in the transmission of data led also to an increase of battery life in mobile handsets, which was definitely a plus, especially in respect of 1G.

Nonetheless, a few technical challenges had to be solved before making CDMA successful for commercialisation. Among them, the near/far power challenge was one of the most relevant. When the user was close to the base station, the tower overpowered the uplink channel, thus significantly reducing the band available to other users located farther away from the cell infrastructure. This issue was circumvented by introducing continuous transmitted power control on the handset side, depending on the strength of the link signal to the base station. In addition, interference between adjacent base stations when the user was at the boundary between cells was also a relevant issue. This challenge was addressed by encompassing the possibility for the handset to communicate simultaneously with both towers, thus smoothing the handoff process. After addressing the mentioned problems, CDMA was released commercially in the mid-1990s, under the cdmaOne (IS-95) standard.

Despite the fact that CDMA was not already employed at a commercial level, 2G evolved further based upon TDMA, by introducing protocols with enhanced transmission data rates and with more pronounced use of packet switching besides traditional circuit switching. This brought advanced (sub-)generations of 2G, commonly referred to as 2.5G and 2.75G [5]. In particular, in 2.5G the General Packet Radio Services (GPRS) was launched (data rates up to 115 kbps), while in 2.75G the Enhanced Data Rate for GSM Evolution (EDGE) was also employed (data rates up to 384 kbps).

3.2.3 The 3rd Generation—3G

Capitalising on the important advancements introduced by 2G, 3G was launched in 2000. Before listing the achievements at service and implementation level brought by 3G, a couple of brief, general considerations are necessary. At the turn of the millennium, people were getting used to the Internet and having broadband access, both in the office and home. This scenario was also driving an increasing demand for Internet access via mobile phones. However, handsets' technologies were taking significant steps forward, introducing relevant innovations, not only in RF systems and transmission data capabilities, but also in relation to other components integrated in the device, such as cameras and screens with improved characteristics. Also, after the widespread distribution of 2G, more and more people had a mobile subscription, and most of them were expecting more capacity in terms of data access on the move. Therefore, it was foreseeable that 3G had to deal with all of the mentioned trends and demands.

From the viewpoint of voice transmission, 3G consolidated the exploitation of the CDMA technology by means of the CDMA2000 standard (channel width of 1.25 MHz). Nonetheless, while in 2G the transmission of data (for SMS, e-mail and news services) travelled together with the voice signal, 3G introduced a paired channel entirely dedicated to non-voice services. In particular, the Evolution-Data Optimised (EV-DO) introduced a data-only 1.25 MHz optimised channel for Internet, multimedia and navigation services. This solution ensured access to richer content and provided the main pillars of mobile broadband. In this regard, the EV-DO introduced the concept of adaptive modulation, which means using a higher order of modulation when the signal strength and quality are good, such as when the handset is close to the base station. Increasing the modulation order means improving the spectral efficiency in terms of bits per second per Hertz (bps/Hz) and, in turn, the amount of transmitted/received data, keeping the bandwidth fixed. Moreover, the management of transmission according to time slots, typical of EV-DO, enabled the implementation of opportunistic scheduling concepts. In other words, channel use was optimised, scheduling transmission of data when users established a good link to the base station in terms of signal strength.

The combination of CDMA2000 and EV-DO brought 3G to the further evolution of Wideband Code Division Multiple Access (WCDMA), Universal Mobile Telecommunications System (UMTS) and High Speed Uplink/Downlink Packet Access (HSUPA/HSDPA). In particular, the concurrence of the just-mentioned protocols brought an increase of data rate in the range from 5 Mbps to 30 Mbps, commonly called 3.5G [5]. In particular, the Evolved High Speed Packet Access (HSPA+) and EV-DO Rev. B (Revision B) ventured into the fields of higher-order modulation and Carrier Aggregation (CA). Concerning the former, Quadrature Amplitude Modulation (QAM) with 64 possible combinations (64-QAM) was explored, thus transmitting 6 bits per symbol, which led to an increase of about 50% in terms of bps/Hz. On the other hand, CA enabled one to extend the spectrum by combining diverse carrier signals and putting together different frequency bands and channels. These technologies, among other benefits, enabled reduced operator costs for data services and built a platform triggering continual enhancement of the functionalities provided to the final users. Eventually, further released standards, like Worldwide Interoperability for Microwave Access (WiMAX) and Ultra Mobile Broadband (UMB), led to a sort of seamless transition between 3G and 4G, across the so-called 3.75G.

3.2.4 The 4th Generation—4G

The scenario depicted in the previous pages elucidated the trends that mobile telecommunications have been seamlessly following for several years, stepping from one generation to the next, as well as within the same generation, introducing continual innovations and improvements. Given such a landscape, 4G Long Term Evolution (4G-LTE) is not an exception. Integration of more services (downloading, browsing, streaming, etc) under the umbrella of 4G, posed non-reversible demands in terms of faster and more efficient data-transfer capacity. In addition, another self-sustaining mechanism triggering the former demands should also be borne in mind. When mobile services supported by modern smartphones combine with good network performance and desirable content, the flywheel effect shown in figure 3.8 is established, leading to further increase of data consumed by end users [9].

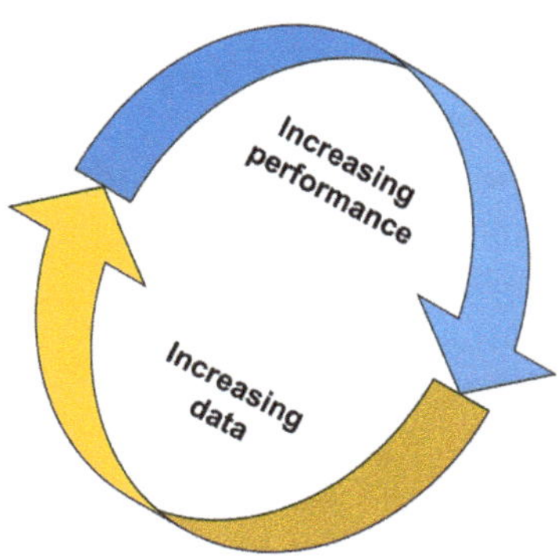

Figure 3.8. Flywheel effect triggering more and more data demand from the end-user side.

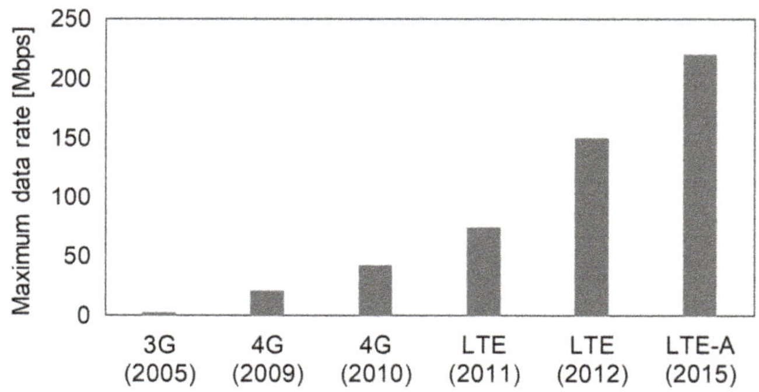

Figure 3.9. Comparison of maximum data rate capacity, in terms of Mbps, from 3G to early releases of 4G-LTE, also including LTE Advanced (LTE-A). The year of release is indicated per each generation.

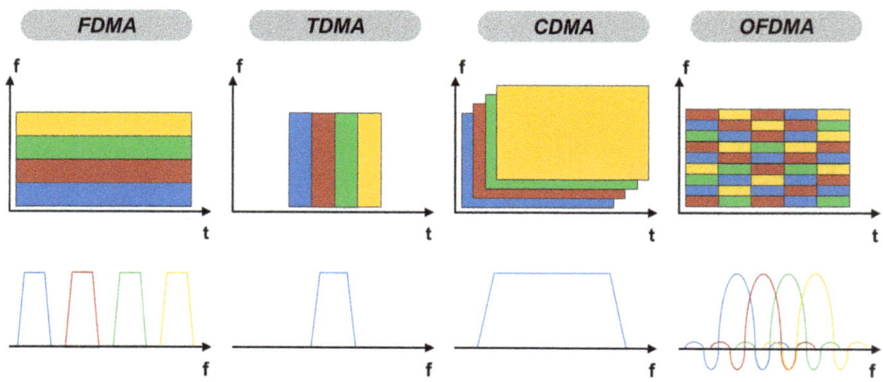

Figure 3.10. Comparison of FDMA, TDMA, CDMA and OFDMA, versus time and frequency. OFDMA exploits numerous subcarriers to build a band that is wider in comparison to other multiple-access strategies.

One can clearly understand the relevance of the abovementioned trend simply by looking at how maximum data rates (in terms of Mbps) increased when stepping from 3G to 4G, and then, within 4G-LTE, from one release to the next. For the benefit of this purpose, the plot in figure 3.9 shows the maximum data rates, also reporting the year of release [10].

Bearing in mind the scenario just depicted, 4G addresses the increasing demands, acting at different levels. From a technical point of view, 4G enables flexible support of channels up to 20 MHz by exploiting Orthogonal Frequency Division Multiple Access (OFDMA) [11]. The latter concept can be easily explained if compared to the other multiple-access techniques, previously discussed when listing former generations of mobile communication standards. To this purpose, figure 3.10 shows the time and frequency characteristics of FDMA, TDMA, CDMA and OFDMA. Each colour corresponds to the band allocated to a different user.

In the FDMA scenario, the band occupied by each user is a horizontal strip on the frequency versus time plot (top of figure 3.10), corresponding to separated bands on the frequency axis (bottom of figure 3.10). In the TDMA case, instead, each user has a vertical strip on the frequency versus time plot, yielding a unique frequency band on the bottom plot (as its occupation is split over time). However, CDMA exploits unique encoding/decoding to transmit, identify and reconstruct the signal of a certain user (see figure 3.7). Therefore, data of all users are broadcasted together over a unique frequency band (wider than in FDMA and TDMA). Instead, a radically different approach is pursued by OFDMA. The spectrum is fractioned down in a variable number of narrow bands (15 kHz wide in LTE) built around multiple subcarriers. The latter are orthogonal, meaning that the centre frequency of each narrow band intercepts nulls of adjacent subcarriers [11]. Each narrow band carries a limited amount of information, albeit the whole throughput is given by the sum of all the employed subcarriers, leading to a rather wide effective band. The significant advantage of OFDMA is that the width of the band associated to each user can be widened or narrowed depending on the amount of data to be transferred at a given time, thus making part of it available to other users when not utilised.

Returning to the technical innovations introduced by 4G-LTE, the coexistence of paired and unpaired spectra for Downlink (DL) and Uplink (UL) is also supported. In more detail, both Frequency Division Duplex (FDD) and Time Division Duplex (TDD) are possible. In FDD mode, the DL and UL spectra are paired and lie on different frequency bands. In contrast, in TDD mode, the DL and UL spectra are unpaired, lie on the same frequency band and are assigned different time slots [12]. Both FDD and TDD have pros and cons, also including space coverage, as reported in figure 3.11.

In particular, FDD enables coverage of large areas, while TDD makes asymmetry of DL and UL possible, thus extending the DL capacity when needed. However, TDD is able to cover reduced areas, mainly because the UL device power is used part of the time in TDD mode, but continuously in FDD mode.

To conclude the overview of technical trends followed by 4G-LTE, another two aspects must be mentioned. On one hand, further advancement of Carrier Aggregation (CA) techniques, previously introduced when discussing 3G, is leading and will lead to wider data pipes and, in turn, increased data rates. Also relevant, 4G-LTE capitalises on techniques aimed to sustain spatially separated multiple

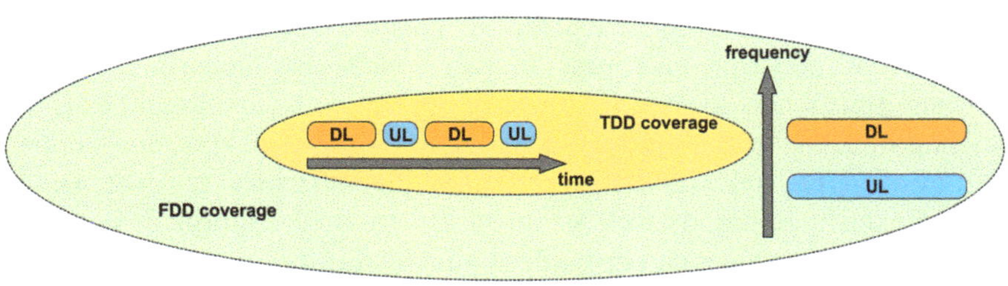

Figure 3.11. Schematic of Frequency Division Duplex (FDD) and Time Division Duplex (TDD) related to Downlink (DL) and Uplink (UL), also including space coverage.

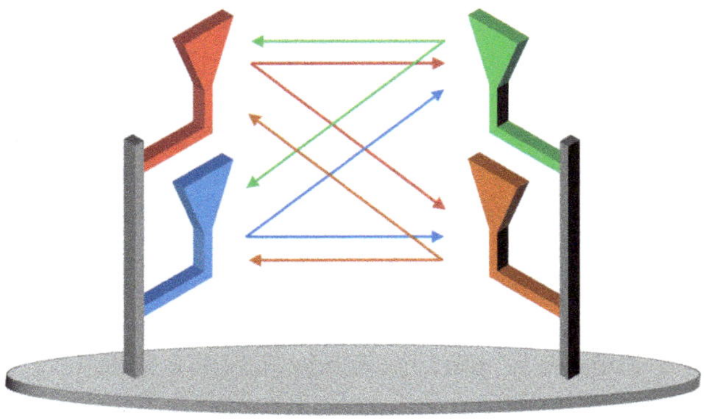

Figure 3.12. Schematic example of 2×2 Multiple Input Multiple Output (MIMO) solution, featuring two antennas both on the transmitter and receiver ends.

transmission paths. In this regard, Multiple Input Multiple Output (MIMO) techniques are massively pursued to enable multi-path broadcasting [13, 14]. They consist in having multiple antennas, both on the transmitter and receiver end, in order to transmit/receive in parallel through distinct and frequency diverse channels. The schematic in figure 3.12 shows the case of 2×2 MIMO, in which two antennas are featured at both ends.

The advantages introduced by employing MIMO solutions are multiple. As first instance, the overall data rate can be significantly increased. Moreover, interferences and/or non-optimal transmission/reception, due to the presence of obstacles or weak signals, can be significantly mitigated by exploiting different frequencies. In addition, handoffs at the cells' boundary could be made even smoother. Of course, these improvements can be further enhanced acting on the solution complexity. This is the reason why massive MIMO solutions are being investigated, featuring more than just two antennas for each end [9].

For the sake of completeness, it must be mentioned that 4G-LTE is pushing forward significant innovation in terms of infrastructure and network settings. By summarising in just three keywords the trends followed at high-level, it can be stated that 4G-LTE pursues *virtualisation*, *densification* and *diversification* of the network.

As demand and configuration requirements for terrestrial infrastructures are increasing in terms of flexibility and adaptability, the overall architecture is following a flattening trend that complies with the latter requests and simplifies, at the same time, the hardware architecture. This important target is addressed by virtualisation of network functions, together with the increasing adoption of Internet-based protocols to manage and reconfigure such functionalities [15, 16].

Recalling once again the increasing trends in data throughput and connectivity followed by 4G-LTE, network densification and diversification are other key points, intimately linked together, as well as correlated to virtualisation. It is straightforward that improved services can be made available only if the network is deployed in a capillary fashion (i.e. network densification) [17–19]. However, this target cannot be reached by simply increasing the number of base stations available, but rather by

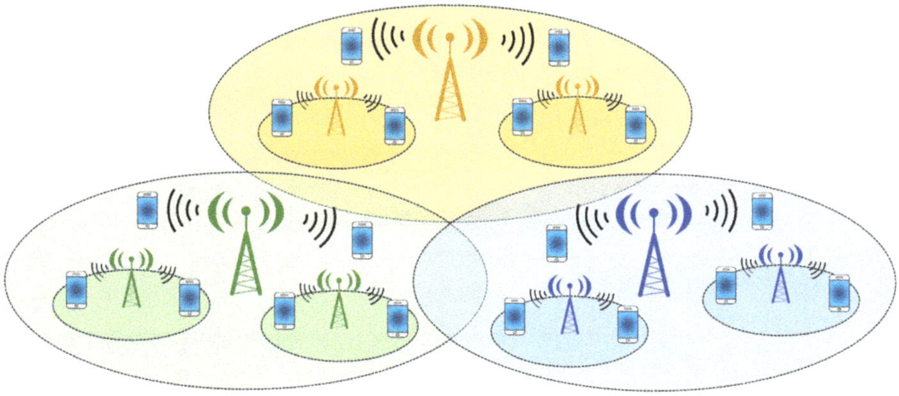

Figure 3.13. Schematic representation of the network diversification concept, including cells with different area coverage.

pursuing diversification of the network (i.e. heterogeneous network) at the same time. In practical terms, this means extending the backhaul with diversified radio towers, in terms of delivered power and covered area, and being able to provide higher throughputs and pronounced connectivity where more is needed (e.g. metropolitan versus rural areas) [20–23]. Network diversification led to the introduction of concepts like macro-, micro-, pico- and femtocells: addressing a variety of radio towers with scaled power, coverage and data volume capacities, realising the hierarchical arrangement in figure 3.13.

In order to conclude this section on 4G-LTE, a few relevant aspects addressed by its evolution are going to be reported in brief, by discussing some of the most relevant advancements at the technical level. Starting from the CA technique, introduced for the first time in Release 10 (in 2011) with two DL carriers, in the subsequent Release 14 it featured up to five DL and three UL carriers [9]. Concerning the order of modulation, 64-QAM in DL and 16-QAM in UL were initially introduced. In the following releases, they were increased to 256-QAM in DL and 64-QAM in UL. The Release 14 was aiming to reach 256-QAM in UL, as well. Furthermore, dealing with multi-paths, 4×4 MIMO have been implemented in the most advanced networks where, in combination with CA with three DL components and 256-QAM, peak data rates quite close to 1 Gbps were demonstrated.

3.2.5 Wrap up of mobile standards from 1G to 4G

This concluding part aims to collect the most salient features that were mentioned in previous pages, in order to enable an easier and more direct understanding of the differences and advancements, in the seamless evolution that brought, across more than four decades, 1G to 4G-LTE releases.

In the first place, it is interesting to report on the lifecycle of mobile generations across the whole timespan including research and technology development, standard

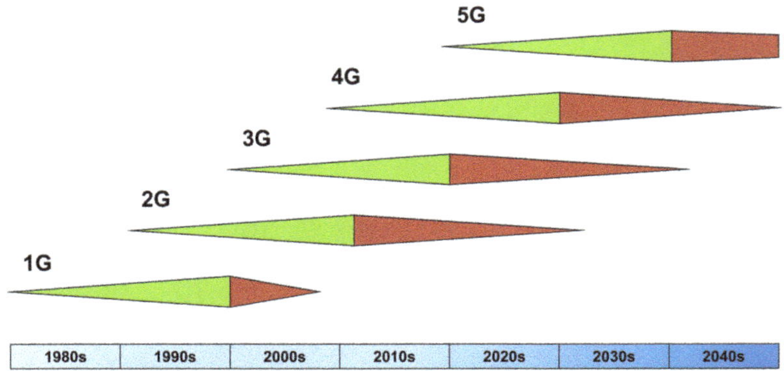

Figure 3.14. Expansion and contraction phase over time of mobile generations from 1G to 5G.

definition, deployment, commercial spread and, finally, decrease and fading-out from the market scenario. In this regard, an effective interpretation is provided in [24] and graphically sketched in figure 3.14.

The scheme reports a kite-like bar over time per each generation. The widening (green) part refers to the timespan from the beginning of development to the maximum spread in terms of market volumes and number of users. On the other hand, the narrowing (red) part provides a time indication concerning the fading-out of a certain generation, until its complete obsolescence and disappearance from the market. Looking at the graph in figure 3.14 a few considerations emerge quite evidently. First, the time necessary for a certain technology to reach its maximum commercial spread, starting from early experimentation and deployments, is as long as two decades. Also, surprisingly, although we are now fully immersed in the 4G-LTE and 5G roll-out has been ongoing for more than two years, the widest worldwide market of 2G started declining about one decade ago, while the decline of 3G commenced just a couple of years ago.

Finally, a comprehensive summary [5] of the main technical characteristics discussed up to now is reported in table 3.1, thus concluding the section.

3.3 The 5th Generation—5G

The ample overview provided on previous and current generations of mobile communications standards helps one to identify the main underlying trends that, despite the interchange between technical solutions and technologies, will continue on being pursued. From this perspective, it is straightforward that the current and future 5th Generation (5G) will be required to provide more data-transfer capacity and increased coverage. Nevertheless, in order to aggregate a more pragmatic vision around the needs that 5G will have to address, one should wonder about the metamorphosis the Internet has been going through in recent years.

We are living in a world where the surrounding environment and objects supporting us in daily life are seamlessly becoming more adaptive and interconnected. Services are broader, more capillary, linked together and easy to access. Experiences are tailored to our preferences and specific (as well as special) needs.

Table 3.1. Summary of the most relevant technical characteristics previously discussed for mobile generations from 1G to 4G-LTE.

Generation	Standard/technology	Data rate	Band	Bandwidth	Switching	Services
1G	AMPS; TACS; NMT—FDMA	2.4 kbps	800 MHz	30 kHz	Circuit	Voice
2G	GSM—TDMA; CDMA	10 kbps	850 MHz; 900 MHz;	200 kHz; 1.25 MHz	Circuit	Voice/data
2.5G	GPRS—TDMA; EDGE—TDMA	50 kbps; 200 kbps	1.8 GHz; 1.9 GHz	200 kHz; 200 kHz	Circuit/Packet	
3G	WCDMA; UMTS; CDMA2000	384 kbps; 384 kbps	800 MHz; 850 MHz; 900 MHz;	5 MHz; 1.25 MHz	Circuit/Packet Circuit/Packet	Voice/data/video calling
3.5G	HSUPA/HSDPA; EV-DO	5–30 Mbps; 5–30 Mbps	1.8 GHz; 1.9 GHz; 2.1 GHz	5 MHz; 1.25 MHz	PacketPacket	
3.75G	LTE—OFDMA/SC-FDMA[1]; Fixed WiMAX—SOFDMA[2]	100–200 Mbps; 100–200 Mbps	1.8 GHz; 2.6 GHz; 3.5 GHz; 5.8 GHz (initially)	From 1.4 MHz to 20 MHz; 3.5 MHz; 7 MHz in 3.5 GHz band; 10 MHz in 5.8 GHz	Packet	Voice/data/video calling/online gaming/high-definition television
4G	LTE-A[3]—OFDMA/SC-FDMA[1]; Mobile WiMAX—SOFDMA[2]	DL 3 Gbps, UL 1.5 Gbps; 100–200 Mbps	1.8 GHz; 2.6 GHz; 2.3 GHz; 2.5 GHz; 3.5 GHz (initially)	From 1.4 MHz to 20 MHz; 3.5 MHz; 5 MHz; 7 MHz; 8.75 MHz; 10 MHz	Packet	Voice/data/video calling/online gaming/high-definition television

[1] Single Carrier Frequency Division Multiple Access.
[2] Scalable Orthogonal Frequency Division Multiple Access.
[3] Advanced Long Term Evolution.

The attribute *smart* is now frequently associated with a diverse application of fields in science and technology, giving rise to familiar sobriquets like Smart Cities, Smart Homes, Smart Objects, etc The Internet of Things (IoT) paradigm portrays an ongoing technology development path through which any object and environment belonging to our daily life experience earns its own identity in the digital world by means of the Internet. The IoT is based on smart objects/environments with one or more of the following functionalities [25–27]:

- Self-awareness, i.e. identification, localisation and self-diagnosis of the object/environment.
- Interaction with the surrounding environment, i.e. data acquisition (sensing and metering) and actuation.
- Data elaboration, both basic (primitive data aggregation) and advanced (statistics, forecasts).

Regardless of the set of functionalities, all smart objects/environments must feature data-transfer capabilities (wired or wireless) in order to be networked and framed within the IoT. Despite the fact that the fields of application are widely diverse, it is possible to identify macro-areas for which IoT represents the common denominator, as follows: (1) Smart City/Environment; (2) Smart Home; (3) Smart Metering/Smart Grid; (4) Smart Building; (5) eHealth; (6) Smart Logistics; (7) Smart Factory; (8) Smart Asset Management; (9) Smart Agriculture; (10) Smart Car.

On a higher-level reference plane, the concept of IoT is better generalised by the paradigm of the Internet of Everything (IoE) [28], meaning that connectivity must be provided also without the perception of a physical object (i.e. a *thing*) as intermediate means. In other words, the IoE encompasses application domains such as Virtual Reality (VR), Augmented Reality (AR), remote (cloud) computing, Machine-to-Machine (M2M) communication and the Tactile Internet [29].

As the IoT/IoE weigh on the shoulders of 5G, it is clear that the demand in terms of data throughput will be massive. Many early predictions appealed for an enhancement in 5G transmission capacity as big as 1000 times with respect to 4G-LTE, delivering 10 Gbps to each individual user [30]. In addition, latency had been marked as needing drastic reduction to the millisecond level. In order to embrace the importance of the latter requirement, one can simply wonder about how low latency can be crucial for applications like Vehicle-to-Vehicle (V2V) communications. Finally, yet importantly, with leverage on M2M applications, cloud computing, IoT and so on, more symmetry between DL and UL data transmission capacity will be demanded for the 5G standard.

Beside speculations ignited by early visions of what the 5G was supposed to be, it is worth developing a comparison with what the 5G is starting to be at present time, in the early phase of its deployment. In order to keep the overview brief, the discussion is restricted to the geographic areas of the United States, looking at what the main operators have undertaken recently, i.e. AT&T (www.att.com), T-Mobile (www.t-mobile.com) and Verizon (www.verizon.com) [31]. In late 2019, AT&T launched its first small-footprint network, followed shortly after by T-Mobile with a 600 MHz 5G network. T-Mobile itself followed up with the first standalone 5G network in August 2020 [32], while Verizon announced a millimetre-wave (mm-wave) footprint at nearly the same time. In particular, thanks to the acquisition of Sprint, T-Mobile managed to provide a standalone 5G network working at 600 MHz and 2.5 GHz. The AT&T was tested for speed in the range of 200–300 Mbps, while the Verizon mm-wave network crossed the 1 Gbps threshold for distances up to 150 m [31]. Given this context, what is being currently pursued, and will persist in the months and years to come, is an increase of effort on the side of operators in getting standalone 5G networks and infrastructures.

From a different perspective, fundamental trends in 5G specifications still remain valid as originally indicated by the International Telecommunication Union (ITU), with reference to the International Mobile Telecommunications-2020 (IMT-2020) standard [33, 34], in light of which the main vectors defining the success of 5G have been defined as follows [35–37]:

- **Enhanced Mobile Broadband (EMBB)** is related to the system capacity growth, mentioned above, with the target of a 1000 times increase, leading to 10 Gbps peak and a minimum of 10 Mbps per each user. This aspect is also linked to the used spectrum, i.e. below 6 GHz and above 6 GHz 5G New Radio (NR) communications.
- **Ultra-Reliable Low-Latency Communications (URLLC)** is demanded by critical applications, like the aforementioned V2V communications and will aim towards the millisecond latency range.
- **Massive Machine-Type Communications (MMTC)** will lead to a very large increase of new connections to be supported by 5G services.

The way that these fundamental drivers are pursued cannot be unfolded in unsophisticated terms, as the complexity of the whole frame of reference is significantly pronounced. It is more convenient to decompose the overall scenario into the fundamental challenges and facilitating elements that can support them across the processes through which they are distilled in design fundamentals to be complied with, as proposed in [38, 39]. The plot of cross-relationships reported in figure 3.15 provides a graphical interpretation of the aforementioned concept.

In order to complete the information shown in figure 3.15, a summary of all the acronyms reported in the scheme is presented in table 3.2.

Pursuing the aforementioned main vectors through the strategic plot shown in figure 3.15 will be unavoidable in pushing forward the vision of the all-communicating world, already undertaken by currently used mobile services. For the sake of clarity, the six main challenges listed in figure 3.15, from the point of view of network-requested performance and characteristics, could be inflected as follows [40]:

1) Data volume increased up to 1000 times with a growth of indoor data traffic up to 70%.
2) Connected devices increased from 10 to 100 times.
3) Typical user data rate increased from 10 to 100 times.
4) Extended battery life up to 10 times for Massive Machine Communication (MMC) devices.
5) E2E latency reduced by 5 times.

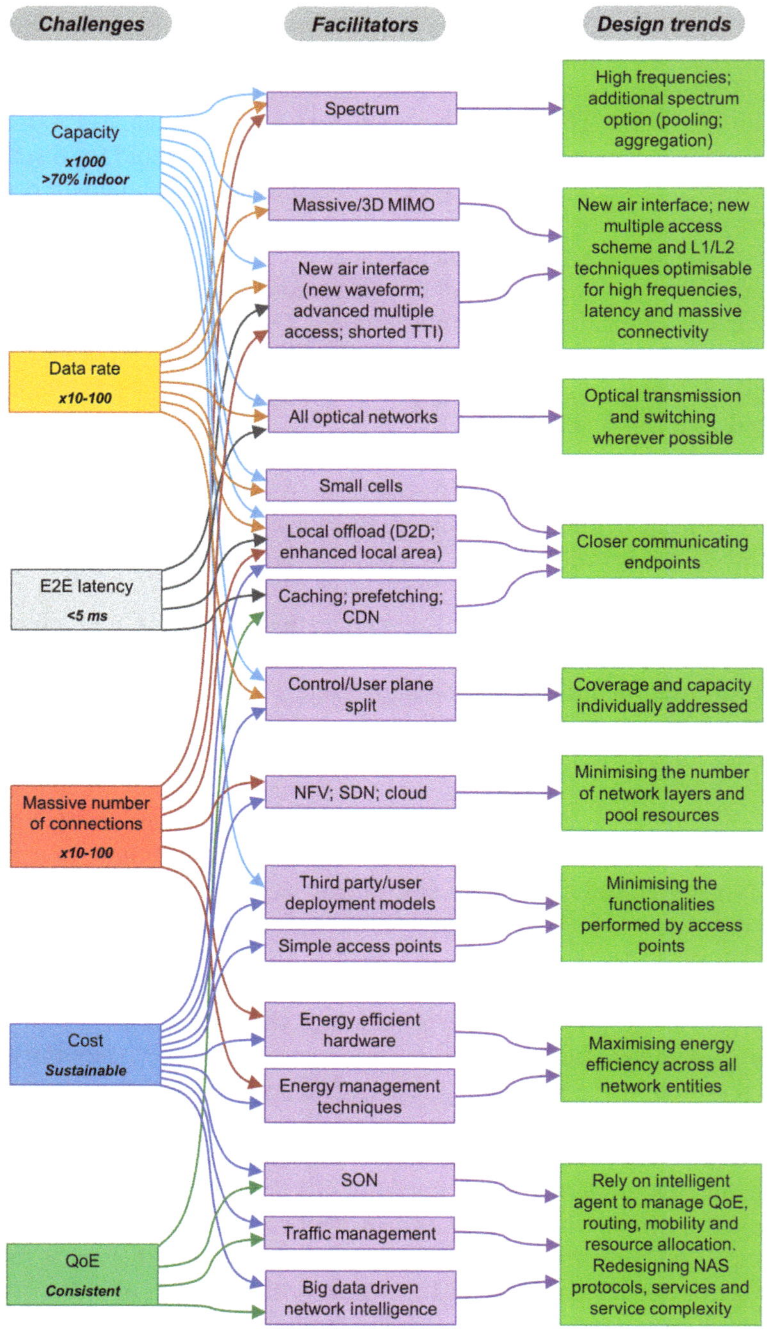

Figure 3.15. Plot of cross-relationships among 5G challenges, facilitators and design trends. © 2015 IEEE. Reprinted, with permission, from [5].

Table 3.2. List of all the acronyms appearing in figure 3.15.

Acronym	Extended form
CDN	Content Delivery Network
D2D	Device-to-Device (communication)
E2E	End-to-End
L1/L2	Layer 1/Layer 2
MIMO	Multiple Input Multiple Output
NAS	Non-Access Stratum
NFV	Network Function Virtualisation
QoE	Quality of Experience
SDN	Software Defined Network
SON	Self-Organising Network
TTI	Transmission Time Interval

The challenge of providing a common connected platform that is able to reach the goals listed will require acting at diverse technical levels, among which the most relevant ones can be identified as reported in the following [40]:

- *Radio links.* New transmission waveforms and new approaches for multiple-access control and resource management.
- *Multi-node/multi-antenna transmissions.* Design of multi-antenna transmission and reception technologies, capitalising on massive antenna configurations and enabling advanced inter-node coordination schemes and multi-hop technologies.
- *Network dimension.* Novel approaches for efficient interference management in complex heterogeneous deployments.
- *Spectrum usage.* Operation in extended spectrum band and in new spectrum regimes, to provide a complete system concept able to address specific needs of diverse usage scenarios.

To complete the discussion, a visual representation of how the 5G network could be architected is shown schematically in figure 3.16.

In order to complement the information shown in figure 3.16, a summary of all the acronyms reported in the scheme is presented in table 3.3. As visible, the previously mentioned concepts of network virtualisation, densification and diversification, already being addressed by 4G-LTE, are going to be further developed in the expanding 5G scenario. In particular, high-data-rate services are forecasted to be made available to end users by massive MIMO solutions and capillary deployment of small cells, implementing the concepts of Moving Networks (MNs), Ultra-Dense Networks (UDNs) and Ultra-Reliable Networks (URNs). Small cells, in particular, will implement the actual 5G NR paradigm, since they are envisaged to work in the millimetre-wave (mm-wave) frequency range, as mentioned below.

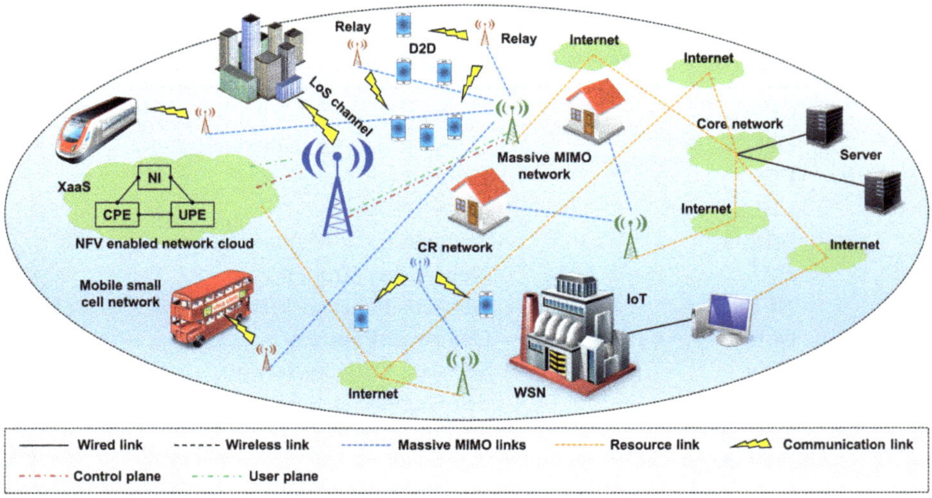

Figure 3.16. A general vision of how a 5G cellular network could look. © 2015 IEEE. Reprinted, with permission, from [5].

Table 3.3. List of all the acronyms appearing in figure 3.16.

Acronym	Extended form
CPE	Control Plane Entity
CR	Cognitive Radio
D2D	Device-to-Device (communication)
IoT	Internet of Things
LoS	Line of Sight
MIMO	Multiple Input Multiple Output
NFV	Network Function Virtualisation
NI	Network Intelligence
UPE	User Plane Entity
WSN	Wireless Sensor Network
XaaS	Network Functionalities as a Service

Bearing in mind the complex scenario depicted up to now, a few critical enabling technologies are going to be briefly discussed in the following subsections.

3.3.1 Massive MIMO

MIMO solutions rely on multiple paths for transmission/reception of signals by means of duplicated antennas arranged in array or matrix fashion. The reasons why

MIMO technology is considered a key enabler of 5G are multiple and act at diverse levels. Massive MIMO has the capability of improving the radiated energy efficiency up to 100 times and to increase, at the same time, the complexity order capacity of 10 times and more [41]. With the increase of antennas, the energy can be concentrated in small regions and the principle of coherent waveforms' superposition can be exploited. Ideally, the wave fronts emitted from the source antenna could add constructively at the intended terminal location and destructively elsewhere, thus reducing interference and energy losses. Moreover, massive MIMO systems can be arranged together with low-power and less costly components. In this regard, MIMO devices require hundreds of inexpensive Power Amplifiers (PAs) in the mW range, rather than ultra-linear, very expensive 30–50 W range PAs, currently in use in base stations [41]. In addition, having channels that individually require low-power levels opens the floor to the exploitation of renewable sources and of environmental Energy Harvesting (EH) techniques [42–46]. From a different point of view, massive MIMO can enable latency decrease, through beamforming and fading dips avoidance, as well as simplifying multiple access, due to the combination of OFDM and MIMO [41]. Eventually, MIMO solutions can increase the strength both against unintended interference as well as against intended jamming. If redundant degrees of freedom are available, they can be exploited for cancelling interference or signals generated by intended jammers.

3.3.2 Interference management

As can be easily pictured, with the increase of traffic capacity, network densification and co-use of the same hardware platforms, interference can turn out to be a critical and even impairing aspect for the whole 5G scenario. Therefore, efficient interference management schemes are of vital importance. Two (among other) techniques are briefly mentioned in the following. One solution relies on advance interference management at the receiver [47]. If advanced receivers are available, it could be possible to detect and even decode the symbols belonging to interference within the constellation and the coding scheme of the actual signal to be received. Therefore, the interference signal could be reconstructed and cancelled by the receiver. Another solution to manage interference is the so-called joint scheduling [47]. Basically, additional performance improvements can be enabled by synchronised broadcasting among multiple transmitters scattered in different locations.

3.3.3 Spectrum sharing

Increasing demands driven by 5G in terms of wideband availability will also reflect in spectrum sharing appropriate techniques in order to be fulfilled. In fact, spectrum sharing can take place according to horizontal or vertical approaches, and 5G gains from both options. Horizontal spectrum sharing refers to the access of several systems with similar access rights to the same resource, such as, for instance, a Wireless Local Area Network (WLAN). However, vertical spectrum sharing involves several systems with hierarchical access rights [48, 49]. Given such a background, there exist two fundamental spectrum-sharing techniques of interest for

5G, namely, distributed and centralised solutions [49]. Distributed solutions are meant to exclusively manage transmissions that generate interference and are based on the frequent transmission of generally understood signals across multiple devices, in a listen-before-talk fashion, with the aim of avoiding conflicting activities on the shared spectrum [50]. But centralised spectrum-sharing techniques rely on access management operated at a higher hierarchical level, and are more suitable for complex and highly granular networks. An example is the geo-localisation database method, in which a database has to be checked in order to obtain information about the resources available at a certain location [51, 52]. Another centralised solution is that of the spectrum broker, in which horizontally sharing systems negotiate short-term grants to access and use spectrum resources with a central management unit [53, 54]. Bearing in mind the spectrum-sharing issue from a general perspective, a key enabling technology for its proper solution is Cognitive Radio (CR), discussed in more details below. It can sense the environment and perform, in dynamic ways, adaption of networking protocols, of spectrum utilisation, channel-access methods and transmission waveform [48].

3.3.4 Device-to-device communication

D2D communications are categorised at the macro-cell level and device level. In the former case, a link is established between a base station and one or more remote devices while, in the latter circumstance, communication takes place directly across devices, i.e. without the involvement/mediation of a base station. 5G will be based upon macro-cell-level D2D communication protocols, complementing them with device-level communication solutions, boosting the overall performance in critical contexts, such as congested areas and at the cell edges [55, 56]. Bearing in mind the macro-cell level and device-level scenarios, four main strategies for D2D communication can be envisaged, taking advantage of different mixing of the just-listed categories, as follows:

1) *Device relaying with base-station-controlled link formation.* Let us imagine two mobile terminals laying within the same cell, with one device close to the base station and another at the cell edge. The first device has, of course, a strong link to the radio tower, while the other does not. Both terminals are exchanging data with the base station, but the one at the cell edge is unable to do it efficiently because of the poor link. Given this scenario, D2D communication can take place directly between the two mobile terminals, with the one at the cell edge exploiting the other close to the base station as a relay (i.e. bridge) to communicate more efficiently with the radio tower. In this context, both direct D2D and device-to-base station communications are managed and supervised by the radio tower itself (i.e. base station operated control link).
2) *Direct D2D communication with base-station-controlled link formation.* In this context, direct D2D communication takes place between two terminals anywhere located within the same cell, with no need to exchange data with

the base station. Nonetheless, direct D2D communication is supervised by the base station that manages the control link.

3) *Device relaying with device-controlled link formation.* In this scenario, several mobile terminals, randomly located in space, communicate among themselves. In particular, if one device has to establish a direct link to another terminal, but they are not in sight or the quality of signal is poor, other terminals physically located among them can serve as multiple relays (i.e. bridges) to establish a good connection between the sender and the receiver. In this type of direct D2D communication the control link is also managed by the remote devices themselves. In other words, the base station is not involved in this type of communication.
4) *Direct D2D communication with device-controlled link formation.* Finally, this scenario can be interpreted as a particular (and simplified) case of the context discussed in the previous point. Direct D2D communication takes place between two mobile devices (sender and receiver) that also manage the control link (i.e. the base station is not involved).

Massive exploitation of direct D2D communications bears the potential to strengthen the quality of service coverage, virtually making it excellent regardless of the location. As a matter of fact, wherever the link to the base station is poor, for instance because of physical obstacles, the distance to the radio tower or interference, there will be other mobile terminals around that can work as relays. Nevertheless, this poses significant concerns in terms of privacy and security, as user data will travel across devices belonging to other users. In any case, research in the field of M2M communication security will be able to provide effective solutions to mitigate this issue [57–61].

3.3.5 Millimetre-wave solutions

The 5G requirements in terms of increased capacity, data rates in the Gbps range, ultra-low latency, and so on, together with the crowded usage of the sub-6 GHz frequency range, are drawing attention towards the exploitation of solutions in the mm-wave range [62, 63]. From the technology point of view, the feasibility of mm-wave wireless communications is supported and enabled by significant advances in Complementary Metal Oxide Semiconductor (CMOS) [64–66] and MEMS-based solutions [67–69]; also concerning high-gain and steerable antennas. On the other hand, 5G's increasing need for pronounced directivity of wireless streams, also to mitigate interference and optimise power consumption as discussed earlier, is pushing in the direction of advanced beamforming capabilities. The exploitation of mm-wave carrier frequencies will enable larger bandwidth allocations, overcoming the previously used 20 MHz channels of 4G-LTE, ensuring at the same time significant reduction of latency. Also, with the increase of frequency, the wavelength reduces, making the scenario of arrays of antennas for massive MIMO with small area occupation possible [70]. As a matter of fact, the mm-wave range also exhibits characteristics making it poorly indicated or unsuitable for cellular communications.

Among them, the most critical ones are unquestionably the path loss effect [71–73], e.g. due to foliage loss, absorption caused by atmosphere and rain [74, 75], poor penetration across obstacles [76, 77], objects and barriers like walls. However, the mm-wave spectrum is appropriate for short-range transmissions [78], thus enabling all the aforementioned pronounced characteristics that 5G will have to meet. Bearing in mind this scenario, a trend to the way of consolidation is the frequency diversity across the backhaul portion of the network hierarchy. In this regard, a clear frequency divide will characterise 5G networks. Classical macro cells, covering rather extended areas, will mainly work in the sub-6 GHz range. On the other hand, remarkable data throughput, capacity and low latency, mentioned above, will be enabled via significant network densification. To this end, small cells will be deployed, covering very limited spaces, like a single building or small metropolitan areas (e.g. the lobby of a train station or of a shopping mall). Such small cells will enable massive data transfer working in the mm-wave range, i.e. well above 6 GHz (up to 60–70 GHz), exploiting massive MIMO solutions and advanced beamforming capabilities [79–82].

3.3.6 Cognitive radio

The issue of spectrum scarcity is relentlessly emerging, due to the spread of more and more bandwidth-hungry applications and services. Such a trend is in full contrast to the classical approach of assigning a certain frequency band portion to users and organisations in a static fashion, for a long period of time and over wide geographical areas. Given this frame of reference, the concept of Software Defined Radio (SDR), i.e. of radio systems which can be agilely and widely reconfigured via software [83, 84], has been evolving in that of Cognitive Radio (CR). The latter one is the idea of a radio that holds the capacity of varying its parameters on the basis of interaction with the surrounding environment in which it operates, thus pursuing the aim of maximising the spectrum utilisation [85, 86]. The mentioned interaction can involve, for instance, active communication and negotiation of spectrum usage with other users, along with passive sensing and decision making within the radio itself, when all the needed information is collected [87]. An important aspect to be stressed about CRs is that typically such systems operate in networks where the users exploiting a CR are identified as Secondary Users (SUs), since Primary Users (PUs) are those using the traditional fixed allocation of spectrum. As it is straightforward to envisage, PUs have priority in spectrum exploitation with respect to SUs. This establishes how critical the ability to negotiate proper time slots and modes for accessing unused spectrum can be in conceiving CR systems.

3.3.7 Massive centralised RAN

The way mobile networks have been developed up to this day has hinged around the fundamental concept of distributed Radio Access Network (RAN) architecture [88]. Within it, each Remote Radio Unit (RRU), typically placed at the top of a radio tower, is connected to a unique and dedicated Baseband Unit (BBU), accommodated at the base of the tower. The BBU and RRU are connected with a point-by-point cable (typically an optical fiber or a coaxial link) [89]. Despite representing a robust and consolidated solution, the distributed RAN approach exhibits intrinsic limitations, both in terms of redundancy (and cost), as well as of energy efficiency. The opposite strategy is that of the so-called Centralised RAN (C-RAN), also referred to as Cloud RAN, in which the number of BBUs is significantly decreased, as each BBU is no longer allocated exclusively to one RRU but serves more RRUs [90, 91]. In a nutshell, the C-RAN approach, among the other advantages, allows significant room for Network Function Virtualisation (NFV). Therefore, it is straightforward that evolution of 5G is pushing towards the employment of partial C-RAN [92–94], in which non-static links are established between RRUs and BBUs. However, the main pros and cons of C-RAN versus a distributed RAN approach deserve a more insightful discussion to be properly framed. Starting from the advantages of centralised over distributed RAN, the former is certainly a key to energy-efficient infrastructures. In C-RAN, BBUs are a shared resource, therefore enabling optimised load allocation, also including the option of switching off certain BBUs when their operation is redundant [95, 96]. Given such preconditions, it is straightforward to grasp how C-RAN, capitalising on pooling of BBUs and optimised scheduling, leads to cost savings [97, 98] and enhanced utilisation of the spectrum [99], along with the possibility, on a more speculative reference plane, of enabling novel business models. On the other hand, a few critical aspects exist in the C-RAN approach. First, transport becomes a non-trivial matter, as the distance of high-bandwidth links between RRUs and BBUs (that in distributed RAN is from the base to the top of the tower) can increase up to tens of kilometres. This might require the presence of repeater sites [100], placed between the pool of BBUs and each RRU, to compensate attenuation. In any case, a combination of wired/free-space optical and wireless RF links has to be regarded as the most viable solution, in order to be more flexible in facing constraints that might arise from the shape of the terrain, from the presence of natural/artificial obstacles, along with cost considerations [101–104]. In particular, with reference to RF links, an interesting opportunity is represented by the exploitation of MIMO technology, also in combination with Line Of Sight (LOS) solutions [105–107]. Other issues linked to the C-RAN approach are connected to real-time baseband processing algorithms, as well as with the dynamic processing capacities necessary to manage the dynamic load of cells [108]. In addition, the fact of having a pool of BBUs physically located close to each other raises issues in terms of security and demands for effective recovery strategies [109, 110].

3.3.8 Vehicular communications

The technology area of Vehicular Communications (VC) employs 5G as a pillar for whatever is concerned in the transmission of data and information. Breaking down the general scenario of VC, four different types of communications are possible, them being Vehicle-to-Vehicle (V2V), Vehicle-to-Network (V2N), Vehicle-to-Infrastructure (V2I) and Vehicle-to-Pedestrian (V2P) communications, and generally grouped under the label of Vehicle-to-Everything (V2X) communications [111, 112].

The evolution of VC technologies, which have been ongoing for two decades, has led to the development of a standard, named Dedicated Short-Range Communication (DSRC) that is able of addressing the V2V and V2I cases [113–115]. From a different perspective, Cellular V2X (C-V2X) solutions are also emerging [116–118], massively relying on cellular communication technologies, both for what concerns direct communications (V2V, V2I, V2P) and network-based communications (V2N), with the latter particularly centred around 5G [119–121]. Given such a scenario, it is worth stressing that VC applications, for which low-latency and high-reliability of communication are critical, like Automatic Driving and Advanced Driver Assistance Systems (ADAS), can more significantly benefit from C-V2X solutions, especially featuring 5G, rather than DSRC protocols [122]. From a more general point of view, VC pose a set of articulated challenges that still need to be fully addressed. Among them, robust and resilient security mechanisms are necessary in order to prevent faulty conditions, like false or missed warnings, as well as external attacks. Moreover, effective and reliable vehicle identification methods have also to be developed, along with proper handover approaches, when switching from cellular infrastructure to Roadside Units (RSUs) [123, 124].

3.3.9 Network slicing

As already pointed out in the previous pages, network softwarisation is a consolidating trend that pushes an evolution of design, implementation and operation of networks, leveraging software programming [125, 126]. It empowers flexibility and modularity in order to create logical networks through technologies like NFV and Software Defined Networking (SDN), effectively defined as network slices [127–130]. Stepping into a more detailed definition, network slicing identifies logical entities exhibiting a certain number of attributes, among which the most relevant can be listed as follows:

- Operating on a physical or virtual network.
- Optimising the network utilisation with respect of each identified application scenario.
- Exploiting a certain amount of access and core network functions.
- Being independently managed and controlled.
- Being generated on demand.
- Avoiding interferences and/or conflicts with other coexisting slices, services and functions.

Bearing in mind such features, network slicing falls way beyond the elemental concept of network sharing. Architecting a network slice means enabling additional functionalities in the control plane (C-plane) and user plane (U-plane), aiming at proper provision of resources over the infrastructure to run the virtual network properly [131, 132]. To this end, a more pronounced separation between the C-plane and the U-plane is also necessary, ensuring the avoidance of conflicts in terms of resources for multiple slices running over the same infrastructure [133, 134]. The same philosophy of increased separation should be implemented between the software and hardware composing the infrastructure, leading to augmented resilience of the network and of the implemented functions, in case of malfunctioning or failure of a physical component, or sub-portion of the whole infrastructure.

3.3.10 Other technologies

In order to conclude this section, a few additional technologies regarded as 5G key enablers and not mentioned in the previous subsections are going to be briefly listed. The items reported below are not to be intended as solutions, bearing limited value with respect to those discussed up to now. On the other hand, they are mentioned here briefly, as an in-depth discussion falls out of the scope of this work. Such technologies can be categorised as follows:

- *Ultra-Dense Networks (UDNs)*. In light of the previously mentioned densification and diversification trends of the 5G network, the characteristics in terms of data rate, accessibility, throughput, and so on, will be relevantly boosted. However, dense and dynamic heterogeneous networks will give rise to challenges related to interference, mobility and backhauling. In order to overcome these issues, the design of new network layer functionalities will be necessary, besides the classical design of physical layers. Such software/algorithm-based functions will help smart devices decide how to manage connectivity on the basis of context information [135, 136].
- *Multi-radio access technology association*. Still referring to network diversification, 5G will unavoidably lead to integration among different radio access technologies. Apart from sub-6 GHz and mm-wave services, future devices will have to support 3G services, various releases of 4G-LTE, different WiFi standards, as well as D2D communications [137]. In this regard, several ongoing studies aim to address in combined ways the issues of user association and resource allocation [138, 139].
- *Full duplex radio*. An intricate challenge at the technology level is that of enabling full duplex systems, i.e. radios able to transmit and receive simultaneously on the same channel. The availability of such a technology would definitely boost the throughput, but poses at the same time challenging issues in minimising and possibly removing self-interference between the receiver and the signal generated and transmitted by the same device [140–142].
- *Cloud technologies for flexible 5G radio access networks*. Eventually, cloud computing and technologies will be increasingly integrated with mobile devices and services, also transcending the classical meaning of the cloud. For instance, virtual cloud resources could be aggregated in limited areas by exploiting other mobile devices in the vicinity [143]. Moreover, local 'cloudlets' could be installed in public areas, in order to provide access to mobile terminals in the vicinity with lower latency and wider bandwidth, with respect to classical remote cloud servers [144]. Besides, more and more functionalities that now are executed at the base station level will be virtualised and centrally executed at cloud level, thus boosting flexibility and easing access to the 5G network [145].

3.4 Beyond 5G—an outlook on the emerging trends and early visions of the Artificial Intelligence (AI)-driven 6G of the future

Wrapping up the discussion previously developed in this chapter, it is out of the question that the currently incipient 5G is not lacking in terms of challenges and demands, and that it is a bearer of disruptive potential, as well. And yet, it does not represent the ultimate frontier, as further milestones are already being set at the time of writing. In a nutshell, in about one decade from now, the 6th Generation (6G) will take shape, acting as a pillar of the Super-Internet of Things (Super-IoT), also referred to with the name of the Tactile Internet (TI) [146–149].

3.4.1 A brief overview of 6G and Super-IoT/TI current visions

Before stepping into the description of 6G, it is worth spending a few words on the Super-IoT and TI scenarios, as they are emerging as driving applications for future telecommunications. The Super-IoT covers a plethora of applications leveraging connected robotics [150], like remote surgery [151, 152], for which ultra-low latency in the transmission of data and very high reliability are critical. According to some definitions, future applications involving robotics define a subset of the Super-IoT identified by the TI name. From a different perspective, the TI can be defined with a much broader meaning, making it equivalent to the Super-IoT, capitalising on the fact that the pursued characteristics of latency and reliability are so pushed forth that the interaction between the end user and the applications (as Virtual/Augmented Reality, or VR/AR) will resemble the physical sense of touch. Beyond a matter of mere definitions, complying with the challenging demands of Super-IoT and TI in terms of data transmission will push towards operation at high frequencies, like mm-wave and even sub-THz ranges, up to 100–300 GHz [153–155]. This scenario already looks quite challenging for 6G, despite it being envisaged to be much more than just a faster 5G.

The backbone of 6G's rationale is perfectly framed in [156], in which it is depicted as an evolution from the present of 'connected things' to a future of 'connected intelligence'. Such a valuable piece of literature stresses how 6G will step way beyond mobile Internet, thanks to the massive employment of Artificial Intelligence (AI) to trigger ubiquitous services, from the core to the periphery of the network, i.e. from the cloud to the edge.

It is straightforward that the central role played by AI in the framework of 6G is unavoidably connected to the services that 6G itself will provide to the end users. Nonetheless, the unprecedented step forward is that AI is predicted to play a novel role in the design and optimisation of portions of 6G network architectures, protocols and functionalities, across their operation, that is, in compliance to a runtime/real-time approach [157–161]. In other terms, AI will lead the evolution both of Super-IoT/TI and 6G functionalities (front end or services level), as well as of the network infrastructure (back end or operations level), following a twofold approach [162].

Table 3.4. Summary of key performance specifications in the transition from 5G, to Beyond 5G and 6G, as reported (more extensively) in [154].

	5G	Beyond 5G	6G
Reference devices	Smartphones, Wearables, Drones	Smartphones, Wearables, Drones, XR devices	CRAS[a], XR[b], and BCI[c] devices, Body implants
Data rate	1 Gbps	100 Gbps	1 Tbps
E2E[d] delay requirement	5 ms	1 ms	Below 1 ms
Radio delay requirement	100 ns	100 ns	10 ns
Processing delay	100 ns	50 ns	10 ns
E2E reliability requirement	99.999%	99.9999%	99.99999%
Frequency bands	Sub-6 GHz, mm-wave (fixed access)	Sub-6 GHz, mm-wave (fixed access)	Sub-6 GHz, mm-wave (mobile access), THz (up to 300 GHz), Non-RF (e.g. optical, VLC[e], etc)

[a] Connected robotics and autonomous systems.
[b] Extended reality.
[c] Brain–computer interaction.
[d] End-to-end.
[e] Visible light communication.

Wrapping together the cornerstone concepts developed above, the future 6G will build upon novel pillars, such as (self-)adaptivity and (self-)evolution of the network, never experienced before, through the whole six-decade span of the history of telecommunications. It is foreseeable that the impact on the technologies needed to sustain the mentioned trends is of paramount importance. In this regard, it has to be stressed that a seamless development is already ongoing in relation to software-centric Information and Telecommunication Technologies (ICTs). Nonetheless, given the extent of future 6G, a more radical and disruptive leap will have to be taken concerning Hardware (HW) technologies, very likely leading also to a partial reformulation of the concept of HW itself. This latter item will be better framed in the following pages. To conclude this section, a comparison of the main performance indicators of 5G, Beyond 5G and 6G, as reported in [154], are listed in table 3.4.

3.4.2 Emerging technology trends driving 6G

The emerging visions of what 6G will be, scatter along the way so many challenges in terms of Technologies, Applications, Specifications and Services (TASS), that thinking to address them utilising the classical approaches explored up to date is simply not an option. As already pointed out before, the extensive employment of AI in the frame of 6G networks will trigger an unprecedented transition from connected things to connected intelligence [156]. Bearing this in mind, it is worth reviewing a few relevant emerging trends and performance metrics of 6G, leveraging the comprehensive discussion provided in [154]:

(1) *Increase of the Bits, Spectrum, Reliability (BSR) multi-dimensional space.* The first visions of 5G used to set the reference data rate target to 1 Gbps (per user) [163], while now, i.e. a few years later, 6G aims at the 1 Tbps frontier, which is 1000 times higher [154, 164]. At a glance, such a number seems to be excessively large. Yet, application frameworks, such as multisensory Extended Reality (XR) [165] and Brain–Computer Interaction (BCI) [166, 167], will demand for such a massive data rate, as well as for transmissions in the mm-wave range, and possibly in the sub-THz range.

(2) *Transition to Volumetric Spectral and Energy Efficiency (3D-SEE) and diffusion of Large Intelligent Surfaces (LISs).* The increase of BSR, along with the exploitation of higher frequencies, that are by nature less isotropic in propagation, will lead the evolution towards the concept of volumetric (3D) bandwidth distribution [154, 168, 169]. From the infrastructural point of view, a change of paradigm is envisaged from classical radio towers/stations to electromagnetic active distributed surfaces [170, 171], featuring common objects, like walls and posts, able to interact with mobile devices (i.e. LISs) [172–174].

(3) *Convergence of services and functionalities.* Already 5G is introducing the founding concept of a standard that covers services falling beyond the communication of mobile devices. Having stated that, 6G will take significant steps forward along this direction, pushing forth the Convergence of Communications, Computing, Control, Localisation and Sensing (3CLS) functionalities [154, 175–177]. Such a convergence will be led in particular by applications like the previously mentioned XR and Connected Robotics and Autonomous Systems (CRAS) [178–181].

(4) *Transition from Self-Organising Networks (SONs) to Self-Sustaining Networks (SSNs).* As already pointed out before, the extensive employment of AI in the services and, most importantly, in the network arrangement and management within the 6G frame of reference, will naturally demand for the development of self-evolutionary capacities. Bearing this concept in mind, 4G and 5G already introduced the concept of SONs [182–184], which are portions of the network (i.e. subnetworks) able to optimise their operation on the basis of real-time conditions, like the number of users, data traffic, and so on. Given such a context, 6G aims at raising the bar, taking the concept of SONs to that of SSNs. While SONs adapt, that is limit, their operation on the basis of surrounding constraints, SSNs will have the capacity of gathering, reaching and collecting the necessary resources, to counteract real-time constraints and ensure the maintenance of networks' and services' Key Performance Indicators (KPIs) [185–187].

(5) *Massive amounts of small pieces of intelligence.* When putting together the self-sustenance of 6G supported by AI, with the upcoming applications previously mentioned, like Connected Robotics and Autonomous Systems (CRAS), Brain–Computer Interaction (BCI) and Extended Reality (XR), it emerges in a rather straightforward way that, differing from what has taken place up to now, the network cannot be structured according to strictly centralised approaches. More reasonably, computational capacities, along with small sets of data, that is, in one word, intelligence, have to be imagined to be distributed through and across the network edge [188, 189]. Given this frame, 6G will have to play with the opposing levers of cloud computing (centralised) and edge computing (peripheral) [190–192], aiming at the threefold target of allowing proper room for local self-adaptation, keeping the overall network functioning as a whole, while reducing latency and optimising the data flows.

(6) *Facing physical atomisation.* Bearing in mind the wide context depicted in the previous points, it is straightforward to figure out that such radical changes and advancements will be possible only if the Hardware (HW) infrastructure, intended with a very broad meaning (i.e. from end-user devices to radio towers), will be reconceived and developed along with the 6G applications and services it has to support. In light of that, the ongoing trend in the realisation of smart interconnected wearable objects will very likely proceed further in the future. The relentlessly increasing needs for multiple-sensing functionalities, coupled with the demands for local computation, data storage and transmission, will orient HW development towards massive integration and miniaturisation. One of the main implications of such trends at end-user level will probably be a perceived increasing separation of the services exploited from the physical devices implementing them, thus unleashing the core concept of ubiquity, also boosted by body implants (not only for medical purposes) [193–197] and direct interaction with human senses. The mentioned trend towards HW miniaturisation is labelled here as physical atomisation, and one of its implications will be the 'Sunset' of smartphones [154].

(7) *Facing Energy-Awareness-Driven Operation (EADO).* Putting together the capillarisation of services, the AI-driven self-evolution of the network, along with the physical atomisation of the HW infrastructure at the edge, energy availability emerges as one of the most crucial aspects to be handled [155, 156]. The self-sustenance of network portions has to feature the capacities to store, convert and transfer proper amounts of energy, aiming at resilience of operation and maintenance of KPIs. In light of these considerations, technologies like EH, i.e. the conversion of part of the energy scattered in the environment into electricity (e.g. from mechanical vibrations, thermal gradients, light) [198–202], Wireless Power Transfer (WPT) [203–209] and innovative energy storage units [210–212] will be of paramount importance and centrality in the 6G scenario.

3.4.3 Potential limiting factors to the full disruption of 6G

Electronics, telecommunications and computer science have marked out a relentless evolutionary process, which started with the spread of technologies based on semiconductors and is still ongoing today, after more than half a century. To this end, Moore's law sketched, with remarkable accuracy, the path followed by silicon-based HW technologies, built on the relationship between size down-scaling (due to the progress of micro-fabrication technologies) and the increase of computation capacities per unit area [213–216].

From a diverse perspective, the development of HW technologies also followed paths that are different from Moore's law. Such additional directions are grouped and labelled as More than Moore technologies [217–220]. The most relevant items falling under the More than Moore classification are Micro/Nanosystems and Micro/Nanotechnologies (MEMS/NEMS) [221–224], novel materials [225, 226], heterostructures-based transistors [227], nanoelectronics [228], as well as packaging [229–232], 3D stacking and integration solutions and methodologies [233–235].

The frame defined by the *More Moore* and *More then Moore* [236] drivers ruled crucial trends in electronics, like integration, reduction in power consumption and miniaturisation, and went on for decades. Within such a scenario, some crucial aspects remained unchanged, like the marked asymmetry between the HW and Software (SW) parts in each system, subsystem or building block.

The typifying immateriality of SW brings attributes like the ease of fixing and upgrading it. On the other hand, HW has always been strictly nailed to its physical dimension, as fixing or upgrading it means most of time replacing or adding parts to the system. It is straightforward that HW can be widely reconfigured, as well as can ensure seamless operation. However, to do so, redundancy has to be significantly leveraged. In other terms, differently from what happens for SW, the flexibility and adaptivity of HW has to be *a priori* accounted for at the design and deployment level, but certainly not in the system operation phase.

In light of such considerations, HW-SW systems and subsystems are conceived according to a codesign approach [237–240]. This means that the physical architecture and the algorithms of a physical device are developed with the target of optimising the operation of the system as a whole. Therefore, the HW is tailored with respect to the functions and applications offered by the SW, aiming at reasonable compromises in terms of performance, compliance of specifications, complexity, reliability and, last but not least, in terms of the costs of the final product.

The just-described asymmetry is addressed in this work as the *HW-SW divide* [241]. Based on this principle, telecommunication standards evolved, across several decades, from 1G to 5G, without any substantial need for its revision, modification or extension.

From a different perspective, however, 5G is already today expected not to fully meet the challenge of acting as the technical enabler of IoE and Super-IoT [153]. This work speculates that a likely reason for such an early failure is the persistence of the HW-SW divide philosophy in the current development of 5G services and physical infrastructure.

3.4.4 Possible novel approaches to empower 6G

The current deployment and development of the 5G paradigm addresses the targets of adaptivity and flexibility by means of virtualisation. To this end, crucial technologies like SDN [242–245] and NFV [246–248] are aiming at the establishment of SW-based virtual networks, that, despite persisting in relying on a physical (HW) infrastructure, unleash extended features in terms of reconfigurability and wide usability, among which network slicing is one [249–251].

The crucial statement here at stake is that the just-discussed softwarisation pursued by 5G is not fully appropriate to address the 'making-intelligent' of networks that is going to be realised by 6G, via the wide exploitation of AI [156]. The future 6G will have to support diverse, heterogeneous and often orthogonal functions, like communications (from GHz, to mm-wave, to sub-THz), computing, energy distribution (via, e.g., EH and WPT), as well as sensing, data collection and storage, analytics and their elaboration.

The sketched scenario, as a whole, highlights the intrinsic limitations linked to the HW-SW divide, also including the mentioned HW-SW codesign approach. On a different plane of reference, the need for a partial reformulation of the established set of relationships, between the HW and SW parts, starts emerging and, according to the point supported in this work, it has to pass through an increase of the concepts of *separation* and *symmetry*, detailed as follows:

- *Separation.* The trend in increasing algorithm-HW separation is in full opposition to the HW-SW codesign approaches, yet it will be crucial in sustaining the realisation of the 6G vision. As a matter of fact, increased HW-SW separation will be fundamental in order to ensure the heterogeneous, adaptable and upgradable capacities of the HW implementing the services mentioned above, along with the network self-sustaining features unleashed by AI.
- *Symmetry.* Starting from the reference plane of an enhanced separation between HW and SW, increase of symmetry between them will turn into a cornerstone feature, as well. In other terms, diverging from the current conception, HW will be demanded to gain more flexibility, self-adaptivity and, in general, more intelligence, therefore reducing the currently existing conceptual gap with respect to SW. This means that the 6G will push towards an increased level of abstraction of the HW, somehow loosening the classical tight link to its physical dimension.

A graphical description of the just-discussed concepts of HW-SW separation and symmetry is reported in the following figure 3.17 and figure 3.18, respectively.

Concerning separation, figure 3.17(a) provides a symbolic representation of typical systems designed leveraging the classical HW-SW codesign concept. Diversely, figure 3.17(b) depicts what is envisaged in the future 6G, according to the point supported in this work, capitalising on an increased separation between HW and SW.

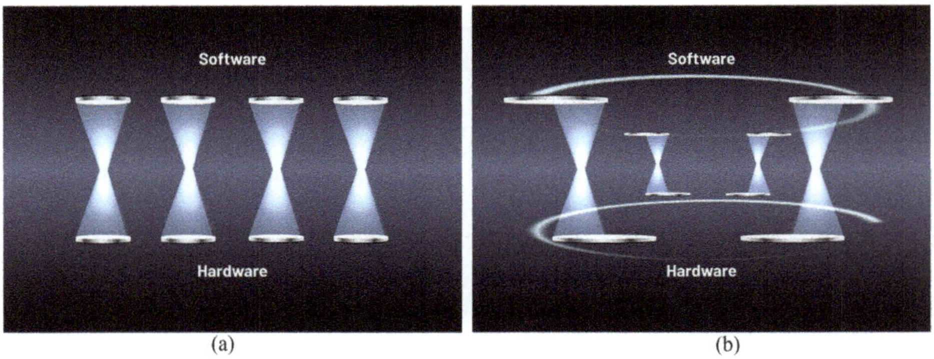

Figure 3.17. (a) Symbolic representation of typical HW-SW systems conceived according to the classical HW-SW codesign approach. (b) Sketch of envisaged HW-SW systems designed capitalising on an increased separation between the two parts. Copyright (2021) IEEE, reproduced with permission. [252].

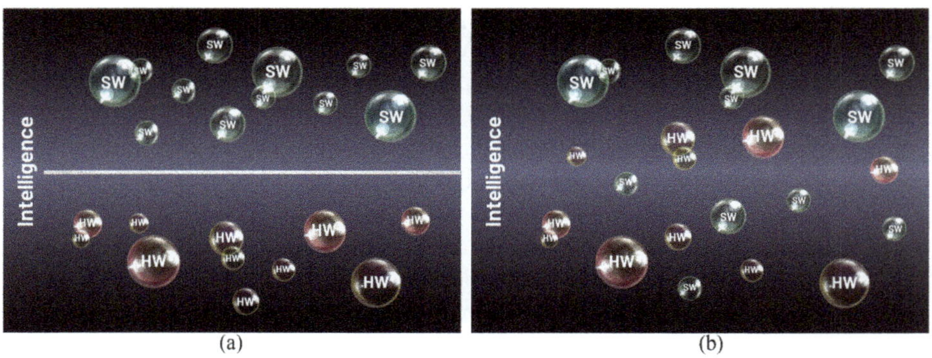

Figure 3.18. Symbolic sketch of the concept of symmetry between HW and SW. (a) Standard and currently in use approaches, like the HW-SW codesign, define a ceiling (upper boundary) to the level of intelligence that the HW can achieve. Whatever lies above such a plane is the floor of the SW domain. (b) The future envisaged scenario supported in this work features a weakening of the mentioned ceiling, letting the HW gain more intelligence. Copyright (2021) IEEE, reproduced with permission. [252].

Reformulated in a different fashion, the content of figure 3.17 can be sketched as follows. The case reported in figure 3.17(a) is that of systems or subsystems in which the HW and SW parts are optimised against each other, in order to implement a set of functionalities and/or services known *ab initio*. Diversely, figure 3.17(b) reports multiple HW and SW parts, able to combine to each other according to dynamic ways, therefore realising a wide range of functionalities and services, some of which could be not known nor scheduled *a priori*.

Stepping forth, figure 3.18 targets the concept of symmetry between HW and SW. To this end, figure 3.18(a) describes in a symbolic way the current state of the HW-SW codesign philosophy, leading to an ideal ceiling (upper boundary) for what concerns the intelligence that the HW can score. It must be stressed that the word *intelligence* is used in this case with a very broad meaning. In fact, it refers to characteristics like reconfigurability, replaceability, flexibility, interoperability, along with self-referred features, like self-healing and self-repairability. Last but not least, in this context, intelligence comprises simple computational, data storage and data elaboration capacities. In light of this, HW-SW systems nowadays are still based on a pronounced asymmetry between the HW and SW parts' intelligence, with a sort of limit that is not surpassed by the HW (as shown in figure 3.18(a)).

Diversely, the novel scenario envisaged in this work is shown in figure 3.18(b), where the boundary to intelligence is weakened, and possibly removed, letting HW raise to higher levels of intelligence, as well as the SW lowering it, if and when necessary.

Wrapping up what has been reported in the previous pages, the deployed scenario has to be intended at the moment as a vision of the implications that the future 6G might exert at HW level. Nonetheless, already today there exist indications in the literature supporting the discussion and the point of view developed in this work. To this end, a few relevant considerations are reported in [156], referring to RF transceivers. It is stated that HW capabilities, like the number of antennas, RF chains, the resolution and sampling rates of ADCs, etc, remained quasi-static in the jointly designed HW-SW systems from 1G to the current 5G.

In addition, the mentioned work highlights that the recent state-of-the-art research in circuits and antennas is relevantly improving the HW capabilities, with the target of enabling 6G base stations and handsets to be diversified and upgradable. Finally, the following sentence in [156] is of particular relevance in light of the scope of this work: '*6G will not be operating under the conventional joint design, which fails in allowing agile adaptation to a diversified and upgradable hardware*'.

3.5 Conclusion

In this chapter, comprehensive discussion around mobile communication standards was developed. In the first place, a general background of cellular mobile networks was provided, focusing on the fundamental working principles and on supplementary considerations related to the mobile services ecosystem, listing the most important stakeholders involved (policy makers and private entities). Subsequently, the evolution of mobile standards' generations was discussed, reporting the most relevant innovations introduced by each of them, along with their strengths and weaknesses.

The 1st Generation (1G), deployed at the turning point of the 1970s and 1980s, brought in the founding innovation of enabling mobile communication among end users. Exclusively voice-based services (i.e. no data) were available and broadcasted as analogue signals. The exploited Frequency Division Multiple Access (FDMA) was limited in terms of spectrum efficiency, mobile terminals were bulky as well as power hungry and broadcasted signals were not encrypted. However, 1G introduced important features like frequency band reuse across cells.

The 2nd Generation (2G), deployed from the early 1990s, stepped into the digital world. Digital voice signals could be more efficiently broadcasted by means of the Time Division Multiple Access (TDMA) and non-voice features, such as Short Message Service (SMS), started to appear on the landscape. Like 1G, 2G was also not efficient in terms of spectrum utilisation and encryption of data was not yet introduced. Nonetheless, a key enabling technology for subsequent generations of mobile services was developed (despite not being widely deployed) under the 2G umbrella, namely, Code Division Multiple Access (CDMA).

The 3rd Generation (3G), deployed from 2000, led to significant enhancements in terms of performance. First, thanks to CDMA, better spectrum exploitation was possible as well as encryption, thus increasing the security of transmitted data. Then, by means of wider channels and of Evolution-Data Optimised (EV-DO), i.e. channels exclusively dedicated to non-voice services, more and more end users could benefit from capillary voice, messaging and Internet access services.

The 4th Generation Long Term Evolution (4G-LTE), deployed from 2010, besides continual increase of performance, started to pursue the concept of flexibility and adaptation of resources and services, based upon end-user demands. In this regard, the possibility of dynamically increasing or decreasing the bandwidth is boosted by Carrier Aggregation (CA) techniques as well as by Orthogonal Frequency Division Multiple Access (OFDMA). 4G-LTE also started to introduce the multi-path transmission concept, through the use of Multiple Input Multiple Output (MIMO) techniques.

Given the aforementioned scenario, the core topic of the future of 5th Generation (5G) mobile communications was then introduced. The main three vectors that 5G will pursue have been mentioned, and they can be listed as follows:

- **Enhanced Mobile Broadband (EMBB)**, related to the system capacity growth, with the target of 1000 times increase with respect to current 4G-LTE, leading to 10 Gbps peak and a minimum of 10 Mbps per each user.
- **Ultra-Reliable Low-Latency Communications (URLLC)**, demanded by critical applications, like Vehicle-to-Vehicle (V2V) communications, aiming at the millisecond latency range.
- **Massive Machine-Type Communications (MMTC)**, leading to a very large increase of new connections to be supported by 5G services.

Given these main drivers, the most relevant key enabling technologies bearing the potential to realise 5G were then listed and discussed in detail. Among them, the most critical seems to be the massive exploitation of MIMO techniques, interference management, spectrum sharing and device-to-device (D2D) communications. Eventually, apart from the listed items, exploration of 5G New Radio (NR) concepts, through the use of millimetre-wave (mm-wave) frequencies and advanced beamforming, appears as a disruptive solution to empower capillary deployment of small cells, bringing impressive network capacity to end users.

Furthermore, the current state of the early 5G deployment and services exploitations was briefly reported, along with a description of technologies and strategies, whose centrality in the frame of 5G has been strengthened in recent times, among which Cognitive Radio (CR), Massive centralised Radio Access Network (RAN), Vehicular Communications (VCs) and network slicing are the most relevant items.

Eventually, the last part of the chapter developed a rather extensive outlook of what the future 6G is envisaged to be, beyond 5G, having as a reference target the year 2030. To this end, according to the current visions, the central role that Artificial Intelligence (AI) will play in the frame of 6G was stressed, focusing on the implications at the performance, services, and, most of all, network infrastructure level. After reviewing the expected key specifications of 6G and how they compare to what 5G is targeting today, the main technology trends that will be triggered by such an AI-driven 6G were analysed. Among them, the transition from Self-Organising Networks (SONs) to Self-Sustaining Networks (SSNs), physical atomisation and Energy-Awareness-Driven Operation (EADO), are certainly worth mentioning.

Bearing in mind the just-depicted scenario, the limitations of the current strategies in the development of Hardware-Software (HW-SW) systems and components in the realisation of future 6G visions were placed under scrutiny, also providing some hints as to how the HW-SW codesign strategies could be re-architected, in order to meet the challenge of an AI-driven network.

References

[1] Tse D and Viswanath P 2005 *Fundamentals of Wireless Communication* 1st edn (Cambridge: Cambridge University Press) p 564

[2] Lee W C Y 2010 *Mobile Communications Design Fundamentals* 2nd edn (Hoboken, NJ: Wiley) p 398

[3] Bertoni H L 1999 *Radio Propagation for Modern Wireless Systems* 1st edn (Upper Saddle River, NJ: Prentice Hall) p 258

[4] Vidojkovic M 2011 Configurable *Circuits and Their Impact on Multi-Standard RF Front-End Architectures* 1st edn (Eindhoven: Technische Universiteit Eindhoven) p 179

[5] Gupta A and Jha R K 2015 A survey of 5G network: architecture and emerging technologies *IEEE Access* **3** 1206–32

[6] Qualcomm *The Evolution of Mobile Technologies: 1G → 2G → 3G → 4G LTE* https://qualcomm.com/documents/evolution-mobile-technologies-1g-2g-3g-4g-lte)

[7] Santhi K R, Srivastava V K, SenthilKumaran G and Butare A 2003 Goals of true broad band's wireless next wave (4G-5G) *Proc. IEEE Vehicular Tech. Conf. VTC (Orlando, FL, Oct. 2003)* pp 2317–21

[8] Halonen T, Romero J and Melero J 2003 *GSM, GPRS and EDGE Performance: Evolution Towards 3G/UMTS* 2nd edn (NewYork: Wiley) p 654

[9] Gammel P, Pehlke D R, Brunel D, Kovacic S J and Walsh K *5G in Perspective: A Pragmatic Guide to What's Next* https://microwavejournal.com/articles/28295-g-in-perspective-a-pragmatic-guide-to-whats-next)

[10] Johansen S *1G, 2G, 3G, 4G, 5G* https://its-wiki.no/images/c/c8/From_1G_to_5G_Simon.pdf)

[11] Sesia S, Toufik I and Baker M 2009 *LTE—The UMTS Long Term Evolution: From Theory to Practice* 1st edn (Hoboken, NJ: Wiley) p 648

[12] Qualcomm LTE *TDD: The Global Solution For Unpaired Spectrum* https://qualcomm.com/documents/lte-tdd-global-solution-unpaired-spectrum)
[13] Ref Ai B, Guan K, He R, Li J, Li G, He D, Zhong Z and Huq K M S 2017 On indoor millimeter wave massive MIMO channels: measurement and simulation *IEEE J. Sel. Areas Commun.* **35** 1678–90
[14] Sipal D, Abegaonkar M P and Koul S K 2017 Easily extendable compact planar UWB MIMO antenna array *IEEE Antennas Wirel. Propag. Lett.* **16** 2328–31
[15] Nguyen V G, Do T X and Kim Y 2016 SDN and virtualization-based LTE mobile network architectures: a comprehensive survey *Wirel. Pers. Commun.* **86** 1401–38
[16] Zaki Y, Zhao L, Goerg C and Timm-Giel A 2011 LTE mobile network virtualization *J. Mobile Netw. Appl.* **16** 424–32
[17] Bhushan N, Li J, Malladi D, Gilmore R, Brenner D R, Damnjanovic A D, Sukhavasi R T, Patel C and Geirhofer S 2014 Network densification: the dominant theme for wireless evolution into 5G *IEEE Commun. Mag.* **52** 82–9
[18] Andrews J G, Zhang X, Durgin G D and Gupta A K 2016 Are we approaching the fundamental limits of wireless network densification? *IEEE Commun. Mag.* **54** 184–90
[19] Nguyen V M and Kountouris M 2017 Performance limits of network densification *IEEE J. Sel. Areas Commun.* **35** 1294–308
[20] Zhao C, Huang L, Zhao Y and Du X 2017 Secure machine-type communications toward LTE heterogeneous networks *IEEE Wirel. Commun.* **24** 82–7
[21] Slamnik N, Okic A and Musovic J 2016 Conceptual radio resource management approach in LTE heterogeneous networks using small cells number variation *Proc. Int. Symp. on Telecommun. BIHTEL (Sarajevo, Oct. 2016)* pp 1–5
[22] Petrut I, Otesteanu M, Balint C and Budura G 2016 On the uplink performance in LTE heterogeneous network *Proc. Int. Conf. on Commun. COMM (Bucharest, June 2016)* p 191–4
[23] Elfadil H E, Ali M A I and Abas M 2015 Performance evaluation of heterogeneous networks schemes in LTE networks *Proc. Int. Conf. on Computing, Control, Networking, Electron. and Embedded Syst. Eng. ICCNEEE (Khartoum, Sept. 2015)* pp 401–8
[24] Rysavy Research *LTE and 5G Innovation: Igniting Mobile Broadband* https://5gamericas.org/wp-content/uploads/2019/07/4G_Americas_Rysavy_LTE_and_5G_Innovation_PPT.pdf
[25] Econocom *How the Internet of Things is revolutionising industry* https://blog.econocom.com/en/blog/how-the-internet-of-things-is-revolutionising-industry/
[26] Vermesan O and Friess P 2014 *Internet of Things Applications—From Research and Innovation to Market Deployment* 1st edn (Aalborg: River Publishers) p 364
[27] Uckelmann D, Harrison M and Michahelles F 2011 *Architecting the Internet of Things* 1st edn (Berlin: Springer) p 353
[28] MachNation *The Development of the Internet of Everything* https://d175jir5ufcz29.cloudfront.net/legacy-mn-wp-media/2014/06/MachNation-Development-of-the-IoE.pdf
[29] ITU *The Tactile Internet* https://itu.int/dms_pub/itu-t/oth/23/01/T23010000230001PDFE.pdf
[30] Xiang W, Zheng K and Shen X S 2017 *5G Mobile Communications* 1st edn (Cham: Springer) p 691
[31] Sag A *The State Of 5G In Early 2021, Part 1* https://forbes.com/sites/moorinsights/2021/03/08/the-state-of-5g-in-early-2021-part-1/?sh=69240be44d98

[32] Sag A *T-Mobile Is The First Carrier Globally To Launch Nationwide Standalone (SA) 5G* https://forbes.com/sites/moorinsights/2020/08/04/t-mobile-is-the-first-carrier-globally-to-launch-nationwide-standalone-sa-5g/?sh=73c334362d0e

[33] ITU IMT *Vision—Framework and Overall Objectives of the Future Development of IMT for 2020 and Beyond* https://itu.int/dms_pubrec/itu-r/rec/m/R-REC-M.2083-0-201509-I!!PDF-E.pdf

[34] ITU *Minimum Requirements Related to Technical Performancefor IMT-2020 Radio Interface(s)* https://itu.int/dms_pub/itu-r/opb/rep/R-REP-M.2410–2017-PDF-E.pdf

[35] Ericsson *Evolving LTE to Fit the 5G Future* https://ericsson.com/en/publications/ericsson-technology-review/archive/2017/evolving-lte-to-fit-the-5g-future

[36] Huawei *5G Network Architecture a High-Level Perspective* http://huawei.com/minisite/hwmbbf16/insights/5G-Nework-Architecture-Whitepaper-en.pdf

[37] 3GPP *The Path to 5G: as much Evolution as Revolution* http://3gpp.org/news-events/3gpp-news/1774-5g_wiseharbour

[38] Agyapong P K, Iwamura M, Staehle D, Kiess W and Benjebbour A 2014 Design considerations for a 5G network architecture *IEEE Commun. Mag.* **52** 65–75

[39] PPP in Horizon 2020 *Advanced 5G Network Infrastructure for the Future Internet*—Creating a Smart Ubiquitous Network for the Future Internet https://5g-ppp.eu/wp-content/uploads/2014/02/Advanced-5G-Network-Infrastructure-PPP-in-H2020_Final_November-2013.pdf

[40] Osseiran A *et al* 2014 Scenarios for 5G mobile and wireless communications: the vision of the METIS project *IEEE Commun. Mag.* **52** 26–35

[41] Larsson E G, Edfors O, Tufvesson F and Marzetta T L 2014 Massive MIMO for next generation wireless systems *IEEE Commun. Mag.* **52** 186–95

[42] Ulukus S, Yener A, Erkip E, Simeone O, Zorzi M, Grover P and Huang K 2015 Energy harvesting wireless communications: a review of recent advances *IEEE J. Sel. Areas Commun.* **33** 360–81

[43] Iannacci J, Sordo G, Serra E and Schmid U 2016 The MEMS four-leaf clover wideband vibration energy harvesting device: design concept and experimental verification *Microsyst. Technol.* **22** 1865–81

[44] Iannacci J and Sordo G 2016 Up-scaled macro-device implementation of a MEMS wideband vibration piezoelectric energy harvester design concept *Microsyst. Technol.* **22** 1639–51

[45] Iannacci J *et al* 2014 Multi-modal vibration based MEMS energy harvesters for ultra-low power wireless functional nodes *Microsyst. Technol.* **20** 627–40

[46] Iannacci J 2017 Microsystem based energy harvesting (EH-MEMS): powering pervasivity of the Internet of Things (IoT)—a review with focus on mechanical vibrations *J. King Saud. Univ.* pp 191–9

[47] Nam W, Bai D, Lee J and Kang I 2014 Advanced interference management for 5G cellular networks *IEEE Commun. Mag.* **52** 52–60

[48] Wyglinski A M, Nekovee M and Hou T 2009 *Cognitive Radio Communications and Networks: Principles and Practice* 1st edn (Amsterdam: Academic) p 736

[49] Irnich T, Kronander J, Selén Y and Li G 2013 Spectrum sharing scenarios and resulting technical requirements for 5G systems *Proc. IEEE Int. Symp. on Personal, Indoor and Mobile Radio Commun. PIMRC (London, Sept. 2013)* pp 127–32

[50] Buchwald G J, Kuffner S L, Ecklund L M, Brown M and Callaway E H Jr 2008 The design and operation of the IEEE 802.22.1 disabling beacon for the protection of TV whitespace

incumbents *Proc. IEEE Symp. on New Frontiers in Dynamic Spectrum Access Networks (Chicago, IL, Oct. 2008)* pp 1–6

[51] Electronic Communications Committee (ECC) *Technical and Operational Requirements for the Possible Operation of Cognitive Radio Systems in the 'White Spaces' of the Frequency Band 470–790 MHz* https://docdb.cept.org/download/be051b35-91e9/ECCREP159.PDF

[52] Federal Communications Commission (FCC) *Third Memorandum Opinion and Order in the Matter of Unlicensed Operation in the TV Broadcast Bands (ET Docket No. 04–186) and Additional Spectrum for Unlicensed Devices Below 900 MHz and in the 3 GHz Band* (ET Docket No. 02–380) https://apps.fcc.gov/edocs_public/attachmatch/FCC-12-36A1_Rcd.pdf

[53] Huang J, Berry R A and Honig M L 2006 Auction-based spectrum sharing *Mob. Netw. Appl.* **11** 405–18

[54] Shen F, Li D, Lin P H and Jorswieck E 2015 Auction based spectrum sharing for hybrid access in macro-femtocell networks under QoS requirements *Proc. IEEE Int. Conf. on Commun. ICC (London, June 2015)* pp 3335–40

[55] Tehrani M N, Uysal M and Yanikomeroglu H 2014 Device-to-device communication in 5G cellular networks: challenges, solutions, and future directions *IEEE Commun. Mag.* **52** 86–92

[56] Orsino A, Gapeyenko M, Militano L, Moltchanov D, Andreev S, Koucheryavy Y and Araniti G 2015 Assisted handover based on device-to-device communications in 3GPP LTE systems *Proc. IEEE Globecom Workshops (San Diego, CA, Dec. 2015)* pp 1–6

[57] Cha I, Shah Y, Schmidt A U, Leicher A and Meyerstein M V 2009 Trust in M2M communication *IEEE Veh. Tech. Mag.* **4** 69–75

[58] Yue J, Ma C, Yu H and Zhou W 2013 Secrecy-based access control for device-to-device communication underlaying cellular networks *IEEE Commun. Lett.* **17** 2068–71

[59] Perrig A, Stankovic J and Wagner D 2004 Security in wireless sensor networks *ACM J. Commun.* **47** 53–7

[60] Zhou Y, Fang Y and Zhang Y 2008 Securing wireless sensor networks: a survey *IEEE Commun. Surv. Tutor* **10** 6–28

[61] Muraleedharan R and Osadciw L 2006 Jamming attack detection and countermeasures in wireless sensor network using ant system *Proc. SPIE* **6248** 1–12

[62] Rappaport T S, Sun S, Mayzus R, Zhao H, Azar Y, Wang K, Wong G N, Schulz J K, Samimi M and Gutierrez F 2013 Millimeter wave mobile communications for 5G cellular: it will work! *IEEE Access* **1** 335–49

[63] Pi Z and Khan F 2011 An introduction to millimeter-wave mobile broadband systems *IEEE Commun. Mag.* **49** 101–7

[64] Gutierrez F, Agarwal S, Parrish K and Rappaport T S 2009 On-chip integrated antenna structures in CMOS for 60 GHz WPAN systems *IEEE J. Sel. Areas Commun.* **27** 1367–78

[65] Rappaport T S, Ben-Dor E, Murdock J N and Qiao Y 2012 38 GHz and 60 GHz angle-dependent propagation for cellular & peer-to-peer wireless communications *Proc. IEEE Int. Conf. on Commun. ICC (Ottawa, June 2012)* pp 4568–573

[66] Rappaport T S, Murdock J N and Gutierrez F 2011 State of the art in 60-GHz integrated circuits and systems for wireless communications *Proc. IEEE* **99** 1390–436

[67] Iannacci J, Tschoban C, Reyes J, Maaß U, Huhn M, Ndip I and Pötter H 2016 RF-MEMS for 5G mobile communications: a basic attenuator module demonstrated up to 50 GHz *Proc. IEEE SENSORS (Orlando, FL, Oct.–Nov. 2016)* pp 1–3

[68] Iannacci J, Huhn M, Tschoban C and Pötter H 2016 RF-MEMS technology for future mobile and high-frequency applications: reconfigurable 8-bit power attenuator tested up to 110 GHz *IEEE Electron Device Lett.* **37** 1646–9

[69] Iannacci J, Huhn M, Tschoban C and Pötter H 2016 RF-MEMS technology for 5G: series and shunt attenuator modules demonstrated up to 110 GHz *IEEE Electron Device Lett.* **37** 1336–9

[70] Rusek F, Persson D, Lau B K, Larsson E G, Marzetta T L, Edfors O and Tufvesson F 2013 Scaling up MIMO: opportunities and challenges with very large arrays *IEEE Signal Process Mag.* **30** 40–60

[71] Al-Samman A M, Rahman T A, Azmi M H, Zulkefly N R and Mataria A M S 2016 Path loss model for outdoor environment at 17 GHz mm-wave band *Proc. IEEE Int. Coll. on Signal Processing & its Applications CSPA (Malacca City, March 2016)* pp 179–82

[72] Sarma A D, Prasad M V S N and Pandit S N N 1999 Modelling of path loss for land mobile cm and mm wave communication systems *Proc. Int. Conf. on Electromagnetic Interference and Compatibility (New Delhi, Dec. 1999)* pp 75–8

[73] Ravindra K and Sarma A D 1999 Modelling path loss in the near field region for cm and mm wave mobile communication *Proc. Int. Conf. on Electromagnetic Interference and Compatibility (New Delhi, Dec. 1999)* pp 79–82

[74] Serov E A, Koshelev M A, Parshin V V and Tretyakov M Y 2010 Atmosphere continuum absorption investigation at MM waves *Proc. Int. Kharkov Symp. on Phys. and Eng. of Microwaves, Millimeter and Submillimeter Waves (Kharkiv, June 2010)* pp 1–3

[75] Lehto A, Tuovinen J and Raisanen A 1991 Proposed method for measurement of absorption loss and beam efficiency of large MM-wave reflector antennas in compact antenna test range by using hot and cold loads *IEE Proc. H—Microw. Antennas Propag.* **138** 13–8

[76] Capone A, Filippini I, Sciancalepore V and Tremolada D 2015 Obstacle avoidance cell discovery using mm-waves directive antennas in 5G networks *Proc. IEEE Annual Int. Symp. on Personal, Indoor, and Mobile Radio Commun. PIMRC (Hong Kong, Aug.–Sept. 2015)* pp 2349–53

[77] Simic L, Panda S, Riihijarvi J and Mahonen P 2017 Coverage and robustness of mm-wave urban cellular networks: multi-frequency HetNets are the 5G future *Proc. Ann. IEEE Int. Conf. on Sensing, Commun., and Networking SECON (San Diego, CA, June 2017)* pp 1–9

[78] Int. Telecommunication Union (ITU) *Estimated Spectrum Bandwidth Requirements for the Future Development of IMT-2000 and IMT-Advanced* http://itu.int/pub/R-REP-M.2078)

[79] Krasko O, Brych M, Masyuk A and Klymash M 2016 Flexible backhaul architecture for densely deployed 5G small cells based on OWTDMA network *Proc. Third Int. Scientific-Practical Conf. Problems of Infocommunications Science and Technology PIC S&T (Kharkiv, Oct. 2016)* pp 33–5

[80] Hou X, Wang X, Jiang H and Kayama H 2016 Investigation of massive MIMO in dense small cell deployment for 5G *Proc. IEEE Vehicular Technol. Conf. VTC-Fall (Montreal, Sept. 2016)*

[81] Behnad A and Wang X 2017 Virtual small cells formation in 5G networks *IEEE Commun. Lett.* **21** 616–9

[82] Nasr A I and Fahmy Y 2016 Millimeter-wave wireless backhauling for 5G small cells: star versus mesh topologies *Proc. Int. Conf. on Microelectron. ICM (Cairo, Dec. 2016)* pp 885–88

[83] He L, Zhong H T and Song D 2018 A software defined radio based airplane communication navigation simulation system *J. Phys.: Conf. Ser.* **960** 012014
[84] Ryabov I V, Nikitin P V, Bazhenov R I, Melnikova A I and Gorokhova R I 2021 Physical principles of work and basic features of universal software radio peripheral *J. Phys.: Conf. Ser.* **1728** 012013
[85] Hassan M, Singh M and Hamid K 2021 Overview of cognitive radio networks *J. Phys. Conf. Ser.* **1831** 012013
[86] Al-Dulaimi O M K, Vlâdeanu C, Martia A, Al-DulaimiMohammed A M K and Al-Dulaim M K H 2021 Cognitive radio using spectrum sensing by cooperative two secondary users *IOP Conf. Ser.: Mater. Sci. Eng.* **1094** 012031
[87] Alhammadi A, Roslee M and Alias M Y 2016 Fuzzy logic based negotiation approach for spectrum handoff in cognitive radio network *2016 IEEE 3rd Int. Symposium on Telecommunication Technologies (ISTT)* pp 120–4
[88] Tang H 2003 Issues of the radio access network with distributed radio network control functions for universal mobile telecommunication system *IEEE Int. Symp. on Personal, Indoor and Mobile Radio Communications* (PIMRC) **1**
[89] Binh L N 2020 Historical overview and digital transmission technologies in cloud networking *Transmission and Processing for Data Center Networking* (Bristol: Institute of Physics Publishing) pp 1–37
[90] Rabia T, Braham O and Pujolle G 2017 Partially centralized C-RAN architecture using TRILL protocol *2016 7th Int. Conference on the Network of the Future (NOF)* pp 1–3
[91] Gulati S, Natarajan B, Kalyanasundaram S and Agrawal R 2016 Performance analysis of centralized ran deployment with non-ideal fronthaul in LTE-advanced networks *IEEE Vehicular Technology Conf.*
[92] Ramantas K, Antonopoulos A, Kartsakli E, Mekikis P V, Vardakas J and Verikoukis C 2018 A C-RAN based 5g platform with a fully virtualized, sdn controlled optical/wireless fronthaul *Int. Conf. on Transparent Optical Networks*
[93] Valastro G C, Panno D and Riolo S 2018 A SDN/NFV based C-RAN architecture for 5G mobile networks *2018 Int. Conference on Selected Topics in Mobile and Wireless Networking (MoWNeT)* pp 1–8
[94] Gavrilovska L, Rakovic V, Ichkov A, Todorovski D and Marinova S 2017 Flexible C-RAN: radio technology for 5G *2017 13th Int. Conf. on Advanced Technologies, Systems and Services in Telecommunications, TELSIKS 2017—Proc.*
[95] Sahu B J R, Dash S, Saxena N and Roy A 2017 Energy-efficient BBU allocation for green C-RAN *IEEE Commun. Lett.* **21** 1637–40
[96] Zhang X, Jia M, Gu X and Guo Q 2019 An energy efficient resource allocation scheme based on cloud-computing in H-CRAN *IEEE Internet Things J.* **6** 4968–76
[97] Checko A, Avramova A P, Berger M S and Christiansen H L 2016 Evaluating C-RAN fronthaul functional splits in terms of network level energy and cost savings *J. Commun. Networks* **18** 162–72
[98] Chaudhary J K, Zou J and Fettweis G 2018 Cost saving analysis in capacity-constrained C-RAN fronthaul *2018 IEEE Globecom Workshops* (GC Wkshps) pp 1–7
[99] Park S H, Simeone O and Shamai S 2018 Multi-tenant C-RAN with spectrum pooling: downlink optimization under privacy constraints *IEEE Trans. Veh. Technol.* **67** 10492–503

[100] Marotta M A, Ahmadi H, Rochol J, Dasilva L and Both C B 2018 Characterizing the relation between processing power and distance between BBU and RRH in a cloud RAN *IEEE Wirel. Commun. Lett.* **7** 472–5
[101] Feng D, Sun W and Hu W 2018 CAPEX optimization with joint allocation of hybrid RF/FSO and optical fibre resources in 5G backhaul *Optics InfoBase Conf. Papers* Part F84-OFC 2018
[102] Almekhlafi M H, Ibrahim A S, Fahmy Y A, Elsayed H M and Khairy M M 2018 Hybrid optical/RF backhauling for 5G networks *IEEE Int. Symp. on Personal, Indoor and Mobile Radio Communications* (PIMRC)
[103] Guerrero L G *et al* 2020 Integrated wireless-optical backhaul and fronthaul provision through multicore fiber *IEEE Access* **8** 146915–22
[104] Douik A, Dahrouj H, Al-Naffouri T Y and Alouini M S 2016 Hybrid radio/free-space optical design for next generation backhaul systems *IEEE Trans. Commun.* **64** 2563–77
[105] Cvetkovski D, Halsig T, Lankl B and Grass E 2016 Next generation mm-wave wireless backhaul based on LOS MIMO links *2016 German Microwave Conference (GeMiC)* pp 69–72
[106] Gagnon F, Dahman G and Poitau G 2019 Tactical Backhaul Range Extension using MIMO: Investigation based on ITU-R P.530 *2019 IEEE Canadian Conference of Electrical and Computer Engineering (CCECE)* pp 1–6
[107] Tabassum H, Sakr A H and Hossain E 2016 Analysis of massive MIMO-enabled downlink wireless backhauling for full-duplex small cells *IEEE Trans. Commun.* **64** 2354–69
[108] Da Paixao E A R, Vieira R F, Araujo W V and Cardoso D L 2018 Optimized load balancing by dynamic BBU-RRH mapping in C-RAN architecture *2018 Third Int. Conference on Fog and Mobile Edge Computing (FMEC)* pp 100–4
[109] Tian F, Zhang P and Yan Z 2017 A survey on C-RAN security *IEEE Access* **5** 13372–86
[110] Jeyakumar D and Rajabhushanam C 2019 Security challenges and solutions for cloud radio access networks *2019 Int. Conference on Smart Systems and Inventive Technology (ICSSIT)* pp 961–65
[111] Varga N, Bokor L, Takacs A, Kovacs J and Virag L 2017 An architecture proposal for V2X communication-centric traffic light controller systems *2017 15th Int. Conference on ITS Telecommunications (ITST)* pp 1–7
[112] Zhou H, Xu W, Chen J and Wang W 2020 Evolutionary V2X technologies toward the internet of vehicles: challenges and opportunities *Proc. IEEE* **108** 308–23
[113] Chang Y Y, Yu C Y and Lee L T 2016 Research of vehicle-mounted detection and collision avoidance system based on dedicated short range communications *2016 Int. Symp. on Computer, Consumer and Control (IS3C)* pp 299–302
[114] Lukin K, Konovalov V, Scherbakov V and Breed R 2006 Dedicated short-range communication system for vehicle-to-vehicle data transmission on the basis of chaotic waveform codes (DSRC-VVDT) *2006 Int. Conf. on Microwaves, Radar & Wireless Communications* pp 442–45
[115] Vershinin Y A and Zhan Y 2020 Vehicle to vehicle communication: dedicated short range communication and safety awareness *2020 Systems of Signals Generating and Processing in the Field of on Board Communications* pp 1–6
[116] Chen S, Hu J, Shi Y, Zhao L and Li W 2020 A vision of C-V2X: technologies, field testing, and challenges with chinese development *IEEE Internet Things J.* **7** 3872–81

[117] Naik G, Choudhury B and Park J M 2019 IEEE 802.11bd 5G NR V2X: evolution of radio access technologies for V2X communications *IEEE Access* **7** 70169–84

[118] Kiela K *et al* 2020 Review of V2X-IoT standards and frameworks for ITS applications *Appl. Sci.* **10** 4314

[119] Campolo C, Molinaro A, Romeo F, Bazzi A and Berthet A O 2019 5G NR V2X: on the impact of a flexible numerology on the autonomous sidelink mode *2019 IEEE 2nd 5G World Forum (5GWF)* pp 102–7

[120] Zhu X, Yuan S and Zhao P 2021 Research and application on key technologies of 5G and C-V2X intelligent converged network based on MEC *2021 IEEE Int. Conf. on Power Electronics, Computer Applications (ICPECA)* pp 175–79

[121] Wang D, Sattiraju R R, Qiu A, Partani S and Schotten H D 2020 Effect of retransmissions on the performance of C-V2X communication for 5G *IEEE Vehicular Technology Conf.*

[122] Sattiraju R, Wang D, Weinand A and Schotten H D 2020 Link level performance comparison of C-V2X and ITS-G5 for vehicular channel models *IEEE Vehicular Technology Conf.*

[123] Qi W, Landfeldt B, Song Q, Guo L and Jamalipour A 2020 Traffic Differentiated Clustering Routing in DSRC and C-V2X Hybrid Vehicular Networks *IEEE Trans. Veh. Technol.* **69** 7223–34

[124] Jayaweera S, Manosha S, Rajatheva N and Latva-aho M 2021 Co-existence of ITS-G5 and C-V2X at an urban road intersection *2021 Joint European Conf. on Networks and Communications & 6G Summit (EuCNC/6G Summit)* pp 253–8

[125] Silva F S D, Neto E, Santos C, Almeida T, Silva I and Neto A V 2020 Evolving fast innovation in next-generation networking through flexible and customized softwarization and slicing capabilities *2020 IEEE Conf. on Network Function Virtualization and Software Defined Networks (NFV-SDN)* pp 188–93

[126] Nguyen K, Le Nguyen P, Zhe-tao L and Sekiya H Empowering 5G mobile devices with network softwarization *IEEE Trans. Netw. Serv. Manage.*

[127] Ordonez-Lucena J, Ameigeiras P, Lopez D, Ramos-Munoz J J, Lorca J and Folgueira J 2017 Network slicing for 5G with SDN/NFV: concepts, architectures, and challenges *IEEE Commun. Mag.* **55** 80–7

[128] Barakabitze A A, Ahmad A, Mijumbi R and Hines A 2020 5G network slicing using SDN and NFV: a survey of taxonomy, architectures and future challenges *Comput. Networks* **167** 106984

[129] Wijethilaka S and Liyanage M 2021 Survey on network slicing for internet of things realization in 5G networks *IEEE Commun. Surv. Tutor.* **23** 957–94

[130] Chochliouros I P, Spiliopoulou A S, Lazaridis P, Dardamanis A, Zaharis Z and Kostopoulos A 2020 Dynamic network slicing: challenges and opportunities *IFIP Advances in Information and Communication Technology* **vol 585** (Cham: Springer)

[131] Yan L, Fang X and Fang Y 2017 A novel network architecture for C/U-plane staggered handover in 5G decoupled heterogeneous railway wireless systems *IEEE Trans. Intell. Transp. Syst.* **18** 3350–62

[132] Ishii H, Kishiyama Y and Takahashi H 2012 A novel architecture for LTE-B :C-plane/U-plane split and Phantom Cell concept *2012 IEEE Globecom Workshops* pp 624–30

[133] Wu Y, Fang X and Wang X 2019 Mobility management through scalable C/U-plane decoupling in iov networks *IEEE Commun. Mag.* **57** 122–9

[134] Zhao H, Zhao L, Liang K and Pan C 2017 Radio access network slicing based on C/U plane separation *China Commun.* **14** 134–41

[135] Yu G, Xu L, Feng D, Yin R, Li G Y and Jiang Y 2014 Joint mode selection and resource allocation for device-to-device communications *IEEE Trans. Commun.* **62** 3814–24

[136] Wang L and Wu H 2014 Fast pairing of device-to-device link underlay for spectrum sharing with cellular users *IEEE Commun. Lett.* **18** 1803–6

[137] Andrews J G, Buzzi S, Choi W, Hanly S V, Lozano A, Soong A C K and Zhang J C 2014 What will 5G be *IEEE J. Sel. Areas Commun.* **32** 1065–82

[138] Ye Q, Rong B, Chen Y, Al-Shalash M, Caramanis C and Andrews J G 2013 User association for load balancing in heterogeneous cellular networks *IEEE Trans. Wirel. Commun.* **12** 2706–16

[139] Corroy S, Falconetti L and Mathar R 2012 Cell association in small heterogeneous networks: downlink sum rate and min rate maximization *Proc. IEEE Wirel. Commun. and Networking Conf. WCNC (Paris, April 2012)* pp 888–92

[140] Kim K, Jeon S W and Kim D K 2015 The feasibility of interference alignment for full-duplex MIMO cellular networks *IEEE Commun. Lett.* **19** 1500–3

[141] Wang K, Zhang R, Zhong Z, Zhang X and Pang X 2017 Measurement of self-interference channels for full-duplex relay in an urban scenario *Proc. IEEE Int. Conf. on Commun. ICC (Paris, May 2017)* pp 1153–8

[142] Shojaeifard A, Wong K K, Di Renzo M, Zheng G, Hamdi K A and Tang J 2017 Full-duplex versus half-duplex large scale antenna system *Proc. IEEE Int. Conf. on Commun. ICC (Paris, May 2017)* pp 743–8

[143] Marinelli E E *Hyrax: Cloud Computing on Mobile Devices using MapReduce* http://reports-archive.adm.cs.cmu.edu/anon/2009/CMU-CS-09–164.pdf

[144] Satyanarayanan M, Bahl P, Caceres R and Davies N 2009 The case for VM-based cloudlets in mobile computing *IEEE Pervasive Comput.* **8** 14–23

[145] Rost P, Bernardos C J, De Domenico A, Di Girolamo M, Lalam M, Sabella D and Wübben D 2014 Cloud technologies for flexible 5G radio access networks *IEEE Commun. Mag.* **52** 68–76

[146] Varsha H S and Shashikala K P 2017 The tactile Internet *Proc. Int. Conf. on Innovative Mechanisms for Industry Applications (ICIMIA) (Bangalore)* pp 419–22

[147] Fitzek F H P, Li S-C, Speidel S, Strufe T, Simsek M and Reisslein M (ed) 2021 *Tactile Internet* 1st edn (Cambridge, MA: Academic) pp 1–475

[148] The Tactile Internet—ITU-T Technology Watch Report https://itu.int/dms_pub/itu-t/oth/23/01/T23010000230001PDFE.pdf

[149] Pujolle G 2020 *Software Networks: Virtualization, SDN, 5G, and Security* 2nd edn (Hoboken, NJ: Wiley Telecom) pp 1–300

[150] Yukitake T 2017 Innovative solutions toward future society with AI, Robotics, and IoT *Proc. Symp. on VLSI Circuits, Kyoto* pp C16–9

[151] Su H, Ovur S E, Li Z, Hu Y, Li J, Knoll A, Ferrigno G and De Momi E 2020 Internet of Things (IoT)-based collaborative control of a redundant manipulator for teleoperated minimally invasive surgeries *Proc. IEEE Int. Conf. on Robotics and Automation (ICRA), virtual* 9737–42

[152] Mohanta B, Das P and Patnaik S 2019 Healthcare 5.0: a paradigm shift in digital healthcare system using artificial intelligence, IOT and 5G communication *Proc. Int. Conf. on Applied Machine Learning (ICAML) (Bhubaneswar)* 191–6

[153] Zhang L, Liang Y and Niyato D 2019 6G visions: mobile ultra-broadband, super internet-of-things, and artificial intelligence *China Commun.* **16** 1–14
[154] Saad W, Bennis M and Chen M 2020 A vision of 6G wireless systems: applications, trends, technologies, and open research problems *IEEE Network* **34** 134–42
[155] Katz M, Matinmikko-Blue M and Latva-Aho M 2018 6Genesis flagship program: building the bridges towards 6G-enabled wireless smart society and ecosystem *Proc. IEEE LATINCOM (Guadalajara)* pp 1–9
[156] Letaief K B, Chen W, Shi Y, Zhang J and Zhang Y A 2019 The roadmap to 6G: AI empowered wireless networks *IEEE Commun. Mag.* **57** 84–90
[157] Jiang W, Anton S D and Dieter Schotten H 2019 Intelligence slicing: a unified framework to integrate artificial intelligence into 5G networks *Proc. 12th IFIP Wireless and Mobile Networking Conf. (WMNC) (Paris)* pp 227–32
[158] Zhang C, Ueng Y, Studer C and Burg A 2020 Artificial intelligence for 5G and beyond 5G: implementations, algorithms, and optimizations *IEEE J. Emerg. Sel. Topics Circuits Syst.* **10** 149–63
[159] Vázquez M Á, Pallois J P, Debbah M, Masouros C, Kenyon T, Deng Y, Mekuria F, Pérez-Neira A and Erfanian J 2019 Deploying artificial intelligence in the wireless infrastructure: the challenges ahead *Proc. IEEE 2nd 5G World Forum (5GWF) (Dresden)* pp 458–9
[160] Jiang W, Strufe M and Schotten H D 2017 A SON decision-making framework for intelligent management in 5G mobile networks *Proc. 3rd IEEE Int. Conf. on Computer and Communications (ICCC) (Chengdu)* pp 1158–62
[161] Liu H, Wang Y and Hu B 2019 Application research of artificial intelligence in the fifth generation mobile communication technology *Proc. IEEE 10th Int. Conf. on Software Engineering and Service Science (ICSESS) (Beijing)* pp 498–501
[162] Lawless W, Mittu R, Sofge D, Moskowitz I S and Russell S (ed) 2019 *Artificial Intelligence for the Internet of Everything* 1st edn (Amsterdam: Elsevier Academic Press) pp 1–303
[163] Xiang W, Zheng K and Shen X S (ed) 2017 *5G Mobile Communications* 1st edn (Cham: Springer Int. Publishing Switzerland) pp 1–691
[164] Polese M, Jornet J M, Melodia T and Zorzi M 2020 Toward End-to-end, full-stack 6G terahertz networks *IEEE Commun. Mag.* **58** 48–54
[165] Pereira V, Matos T, Rodrigues R, Nóbrega R and Jacob J 2019 Extended reality framework for remote collaborative interactions in virtual environments *Proc. Int. Conf. on Graphics and Interaction (ICGI) (Faro)* pp 17–24
[166] Thum G E and Gaffar A 2017 The future of brain-computer interaction: how future cars will interact with their passengers *Proc. IEEE SmartWorld, Ubiquitous Intelligence & Comp., Advanced & Trusted Computed, Scalable Computing & Commun., Cloud & Big Data Computing, Internet of People and Smart City Innov. (SmartWorld/SCALCOM/UIC/ATC/CBDCom/IOP/SCI) (San Francisco, CA)* pp 1–5
[167] Pun T, Alecu T I, Chanel G, Kronegg J and Voloshynovskiy S 2006 Brain-computer interaction research at the computer vision and multimedia laboratory, University of Geneva *IEEE Trans. Neural Syst. Rehab. Eng.* **14** 210–3
[168] Lemberg K V, Kosmynin A N, Stupnitsky D A, Grushevsky E O and Podshivalov I V 2020 Tunable meta-surface antenna array with holographic beamforming *Proc. 7th All-Russian Microwave Conf. (RMC) (Moscow)* pp 159–61

[169] Ebadi S, Driscoll T and Smith D 2013 Visual illustrations of microwave holographic beamforming using a modulated surface-impedance metamaterial *Proc. IEEE Antennas and Propagation Society Int. Symp. (APSURSI) (Orlando, FL)* pp 2343–4
[170] Deng F, Xi X, Li J and Ding F 2015 A method of designing a field-controlled active frequency selective surface *IEEE Antennas Wirel. Propag. Lett.* **14** 630–3
[171] Li H, Cao Q, Wang Y, Monorchio A and Costa F 2018 Active frequency selective surface with two independent electromagnetic functions *Proc. IEEE Int. Symp. on Antennas and Propagation & USNC/URSI National Radio Science Meeting (Boston, MA)* pp 2047–8
[172] Hu S and Rusek F 2020 Spherical large intelligent surfaces *Proc. IEEE Int. Conf. on Acoustics, Speech and Signal Processing (ICASSP) (Barcelona)* pp 8673–7
[173] Yan W, Yuan X, He Z and Kuai X 2020 Large intelligent surface aided multiuser MIMO: passive beamforming and information transfer *Proc. IEEE Int. Conf. on Commun. (ICC) (Dublin)* pp 1–7
[174] Hu S, Rusek F and Edfors O 2018 Beyond massive MIMO: the potential of positioning with large intelligent surfaces *IEEE Trans. Signal Process.* **66** 1761–74
[175] Li Y, Liu J, Cao B and Wang C 2018 Joint optimization of radio and virtual machine resources with uncertain user demands in mobile cloud computing *IEEE Trans. Multimedia* **20** 2427–38
[176] Ko B, Lee H and Son S H 2016 GPS-less localization system in vehicular networks using dedicated short range communication *Proc. IEEE 22nd Int. Conf. on Embedded and Real-Time Computing Systems and Applications (RTCSA) (Daegu)* p 106
[177] Makridis E and Charalambous T 2020 Towards robust onboard control for quadrotors via ultra-wideband-based localization *Proc. Int. Wireless Commun. and Mobile Computing (IWCMC) (Limassol)* pp 1630–5
[178] Saxena S, Isukapati I K, Smith S F and Dolan J M 2019 Multiagent sensor fusion for connected & autonomous vehicles to enhance navigation safety *Proc. IEEE Intelligent Transportation Systems Conf. (ITSC) (Aukland)* pp 2490–5
[179] Hu H, Smith S F and Goldstein R 2019 Cooperative schedule-driven intersection control with connected and autonomous vehicles *Proc. IEEE/RSJ Int. Conf. on Intelligent Robots and Systems (IROS) (Macau)* pp 1668–73
[180] Ali M A H, Yusoff W A B W, Hamedon Z B, Yussof Z B M and Mailah M 2016 Mechatronic design and development of an autonomous mobile robotics system for road marks painting *Proc. IEEE Industrial Electronics and Applications Conf. (IEACon) (Kota Kinabalu)* pp 336–41
[181] Pereira J, Premebida C, Asvadi A, Cannata F, Garrote L and Nunes U J 2019 Test and evaluation of connected and autonomous vehicles in real-world scenarios *Proc. IEEE Intelligent Vehicles Symp. (IV) (Paris)* pp 14–9
[182] Premnath K N and Rajavelu S 2011 Challenges in self organizing networks for wireless telecommunications *Proc. Int. Conf. on Recent Trends in Information Technology (ICRTIT) (Chennai)* pp 1331–4
[183] Shuang S, Zhuoming L, Peng W, Wanjun Z, Wei W and Xuanli W 2014 Power control for self-organizing network in LTE-advanced system *Proc. Int. Conf. on Instrumentation and Measurement, Computer, Commun. and Control (Harbin)* pp 836–9
[184] Dou Y, Liu H, Wei L and Chen S 2020 Design and simulation of self-organizing network routing algorithm based on Q-learning *Proc. 21st Asia-Pacific Network Operations and Management Symp. (APNOMS) (Daegu)* pp 357–60

[185] Mei J, Wang X and Zheng K Dec. 2020 An intelligent self-sustained RAN slicing framework for diverse service provisioning in 5G-beyond and 6G networks *Intell. Conv. Netw.* **1** 281–94
[186] Zhang S, Zhang N, Fang X, Yang P and Shen X S 2017 Self-sustaining caching stations: toward cost-effective 5G-enabled vehicular networks *IEEE Commun. Mag.* **55** 202–8
[187] Next generation SON for 5G https://huawei.com/minisite/hwmbbf16/insights/Next_Gene_ation_SON_for_5G.pdf
[188] Miozzo M, Ali Z, Giupponi L and Dini P 2021 Distributed and multi-task learning at the edge for energy efficient radio access networks *IEEE Access* **9** 12491–505
[189] Ioannou I, Vassiliou V, Christophorou C and Pitsillides A 2020 Distributed artificial intelligence solution for D2D communication in 5G networks *IEEE Syst. J.* **14** 4232–41
[190] Arena F and Pau G 2020 When edge computing meets IoT systems: analysis of case studies *China Commun.* **17** 50–63
[191] López Peña M A and Fernández I M 2019 SAT-IoT: an architectural model for a high-performance Fog/Edge/Cloud IoT platform *Proc. IEEE 5th World Forum on Internet of Things (WF-IoT) (Limerick)* pp 633–8
[192] Alrowaily M and Lu Z 2018 secure edge computing in IoT systems: review and case studies *Proc. IEEE/ACM Symp. on Edge Computing (SEC) (Seattle, WA)* pp 440–4
[193] Kaka A O and Toycan M 2017 Design of miniaturized on-body antennas for dual band implant antenna communication *Proc. 25th Signal Proc. and Commun. Applications Conf. (SIU) (Antalya)* pp 1–4
[194] Li M, Song Y, Hou Y, Li N, Jiang Y, Sulaman M and Hao Q 2019 Comparable investigation of characteristics for implant intra-body communication based on galvanic and capacitive coupling *IEEE Trans. Biomed. Circuits Syst.* **13** 1747–58
[195] Agneessens S, Van Torre P, Tanghe E, Vermeeren G, Joseph W and Rogier H 2012 On-body wearable repeater as a data link relay for in-body wireless implants *IEEE Antennas Wirel. Propag. Lett.* **11** 1714–7
[196] Seimiya S, Takabayashi K and Kohno R 2017 A study for the adaptive error correction using QoS-HARQ toward dependable implant body area network *Proc. 11th Int. Symp. on Medical Information and Commun. Technology (ISMICT) (Lisbon)* pp 6–10
[197] Teshome A K, Kibret B and Lai D T H 2019 A review of implant communication technology in WBAN: progress and challenges *IEEE Rev. Biomed. Eng.* **12** 88–99
[198] Mao B, Kawamoto Y and Kato N 2020 AI-based joint optimization of QoS and security for 6G energy harvesting Internet of Things *IEEE Internet Things J.* **7** 7032–42
[199] Siskos S, Gogolou V, Tsamis C, Kerasidou A, Doumenis G, Tsiapali K, Katsikas S and Sakellariou A 2019 Design of a flexible multi-source energy harvesting system for autonomously powered IoT : the PERPS project *Proc. 29th Int. Symp. on Power and Timing Modeling, Optimization and Simulation (PATMOS) (Rhodes)* pp 133–4
[200] Iannacci J, Sordo G, Serra E and Schmid U 2015 A novel MEMS-based piezoelectric multi-modal vibration energy harvester concept to power autonomous remote sensing nodes for Internet of Things (IoT) applications *Proc. IEEE SENSORS (Busan)* pp 1–4
[201] Tsenempis I, Filios G, Katsidimas I and Nikoletseas S 2019 Energy harvesting and smart management platform for low power IoT systems *Proc. 15th Int. Conf. on Distributed Computing in Sensor Systems (DCOSS) (Santorini Island)* pp 339–46

[202] Mukhlif F, Nooridin K A B, AL-Gumaei Y A and AL-Rassas A S 2018 Energy harvesting for efficient 5G networks *Proc. Int. Conf. on Smart Computing and Electronic Enterprise (ICSCEE) (Shah Alam)* pp 1–5
[203] Lee Y-H, Wang A-S, Liao Y-D, Lin T-W, Chi Y-J, Wong C-C, Shinohara N, Yuan Q and Chen Q 2017 Wireless power IoT system using polarization switch antenna as polling protocol for 5G mobile network *Proc. IEEE Wireless Power Transfer Conf. (WPTC) (Taipei)* pp 1–3
[204] Shan S, Hsiao V and Hwang R 2019 Wireless power transfer through low-E glass *Proc. IEEE Wireless Power Transfer Conf. (WPTC) (London)* pp 192–6
[205] Sheng M, Wang L, Wang X, Zhang Y, Xu C and Li J 2016 Energy efficient beamforming in MISO heterogeneous cellular networks with wireless information and power transfer *IEEE J. Sel. Areas Commun.* **34** 954–68
[206] Ponnimbaduge Perera T D, Jayakody D N K, Sharma S K, Chatzinotas S and Li J 2018 Simultaneous wireless information and power transfer (SWIPT): recent advances and future challenges *IEEE Commun. Surv. Tutor.* **20** 264–302
[207] Zhai D, Zhang R, Du J, Ding Z and Yu F R 2019 Simultaneous wireless information and power transfer at 5G new frequencies: channel measurement and network design *IEEE J. Sel. Areas Commun.* **37** 171–86
[208] Qi Q, Chen X, Zhong C and Zhang Z 2020 Integration of energy, computation and communication in 6G cellular Internet of Things *IEEE Commun. Lett.* **24** 1333–7
[209] López O L A, Alves H, Souza R D, Montejo-Sanchez S, Fernandez E M G and Latva-aho M 2021 Massive wireless energy transfer: enabling sustainable IoT towards 6G era *IEEE Internet Things J.* **8** 8816–35
[210] Song B, Chen Y, Moon K and Wong C 2017 Design of miura folding-based micro-supercapacitors as foldable and miniaturized energy storage devices *Proc. IEEE 67th Electronic Components and Technology Conf. (ECTC) (Orlando, FL)* pp 2027–32
[211] Weinreich W, Mart C, Zybell S, Bönhardt S, Czernohorsky M, Kühnel K and Kia A M 2018 Materials for on-chip energy storage *Proc. IEEE 18th Int. Conf. on Nanotechnology (IEEE-NANO) (Cork)* pp 1–2
[212] Choi J, Shin Y, Choi M, Park W and Lee I 2019 Robust control of a microgrid energy storage system using various approaches *IEEE Trans. on Smart Grid* **10** 2702–12
[213] Bondyopadhyay P K 1998 Moore's law governs the silicon revolution *Proc. of the IEEE* **86** 78–81
[214] Adee S 2008 The data: 37 years of Moore's Law *IEEE Spectr.* **45** 56
[215] Kuhn K J 2009 Moore's law past 32nm: future challenges in device scaling *Proc. 3th Int. Workshop on Computational Electronics (Beijing)* pp 1–6
[216] Strawn G and Strawn C 2015 Moore's law at fifty *IT Prof.* **17** 69–72
[217] Kahng A B 2010 Scaling: more than Moore's law *IEEE Des. Test Comput.* **27** 86–7
[218] Brown K H 2000 Moore's law-is there more? *Proc. Int. Microprocesses and Nanotech. Conf. (IEEE Cat. No.00EX387) (Tokyo)* pp 2–3
[219] Borkar S 2000 Obeying Moore's law beyond 0.18 micron [microprocessor design] *Proc. 13th Annual IEEE Int. ASIC/SOC Conf. (Cat. No.00TH8541) (Arlington, VA)* pp 26–31
[220] Lu X 2000 Beyond Moore's law [in photonic technology] *Proc. LEOS Summer Topical Meetings. Electronic-Enhanced Optics. Optical Sensing in Semiconductor Manufacturing. Electro-Optics in Space. Broadband Optical Networks (Cat. No.00TH8497) (Aventura, FL)* IV21

[221] Gad-el-Hak M (ed) *MEMS—Introduction and Fundamentals* 1st edn (Boca Raton, FL: CRC Press) 2005 pp 1–492

[222] Gad-el-Hak M (ed) *MEMS—Design and Fabrication* 1st edn (Boca Raton, FL: CRC Press) 2005 pp 1–664

[223] Gad-el-Hak M (ed) *MEMS—Applications* 1st edn (Boca Raton, FL: CRC Press) 2006 pp 1–586

[224] Nasrollahzadeh M, Issaabadi Z, Sajjadi M, Sajadi S M and Atarod M 2019 Types of nanostructures *Interface Science and Technology* 1st edn ed M Nasrollahzadeh, S M Sajadi, M Sajjadi, Z Issaabadi and M Atarod (Amsterdam: Elsevier) ch 2 pp 29–80

[225] Gogotsi Y (ed) 2017 *Nanomaterials Handbook* 1st edn (Boca Raton, FL: CRC Press) pp 1–712

[226] Vollath D 2013 *Nanomaterials: An Introduction to Synthesis, Properties and Applications* 2nd edn (Weinheim: Wiley-VCH) pp 1–386

[227] Lamberti C and Agostini G 2013 *Characterization of Semiconductor Heterostructures and Nanostructures* 2nd edn (Amsterdam: Elsevier) pp 1–828

[228] Kaushik B K (ed) 2018 *Nanoelectronics. Devices, Circuits and Systems—A volume in Advanced Nanomaterials* 1st edn (Amsterdam: Elsevier) pp 1–476

[229] Tian J, Sosin S, Iannacci J, Gaddi R and Bartek M 008 RF–MEMS wafer-level packaging using through-wafer interconnect *Elseiver Sensors Actuators A Phys.* **142** 442–51

[230] Blackwell G R (ed) 1999 *The Electronic Packaging Handbook* 1st edn (Boca Raton, FL: CRC Press) pp 1–640

[231] Watanabe A O, Ali M, Sayeed S Y B, Tummala R R and Raj P M 2021 A Review of 5G front-end systems package integration *IEEE Trans. Compon. Packaging Manuf. Technol.* **11** 118–33

[232] Tian W, Wang X, Niu J, Cui H, Chen Y and Zhang Y 2020 Research status of wafer level packaging for RF MEMS switches *Proc. 21st Int. Conf. on Electronic Packaging Technology (ICEPT) (Guangzhou)* pp 1–5

[233] Chen G, Ma M, Liu Z, Wei A, Zavabeti A, Hu X, Zhang F, Wang B, Mitchell A and Li Y 2019 Anodic bondable Li–Na–Al–B–Si–O glass-ceramics for Si—ULTCC heterogeneous integration *J. Eur. Ceram. Soc.* **39** 2419–26

[234] Murillo G, Agusti J, López-Suárez M and Gabriel A 2011 Heterogeneous integration of autonomous systems in package for wireless sensor networks *Procedia Eng.* **25** 88–91

[235] Theuss H and Pressel K 2015 3D integration of MEMS *Handbook of Silicon Based MEMS Materials and Technologies* 2nd edn ed M Tilli, T Motooka, V-M Airaksinen, S Franssila, M Paulasto-Kröckel and V Lindroos (Amsterdam: Elsevier) ch 37 pp 678–93

[236] Dan Hutcheson G and Bennett H S More than Moore or more Moore: a SWOT analysis https://tsapps.nist.gov/publication/get_pdf.cfm?pub_id=909225

[237] Kuck D 2013 Keynote talk: a comprehensive approach to HW/SW codesign *Proc. 22nd Int. Conf. on Parallel Architectures and Compilation Techniques (Edinburgh)* pp 1–1

[238] Noguera J and Badia R M 2002 HW/SW codesign techniques for dynamically reconfigurable architectures *IEEE Trans. Very Large Scale Integr. Syst.* **10** 399–415

[239] Yeniçeri R and Hüner Y 2020 HW/SW codesign and implementation of an IMU navigation filter on Zynq SoC with Linux *Proc. 7th Int. Conf. on Electrical and Electronics Engineering (ICEEE) (Antalya)* pp 351–54

[240] Wolff C, Gorrochategui I and Bücker M 2011 Managing large HW/SW codesign projects *Proc. 6th IEEE Int. Conf. on Intelligent Data Acquisition and Advanced Computing Systems (Prague)* pp 919–22

[241] Iannacci J and Poor H V 2022 Review and perspectives of micro/nano technologies as key-enablers of 6G *IEEE Access* **10** 55428–58
[242] Markowski M, Ryba P and Puchała K 2016 Software defined networking research laboratory-experimental topologies and scenarios *Proc. European Network Intelligence Conf. (ENIC) (Wroclaw)* pp 252–6
[243] Theodorou T and Mamatas L 2017 CORAL-SDN: a software-defined networking solution for the Internet of Things *Proc. IEEE Conf. on Network Function Virtualization and Software Defined Networks (NFV-SDN) (Berlin)* pp 1–2
[244] Dahir M H, Alizadeh H and Gözüpek D 2017 Energy efficient virtual network embedding in federated software defined networks *Proc. 25th Signal Processing and Commun. Applications Conf. (SIU) (Antalya)* pp 1–4
[245] Olimjonovich M S 2016 Software defined networking: management of network resources and data flow *Proc. Int. Conf. on Information Science and Commun. Technologies (ICISCT) (Tashkent)* pp 1–3
[246] Al-Quzweeni A, El-Gorashi T E H, Nonde L and Elmirghani J M H 2015 Energy efficient network function virtualization in 5G networks *Proc. 17th Int. Conf. on Transparent Optical Networks (ICTON) (Budapest)* pp 1–4
[247] Casellas R, Vilalta R, Martínez R and Muñoz R 2017 Highly available SDN control of flexi-grid networks with network function virtualization-enabled replication *IEEE/OSA J. Opt. Commun. Netw.* **9** A207–15
[248] Krasko O, Al-Zayadi H, Pashkevych V, Kopets H and Humeniuk B 2018 Network functions virtualization for flexible deployment of converged optical-wireless access infrastructure *Proc. 14th Int. Conf. on Advanced Trends in Radioelecrtronics, Telecommun. and Computer Engineering (TCSET) (Slavske)* pp 1135–8
[249] Guan W, Wen X, Wang L, Lu Z and Shen Y 2018 A Service-oriented deployment policy of end-to-end network slicing based on complex network theory *IEEE Access* **6** 19691–701
[250] Esteves J J A, Boubendir A, Guillemin F and Sens P 2020 Edge-enabled optimized network slicing in large scale networks *Proc. 11th Int. Conf. on Network of the Future (NoF) (Bordeaux)* pp 129–31
[251] Wang W, Guo W and Hu W 2018 Network service slicing supporting ubiquitous access in passive optical networks *Proc. 20th Int. Conf. on Transparent Optical Networks (ICTON) (Bucharest)* pp 1–3
[252] Iannacci J 2021 The WEAF Mnecosystem: a perspective of MEMS/NEMS technologies as pillars of future 6G, tactile internet and super-IoT *Microsyst. Technol.* **27** 4193–207 https://doi.org/10.1007/s00542-021-05230-3

Chapter 4

RF-MEMS passives for 5G applications: case studies and design examples

4.1 An introduction to RF passives' specifications trends in 5G

The insight provided in previous chapters on Radio Frequency Micro-ElectroMechanical Systems (RF-MEMS) technologies is quite exhaustive. First, an achievable performance along with the added value of pronounced reconfigurability/tunability, targetable through Radio Frequency (RF) passive components designed in Micro-ElectroMechanical Systems (MEMS) technology, were reported and discussed in the first chapter. Second, trends in terms of system-level performance, triggered by current 4G-LTE mobile communications and destined to be pursued relentlessly with the advent of 5G standards, have been discussed in a generous amount of detail in the preceding chapter. To complete the picture, an intermediate chapter, between the introduction of RF-MEMS and that of 5G, detailed the main reasons that, in the past, limited concrete market absorption of RF passive products based on MEMS technology, and framed also the significantly changed scenario in which RF-MEMS are on the way to consolidation, especially with reference to 4G-LTE market applications. In light of this landscape, the purpose of this last chapter is to blend together all the elements scattered around, up to here, and build up more technical insight around the possible exploitation of RF-MEMS technology in the 5G frame of reference.

As a matter of fact, the performance indicators and trends discussed throughout the chapter on 5G were mainly based on the system-level plane of abstraction. In addition, the key enabling technologies listed before, like massive Multiple Input Multiple Output (MIMO), millimetre-wave (mm-wave) solutions, and so on, refer to complete systems and subsystems. Therefore, prior to stepping further into the area of RF-MEMS design and simulation technicalities, effort should be oriented

doi:10.1088/978-0-7503-4199-8ch4

towards lowering specifications and characteristics from the system level down to the component level.

Bearing in mind the aforementioned target, a first useful step consists in listing the classes of RF passives that are expected to be beneficial in order to empower the 5G scenario discussed before. Such components, also taking into account the 5G New Radio (NR) concepts that will lie in the mm-wave range, can be grouped as follows:

(1) Very-wideband switches and switching units (e.g. Multiple Pole Multiple Throw, MPMT) with low loss (when ON), high isolation (when OFF) and very low adjacent channels cross-talk, working from 2–3 GHz up to 60–70 GHz (and more).
(2) Reconfigurable filters with pronounced stopband rejection characteristics and very low attenuation of the passed band.
(3) Very-wideband multi-state impedance tuners.
(4) Programmable step attenuators with multiple configurations and very flat characteristics over 60–70 GHz frequency spans.
(5) Very-wideband multi-state/analogue phase shifters.
(6) Hybrid devices with mixed phase shifting and programmable attenuation; functionalities described in points (4) and (5) blended into a unique device.
(7) Miniaturised antennas and arrays of antennas, possibly integrated monolithically with one or more of the devices described in the previous points from (1) to (6).

These listed devices could find application in several parts of the intricate 5G landscape, both with reference to smartphones and end-user terminals, as well as to fixed/mobile infrastructure. For instance, high-performance switches and switching units would be beneficial to MIMO systems, both on the terminal and infrastructure ends. In addition, wideband impedance tuners are critical devices to enable adaptive matching between the antenna and RF Front End (RFFE) of smartphones. From a different perspective, mm-wave small cells will require advanced beamforming capability. To this end, widely reconfigurable phase shifters, as well as multi-state programmable step attenuators, are critical components to ensure proper feeding of the array of antennas. Moreover, it would be desirable to implement both of these functions within the same component (monolithic solution), thus reducing losses, area occupation, hardware complexity and cost. Following the same direction, monolithic realisation of the array of antennas and of the passive front end necessary for advanced beamforming, i.e. phase shifters and attenuators, would mean stepping to a higher level from the point of view of system performance.

Given the just-disclosed background on RF passives of interest for 5G applications, it is now time to move our attention to the expected performance figures and characteristics. In this regard, the *performance bowl* is shown in figure 4.1.

Within the bowl, each *bubble* is a performance indicator and its placement along the vertical axis indicates if it is expected to be high/wide (bowl ceiling), or low/small (bowl floor). For instance, isolation of switching elements (when OFF) is supposed to be very pronounced, while cross-talks between adjacent channels must be kept as

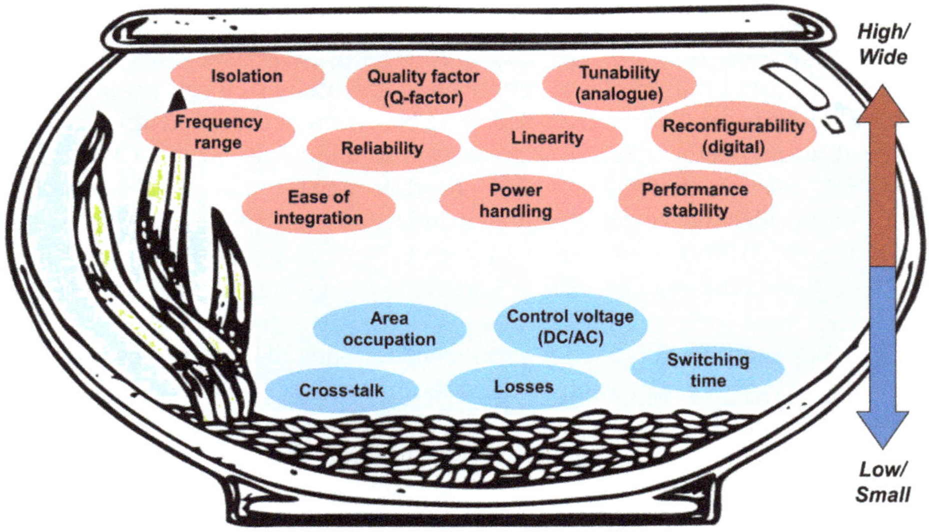

Figure 4.1. Performance bowl including the most critical specifications and characteristics expected from RF passives for 5G applications. The vertical placement of each *bubble* (characteristic) within the bowl depends on how high or low it is expected to be (e.g. high isolation, low losses).

low as possible. These are the reasons why the isolation and cross-talk bubbles are placed to the top and bottom of the bowl height, respectively.

Given the specifications within the performance bowl in figure 4.1, it is not viable to quantify the unique numbers to be targeted, as they strongly depend on multiple factors, such as the frequency range of operation, as well as the requirements imposed by the specific application. In addition, the topmost targetable performance for a certain device also depends on its complexity. For example, if the losses of a single switch (when ON) can reasonably approach a certain number, the same characteristic of a switching unit with multiple cascaded stages will be unavoidably worse than that value. In light of these considerations, it is in any case possible to sketch a quantitative list of reference target numbers for some of the performance indicators reported in figure 4.1, bearing in mind the scenario of 5G.

- *Frequency range.* In light of the discussion developed before, in the 5G scenario there will be significant diversification of the network infrastructure. In this regard, there clearly emerged a frequency divide between the classical backhaul infrastructure, working in the sub-6 GHz range, compared against small cells operating in the mm-wave range. Therefore, RF passives will be required to exhibit good performance in a very wide span, starting from 2–3 GHz and ramping up to 60–70 GHz. This is particularly critical for components to be integrated within smartphones and mobile terminals in general, as such devices will have to work properly when linked both to traditional base stations (sub-6 GHz) and to small cells (mm-wave range).
- *Isolation.* Given the pronounced RFFEs hardware reconfigurability and switching capabilities that need to be addressed within the 5G scenario, a

critical feature of switches and, more in general, switch-based passives, will be isolation (when OFF), which is expected to be as high as possible. Just to provide an indication, isolation better than −30/−40 dB for frequencies as high as possible would be desirable. Of course, if it is a reasonable task to design micro-relays with such an isolation operating up to a few GHz, the target becomes much more challenging when dealing with millimetre wave, because of input/output parasitic coupling effects.

- *Losses.* Referring to the scenario of the previous point, the other side of the coin is that of losses introduced by switches (when ON) and, more in general, RF passives when operating in non-isolating configurations. It is straightforward that losses are to be as limited as possible. It would be desirable to keep losses below −1 dB on the widest possible frequency range. As a matter of fact, however, losses get worse when the frequency of operation increases, and also when the passive is longer/more complex, because of parasitic effects. The challenge of keeping the losses as limited as possible is particularly arduous; as in 5G, both the aforementioned factors will take place, i.e. frequency will rise up to the mm-wave range and passives will have to be widely reconfigurable, which results in having complex, articulated and redundant designs.
- *Cross-talk.* Still bearing in mind the pronounced reconfigurability of 5G, cross-talk emerges as another critical performance figure. It can be defined as the extent to which adjacent channels, ideally conceived to be completely isolated from each other, in fact interfere and couple one another. It is easy to appreciate how critical it is to keep cross-talk low in multi-path scenarios, like the one of massive MIMO technologies and advanced beamforming capability. In this regard, a desirable threshold would keep cross-talks below −50/−60 dB over the widest frequency range possible, although it tends to degrade ramping up to the mm-wave range due to the kicking in of parasitic effects and unwanted couplings.
- *Switching time.* Moving our attention towards performance indicators not directly related to the RF and electromagnetic characteristics of passives, it is worth mentioning the switching time. Given the pronounced reconfigurability demanded of 5G RFFEs and RF hardware, it clearly emerges that idle time in commuting from one state to another must be kept as limited as possible. Just to provide an example, this could be the case of a mobile device hopping from sub-6 GHz coverage to mm-wave small cell services, during an outdoor-to-indoor environment transition. Breaking down the commutation time issue from system to component level, the fundamental critical device is the single switch and the required time for its transition from stable OFF to stable ON configuration, as well as vice versa. Given the millisecond-range latency targeted by 5G services, the basic relay switching time has to be definitely lower than 1 ms, with few fractions of ms (e.g. 200–300 μs) as a reasonable target.
- *Control voltage.* Eventually, still referring to switches and RF passive reconfigurable devices, their operation requires driving them by means of

suitable DC/AC electrical signals, i.e. voltages/currents. It is clear that, if within the reference scenario of ground infrastructures, providing biasing signals is, in principle, not a critical issue, the context changes dramatically in mobile and battery-operated devices. In the latter case, the extent of DC/AC biasing signals, as well as power consumption in driving RF passives, should be kept as low as possible. It would also be desirable to maintain compliance with Complementary Metal Oxide Semiconductor (CMOS) typical driving voltages, in order to avoid incorporation of additional ad hoc voltage pumping circuitry. This means working with control voltages not higher than 1–2 V.

In the following sections, multi-physics electromechanical and electromagnetic modelling of RF-MEMS design case studies is going to be discussed, targeting the just-reported 5G driven performance indicators and characteristics.

4.2 A few notes about the RF-MEMS technology platform

Prior to discussing the technical details of a few case studies and reference designs of RF-MEMS for 5G applications, some references are going to be dispensed concerning the technology fabrication process. In order to keep the understanding of the reported examples simple and coherent, a unique RF-MEMS manufacturing process is chosen, that being the 6-inch silicon wafers RF-MEMS surface micromachining process flow, available at the Center for Sensors and Devices (SD) of the Fondazione Bruno Kessler (FBK) in Trento (Italy). The main fabrication steps of the SD-FBK RF-MEMS technology are schematically depicted by the simplified sequence in figure 4.2.

The technology features two buried conductive layers, namely polycrystalline silicon and aluminium, meant for DC biasing/calibrated resistive loads and RF signals underpasses, respectively. Subsequently, a thin gold layer is evaporated above the opened areas to the aluminium layer underneath, in order to improve the quality of the ohmic contact. Then, a photoresist sacrificial layer is patterned wherever MEMS elevated membranes are supposed to be. Finally, two distinct gold electrodeposition steps define the structural parts of the device, and air gaps are eventually released by removing the sacrificial layer. The most relevant geometrical, mechanical and electrical properties of the layers are reported in table 4.1, while a more detailed discussion around the SD-FBK RF-MEMS technology is available in [1–3].

4.3 Electromechanical simulation of RF-MEMS devices

The target of this section is to describe a fundamental methodology devoted to the modelling and simulation of the coupled electromechanical behaviour of RF-MEMS devices. Before discussing practical cases, however, some considerations must be developed. First, the approach that is going to be shown is neither the only available, nor the best one could think to exploit in any situation, when developing a certain RF-MEMS project. Such a circumstance takes place because it is a specific

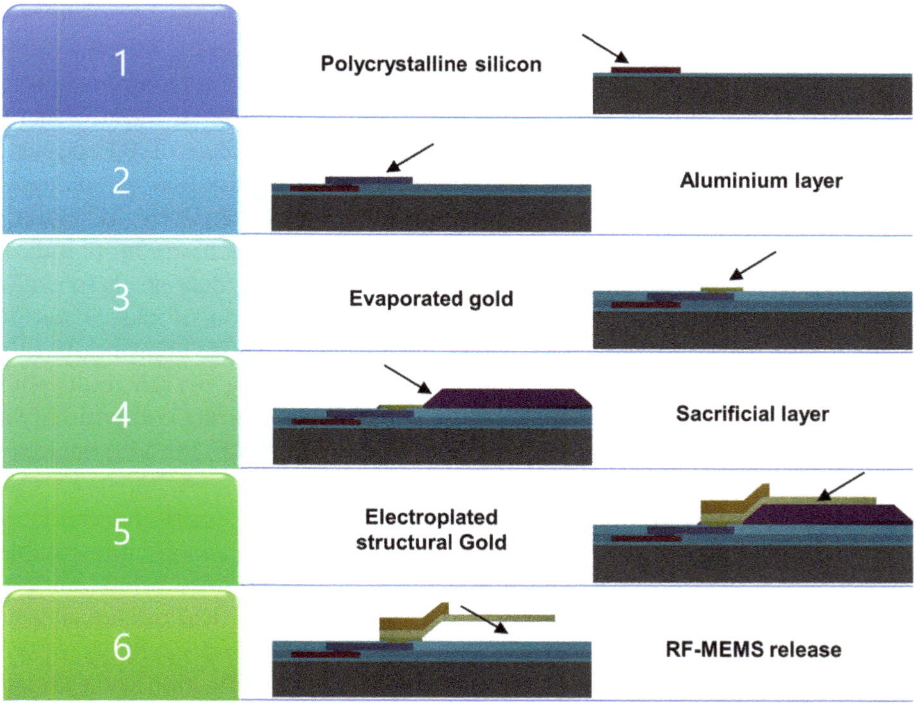

Figure 4.2. Schematic cross-sectional simplified sequence of the main fabrication steps of the SD-FBK RF-MEMS surface micromachining process. Details about the deposited layers are reported in table 4.1.

Table 4.1. Mechanical and electrical fundamental properties of the layers composing the SD-FBK RF-MEMS surface micromachining fabrication flow depicted in figure 4.2.

Physical layer	Fabrication step (see figure 4.2)	Nominal thickness	Sheet resistance [Ω/sq]	Dielectric constant [F/m]	Young's modulus [GPa]
Silicon substrate		625 μm	>5000 Ω.cm	11.9	
Field oxide		1 μm		3.94	
Polycrystalline silicon	1	630 nm	1584		
Insulating silicon oxide		300 nm		3.94	
Aluminium layer	2	630 nm	0.0654		
Insulating silicon oxide		100 nm		3.94	
Evaporated gold	3	150 nm	0.126		
Sacrificial layer	4	3 μm			
Electroplated structural gold (layer 1)	5	2 μm	0.022		98.5
Electroplated structural gold (layer 2)	5	3 μm	0.0055		98.5

problem to be addressed and solves the fundamental driver, defining which simulation approach can be indicated to be more suitable than the others available. For the purposes of clear understanding, let us just mention a reference example. If the RF-MEMS designer/developer/engineer has to develop a more structured comprehension of the materials behaviour against the amount of RF power the device has to stand, or of the metal-to-metal contact degradation versus cycling, simulation tools dedicated to the mechanical/structural analysis of materials, possibly based on the Finite Element Method (FEM) approach, are more indicated than others. On the other hand, if the designer/developer/engineer needs to evaluate device-level characteristics of a certain RF-MEMS component placed within a subsystem also featuring active electronics, it is definitely more appropriate to rely on fast simulation tools. The latter are able to describe the overall multi-physics behaviour of the RF passive in MEMS technology, also ensuring compatibility and proper interface to the CMOS part, so that hybrid (CMOS/RF-MEMS) devices can be simulated as a whole.

Whereas, on one hand, FEM tools address very accurate results and predictions, on the other hand they require more computational capacity and are not suitable to describe the overall electromechanical, RF and electronic coupled behaviour within the same framework. In contrast, fast tools based on analytical and/or simplified (order-reduced) models do not target the same accuracy and flexibility of FEM simulation environments, although they do enable easy identification and evaluation of multiple trade-offs existing in RF-MEMS. Moreover, as mentioned above, they are compatible with standard CMOS design kits and, therefore, can be simulated within the same analysis frameworks suited for electronics development.

Of course, using more accurate or high-level modelling tools is not an exclusive choice, but rather a matter of selecting the proper approach depending on the specific needs arising at different stages of the RF-MEMS design concept development chain. For the purposes of this work, case studies will be reported at device level, in order to emphasise the correlation between changes/choices at layout geometry level and the performance indicators mentioned at the beginning of this chapter. With the aim of maintaining easy understanding of the reported examples to the widest possible extent, specific technical details regarding simulations/simulators settings will be omitted. Nonetheless, references to articles and books dealing with such technicalities will be provided, thus facilitating readers who are interested in carrying out more detailed studies on specific topics of RF-MEMS devices design and development.

4.3.1 Static structural simulation of the pull-in/pull-out characteristic

Bearing in mind the list of critical performances requested of RF passives from high-level 5G specifications outlined in the beginning of this chapter, our attention is now going to be focused on the pull-in voltage of RF-MEMS switching devices. As mentioned in the first chapter, in electrostatically driven MEMS the pull-in voltage is a critical threshold of the imposed DC/AC bias, at which instability between electrostatic attraction force and mechanical restoring force take place [4].

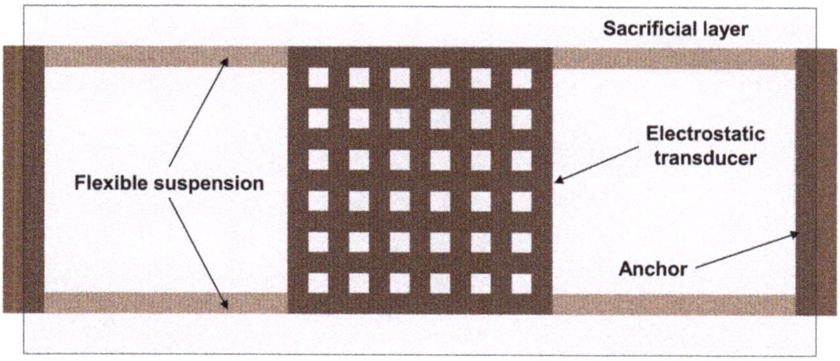

Figure 4.3. Layout of the clamped–clamped MEMS electrostatic micro-relay case study.

As a result, at pull-in voltage the MEMS collapses abruptly on the underlying electrodes, thus establishing metal-to-metal contact between the input/output terminations, if the device is ohmic, or realising a high-capacitance configuration, if the switch is designed to be capacitive.

The reference design is a quite standard configuration of a MEMS movable membrane realised in the SD-FBK RF-MEMS technology platform. It consists of a central square plate for electrostatic transduction, kept suspended above the underlying counter-electrode and input/output contacts by means of four deformable straight beams. Mechanical anchoring areas are placed at both ends of the suspended MEMS, giving rise to a so-called clamped–clamped configuration. The layout of the switch, realised in the L-Edit Layout Editor [5], is shown in figure 4.3.

The central electrostatic transducer is a 130 μm by 130 μm plate, joined to four 120 μm long and 10 μm wide flexible suspensions. The plate has 10 μm square holes (spaced at 10 μm from each other) necessary for the effective removal of the sacrificial layer underneath its surface [2]. Finally, external anchoring overlaps the sacrificial layer, thus realising a gold vertical transition to the substrate underneath that ensures mechanical support to the elevated membrane (air gap). With reference to table 4.1, the whole structure in figure 4.3 is realised with the first 2 μm thick gold metallisation (light colour in figure 4.3). Furthermore, the central plate and anchoring areas also comprise the second 3 μm thick gold metallisation, thus yielding a 5 μm thick membrane (dark colour in figure 4.3). This is done because the central transducer and the anchoring parts are meant to be mechanically stiff. However, flexible suspensions should be as deformable as possible in order to keep the pull-in voltage low. As for studying the pull-in/pull-out characteristics, no other parts of the RF-MEMS device are necessary. In figure 4.3 all the other layers referring to buried metallisations are omitted (see figure 4.2 and table 4.1).

The layout generated in L-Edit is a 2D structure. However, in order to make the model suitable for FEM simulations, it has to be extended to a 3D structure. To do so, the layout in figure 4.3 is exported from L-Edit in Drawing Exchange Format (DXF) [6] and imported into the FreeCAD [7] environment, where a proper thickness is assigned to both layers through an extrude operation. The 3D model of the switch is shown in figure 4.4.

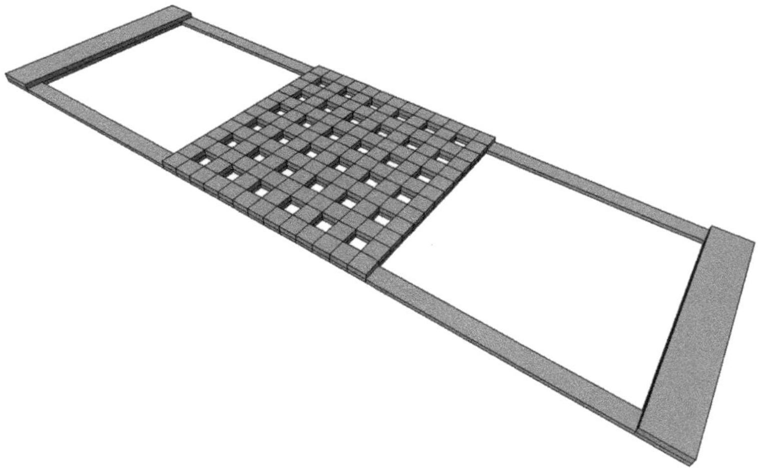

Figure 4.4. 3D model of the MEMS switch after extrusion with proper thickness of both layers performed in a FreeCAD environment.

Once the RF-MEMS switch 3D model is ready, it is exported from FreeCAD in the so-called Standard for the Exchange of Product model data (STEP) format [8], and then imported within the ANSYS Workbench [9] simulation environment; this being the actual tool for carrying out FEM-based analysis. Within Workbench, mechanical properties are applied to the 3D structure according to the information reported in table 4.1. Suitable Boundary Conditions (BCs) are also imposed on the model. In particular, fixed (zero) displacement is set for the anchoring areas, so that the mechanical constraint operated by gold transition from the air-gap to substrate level is properly taken into account. On the other hand, electromechanical transduction is operated by including in the model TRANS126 elements through the Electro-Mechanical Transducers Generation (EMTGEN) macro [10]. In brief, a lumped parallel plate capacitor element is generated per each node of the mesh underneath the transducer plate. An initial air gap (rest position) of 3 μm is set with a minimum allowed gap of 100 nm. The latter is meant to avoid convergence issues when pull-in occurs, as the electrostatic force tends to excessive values when the gap is too small. As dynamic effects are not relevant in observing the pull-in/pull-out characteristic, DC static structural analysis is appropriate for this type of investigation. The voltage imposed across TRANS126 elements is ramped from 0 V up to 60 V, and then back to 0 V. The Workbench 3D schematic of the pulled-in (actuated) model is shown in figure 4.5, where the colour scale refers to the extent of vertical displacement. Deformation was magnified in order to make it visible. The way the 3D model was subdivided into mesh elements is not displayed for the sake of simplicity.

The pull-in/pull-out characteristic, i.e. vertical displacement of the central plate, versus applied DC bias, resulting from the simulation, is reported in figure 4.6. In detail, the pull-in takes place at 51.5 V, while the pull-out happens at 7 V.

Recalling what has been mentioned before concerning 5G specifications, the just-listed values are both quite high. Therefore, efforts at the design level should be

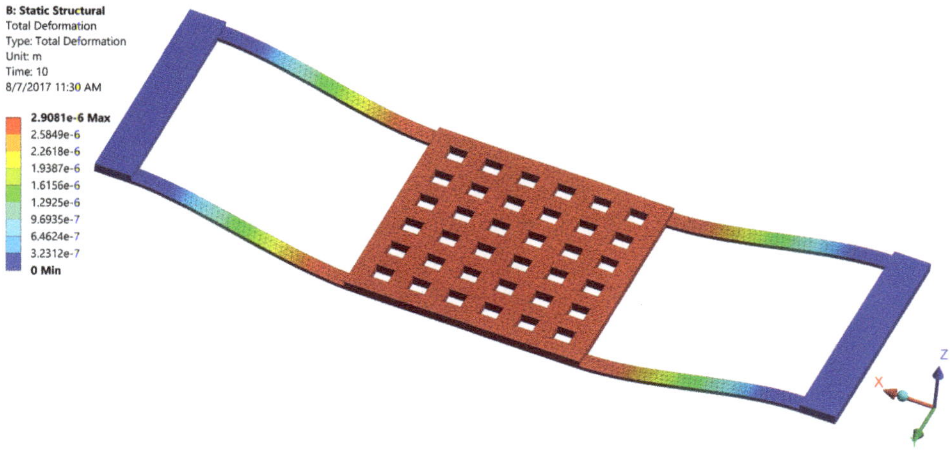

Figure 4.5. Workbench 3D schematic of the pulled-in RF-MEMS switch. The colour scale refers to the extent of vertical deformation. The displacement was magnified to make it easily detectable.

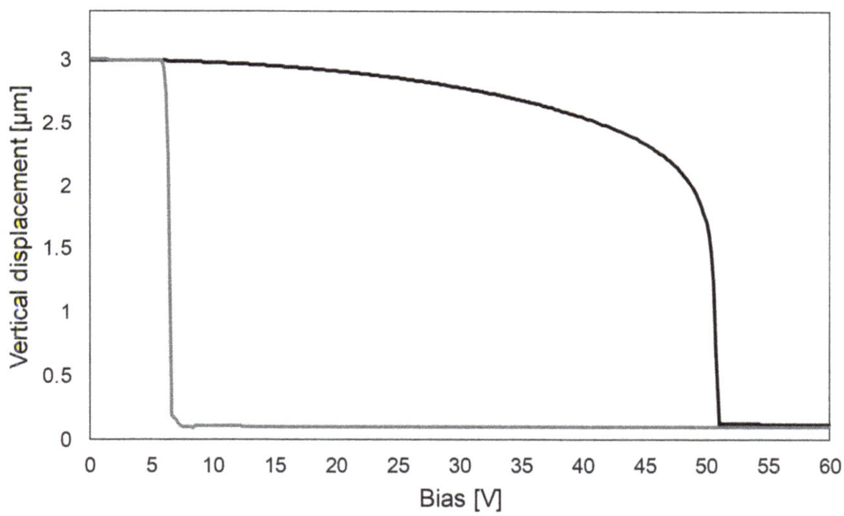

Figure 4.6. Pull-in/pull-out characteristic of the MEMS switch design in figure 4.3 resulting from DC static structural analysis performed within Workbench.

invested with the aim of lowering the activation/deactivation DC levels. One of the Degrees of Freedom (DoFs) available is to reduce the elastic constant of flexible suspensions, e.g. increasing their length. A possible solution to reach this target while keeping the RF-MEMS switch design compact is through designing folded (or meander-/serpentine-like) suspensions [11, 12]. In light of these considerations, the modified layout in figure 4.7 is proposed.

The 2D layout in figure 4.7 is exported/imported following the same procedure previously described for the structure in figure 4.3, and then simulated within Workbench. The pull-in/pull-out characteristic resulting from DC static structural

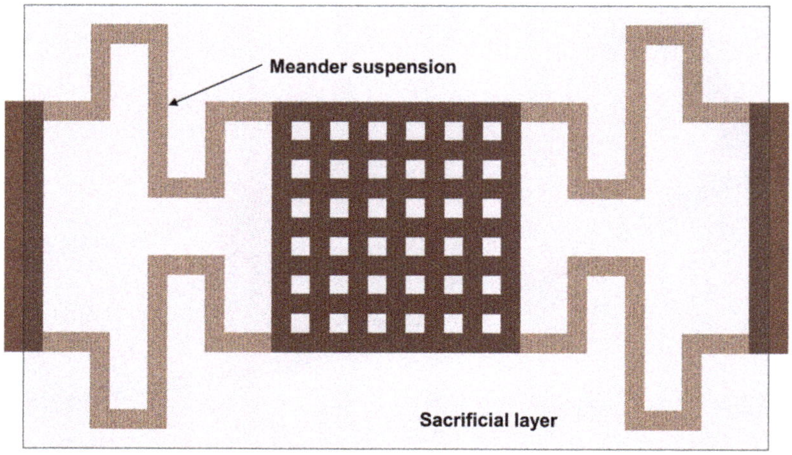

Figure 4.7. Layout of the clamped–clamped MEMS electrostatic micro-relay with meander-like folded flexible suspensions.

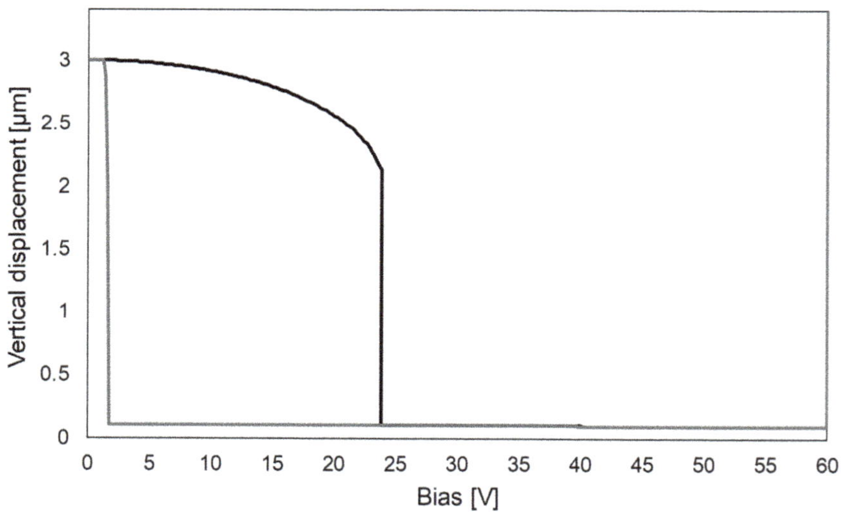

Figure 4.8. Pull-in/pull-out characteristic of the MEMS switch design in figure 4.7 resulting from the DC static structural analysis performed within Workbench.

simulation is reported in figure 4.8, showing activation and deactivation voltages of 24 V and 1.6 V, respectively.

With respect to the original layout with straight beams in figure 4.3, the pull-in and pull-out levels of the solution with folded suspensions are lowered by 47% and 23%, respectively. Of course, the approach of serpentine-shaped suspensions can be further pursued, in order to keep on lowering the pull-in/pull-out threshold levels.

4.3.2 Transient dynamic simulation of the switching (opening/closing) time

Still bearing in mind the specifications derived by the 5G requirements mentioned in the beginning of the chapter, another relevant performance indicator for RF-MEMS micro-relays is the switching time, as it is correlated to the latency of the system. To this purpose, transient structural analysis is performed in Workbench on both MEMS switch layouts previously reported in figure 4.3 and figure 4.7. The electrostatic transduction is implemented once again relying on TRANS126 elements. A 0–60 V square pulse is set as stimulus and is reported in figure 4.9(a).

The analysis is performed over a time interval of 2 ms, with 250 ns step resolution. Rise and fall times are 1 ns and take place at 100 and 500 μs, respectively (pulse length 400 μs; 20% duty cycle). The response in terms of vertical displacement versus

(b)

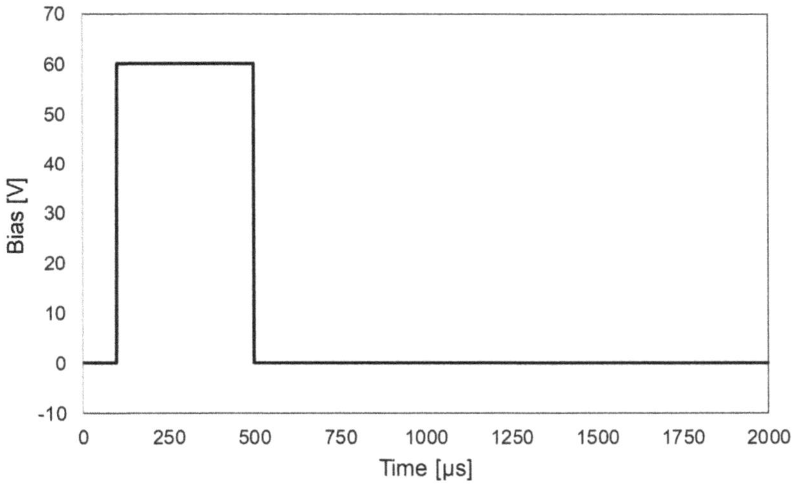

(a)

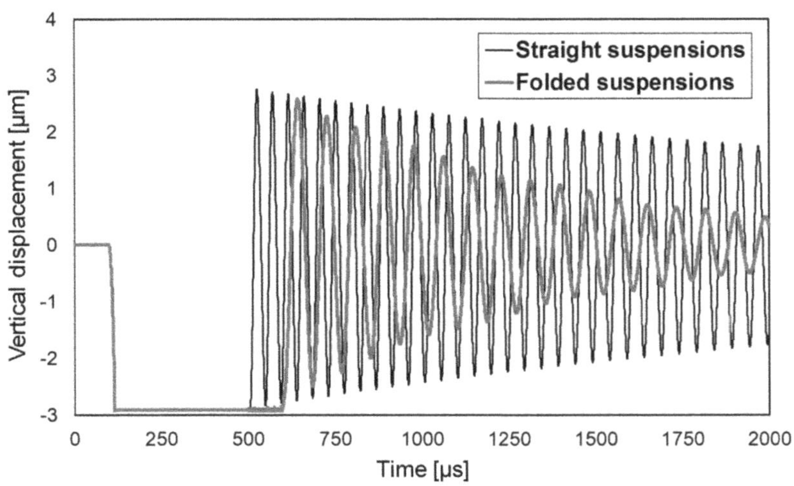

Figure 4.9. (a) Square pulse provided as stimulus in the transient structural analysis. (b) Dynamic vertical displacement response of the MEMS switch layouts discussed in figure 4.3 and figure 4.7.

time is reported in figure 4.9(b). Differing from static pull-in/pull-out characteristics, the rest position level on the vertical axis is set to 0 μm, in order to help visualise and assess overshooting during ringing at release. The pull-in transition is quite steep for both MEMS designs, while after the bias fall edge different dynamic characteristics are evident. First, it should be highlighted that after the bias is zeroed, dynamic evolution of the MEMS structure is free (no stimuli applied). Therefore, its observation provides meaningful insight into the switch's mechanical characteristics. In particular, the frequency of damped ringing, visible in figure 4.9(b), is determined by the elastic constant of flexible suspensions, and corresponds to the fundamental resonant mode of the mass-spring system composed by the rigid central plate and deformable beams [13]. Looking at the oscillations of the MEMS plate, highlighted in the close-up reported in figure 4.10(b), the period measured for the MEMS with

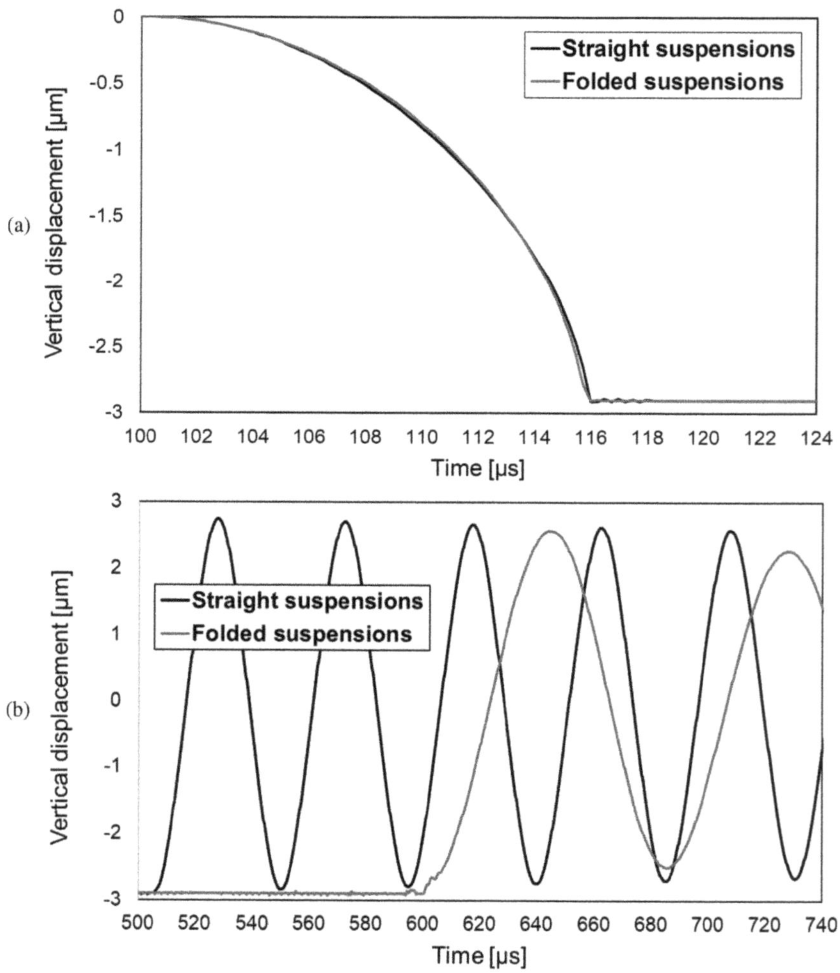

Figure 4.10. Close-ups of the dynamic behaviour of both MEMS switch topologies. (a) Bias rise edge (actuation). (b) Bias fall edge (release).

straight beams and folded suspensions is of 51 and 84 μs, respectively, corresponding to a resonant frequency of 19.6 kHz for the former and 11.9 kHz for the latter. Modal eigenfrequency analysis of both structures was performed in Workbench [14], albeit not reported here for brevity, with the aim to cross-check these values. The results obtained concerning the fundamental resonant mode frequency of the case studies with different suspensions geometry are in close agreement with the numbers reported above, thus confirming the soundness of the modelling approach.

The air-damping effect is not explicitly accounted for in this set of simulations, in order to probe into the dynamic of the MEMS structure itself. Nonetheless, a cumulative damping factor can be set in the transient simulation, to introduce dissipative energy effects that model the main losses of physical systems, like the damping intrinsic to the material (thermoelastic damping) [15–17] and, in fact, dissipative phenomena due to the viscosity of air. In the simulations reported here, the cumulative damping factor is set to a quite low and conservative value, in order to weakly account for the viscosity of air and other dissipative effects. In any case, modelling of viscous damping due to the gas in which the MEMS device is immersed was previously carried out, and it can be fully embodied in simulations, also accounting for the squeeze-damping effect [2, 3] and gas flow across holes of the perforated central transducer [18–20]. Given these statements, damping effects induce a more pronounced attenuation of ringing amplitude in the evolution of the MEMS with folded suspensions (figure 4.7) rather than with straight beams (figure 4.3). This happens because the energy stored by the MEMS device in figure 4.3, when kept actuated, is larger due to the greater spring constant of flexible suspensions [4].

Further considerations are now going to be developed around the switching (closing/opening) times for both the MEMS geometries discussed. To this purpose, close-ups of the dynamic evolution previously shown in figure 4.9(b) are reported for the structures' actuation and release in figure 4.10(a) and figure 4.10(b), respectively. In both plots, the origin of the time axis is placed at the proper edge of the square pulse depicted in figure 4.9(a), i.e. rise edge at 100 μs and fall edge at 500 μs.

Focusing at first on the actuation, the plot in figure 4.10(a) shows that the characteristics of the MEMS devices are very similar to each other. Contact to the underlying plane takes place around 18 μs after bias commutes from 0 to 60 V. Small ripples are visible for a couple of μs after the contact is reached. However, they are too small to be attributed to after-impact mechanical bouncing [21, 22]. More likely, such small fluctuations are due to numerical convergence stabilisation of the simulator in dealing with the abrupt discontinuity of the physical contact, which does not take place smoothly but rather suddenly. In any case, steady pulled-in (actuated) configuration is established in less than 18 μs.

However, the release phase, reported in figure 4.10(b), highlights a different behaviour of the two MEMS geometries. The switch with straight suspensions starts detaching from the underlying contact surface around 6 μs after the square pulse fall edge. Nevertheless, the design variation with folded suspensions commences moving upward around 100 μs after the bias drops from 60 to 0 V. A longer release time can reasonably be expected for mechanical structures with lower elastic constant of suspensions, such as in the case of meander-like designs. As a matter of fact, a lower spring constant leads to less restoring force when the MEMS is actuated and, in turn, to slower release dynamics.

The considerations just developed highlight the existence of conflicting specifications, depicting the typical context within which MEMS designers have to come up with effective trade-offs. If, on one hand, lowering the elastic constant enables significant reduction of the pull-in voltage, on the other hand it makes the device's dynamic response slower. The latter statement holds validity for the actuation phase, as well. One should not be misled by the plot in figure 4.10(a), where the transition time is the same for both MEMS designs. In that case, 60 V was imposed on the switch with meandered suspensions, while its pull-in voltage is around 24 V, as reported in figure 4.8. To this purpose, the plot in figure 4.11 shows the meandered switch dynamic for different bias levels.

Lowering the square pulse peak value from 60 to 26 V, i.e. slightly above pull-in, the transition time steps from 16–18 μs to 47 μs. In addition, reducing the MEMS device restoring force makes it more prone to failure mechanisms, like charge accumulation [23, 24], micro-welding [25, 26], and so on, that can lead to stiction, i.e. missed deactuation when the bias is zeroed.

In conclusion, wrapping up the indications observed up to now, the proposed MEMS switch case study designs scored transition times better than 20 μs for actuation, and better than 100 μs for release, with pull-in voltage around 25 V. Nonetheless, conflicting trends among different specifications emerged, as well, requiring careful investigation in order to identify optimum trade-offs.

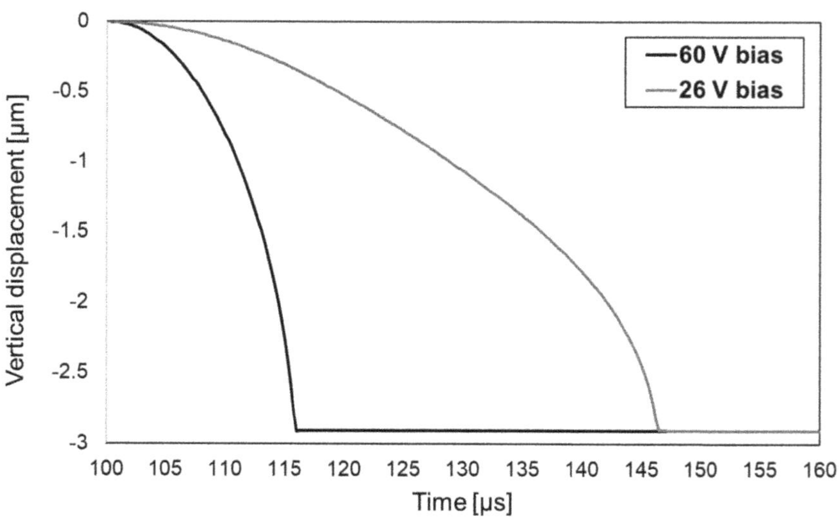

Figure 4.11. Dynamic behaviour of the MEMS switch with meandered suspensions at bias rise edge. (a) Bias peak value of 60 V. (b) Bias peak value of 26 V.

4.4 Electromagnetic simulation of RF-MEMS devices

The focus of this section is the modelling and simulation of RF-MEMS electromagnetic characteristics. Following an approach similar to the one previously developed when dealing with the electromechanical behaviour of microsystems, both FEM analysis and extraction of equivalent lumped-element networks are going to be reviewed and discussed.

4.4.1 Finite Element Method (FEM) analysis

Among the available commercial software tools appropriate for conducting FEM electromagnetic analysis of RF-MEMS, the Ansys High-Frequency Structural Simulator (HFSS) [27] is selected and exploited in the following examples.

4.4.1.1 Validation of the simulation tool

Prior to exploitation of HFSS for design refinement versus target specifications, it is fundamental to build confidence in the accuracy of the simulated results. As a matter of fact, besides the appropriateness of a certain simulator in solving the problem of interest, several settings must be determined in the proper way if accurate, credible and repeatable results are to be achieved. Moreover, reasonable trade-offs must also be scored concerning the accuracy of results and computation complexity. For instance, if an accuracy improvement of 0.3%–0.5% means increasing the simulation time from 20–30 min to 5–7 hr, it is pointless to try.

All these aspects must be solved by carrying out thorough validation of the simulation tool, having a population of a few significant RF-MEMS design concepts and experimental datasets; in this case, scattering parameter (S-parameters) measurements, collected from the physical samples available. In order to perform a validation that is sound and robust to a fair extent, Devices Under Test (DUTs) of different complexity should be selected, such as simple test structures as well as multi-state high-order devices, and the physical specimens should come from different wafers and fabrication batches.

When a certain simulation tool is employed for the first time, a reasonable choice is to start from the nominal technology values, for what concerns the electrical properties of each material as well as for the geometry features, such as layer thickness, planarity, roughness, and so on. Then, by confronting the results of simulations against measurements, the post-layout verification phase (already mentioned in section 2.4.1 and reported in figure 2.6) is performed, leading to a set of actual values describing the specific technology and its spreads. Of course, simulation versus measurement disagreements due to technology non-idealities must not be mixed or confused with those due to the settings of the simulator itself. In fact, several simulator features can influence the quality of results. Among them stand, for example, the mesh type and density, the excitations providing RF power to the 3D structure, the selected BCs, the solver algorithm, and so on.

An effective approach to separate technology-driven inaccuracies from those caused by intrinsic simulator settings is to start the validation with nearly ideal devices. For example, if simple Coplanar Waveguide (CPW) and/or microstrip structures are available, it is reasonable to assume that their S-parameters' behaviour is quite close to ideality. Therefore, benchmarking simulations against such experimental datasets helps determine the most appropriate simulator settings. In addition, or as replacement of physical CPW/microstrip test structures, the simulator can be tuned upon data produced by other (already validated) simulation tools, and/or by trusted analytical models.

For the purposes of this section, the validation of the HFSS FEM tool against S-parameters' measurements is carried out for an RF-MEMS series ohmic switch, whose layout is discussed in [28, 29] and depicted in figure 4.12. In order to simplify the understanding of the proposed geometry, only metal layers defining the elevated MEMS membrane and the sacrificial layer are shown. However, the buried layers, biasing lines, surrounding CPW and protective package [30] are hidden, despite being present in the L-Edit layout file.

Pursuing a similar approach to the one discussed in the section dedicated to electromechanical simulation of RF-MEMS, starting from the layout in figure 4.12, a full 3D model of the RF-MEMS switch is generated (including layers and structures not depicted above) and then imported into the HFSS graphic environment. Material properties and thickness of layers are set according to the information previously reported in table 4.1. HFSS simulation settings are defined as discussed in detail in [31] and not reported here for the sake of brevity. The full 3D model of the RF-MEMS switch imported into the HFSS design environment is shown in figure 4.13. The vertical position of the MEMS elevated membrane was parameterised, in order to easily adapt the 3D model to the ON and OFF switch configurations.

On the experimental side, the S-parameters are characterised on a probe station (on-wafer measurements) by means of a Programmable Network Analyser (PNA) [32]. RF-MEMS physical samples are contacted with Ground-Signal-Ground (GSG) probes (150 μm pitch). The delivered power is 0 dBm and the chosen calibration routine is the Line-Reflect-Reflect-Match (LRRM), performed up to 110 GHz. In order to obtain better accuracy, LRRM calibration and subsequent measurements were performed in two distinct frequency ranges, namely, from 10 MHz to 67 GHz (Low Frequency Range, LFR) and from 67 GHz to 110 GHz

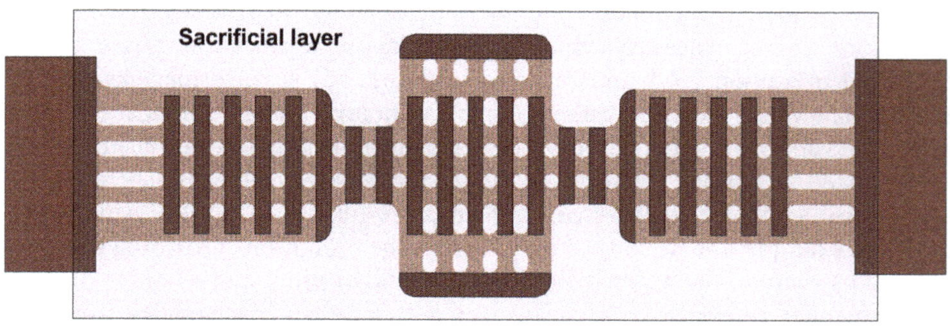

Figure 4.12. Layout of the RF-MEMS series ohmic switch exploited for validation purposes. The sacrificial layer, defining the elevated membrane, is visible.

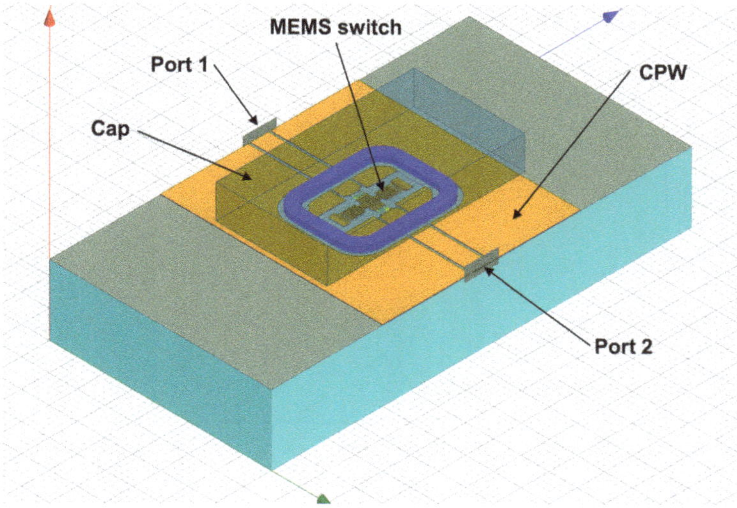

Figure 4.13. HFSS full 3D model of the RF-MEMS series ohmic switch in figure 4.12 generated from the 2D layout. It includes all the layers of the SD-FBK technology, the CPW surrounding frame and the quartz protective cap.

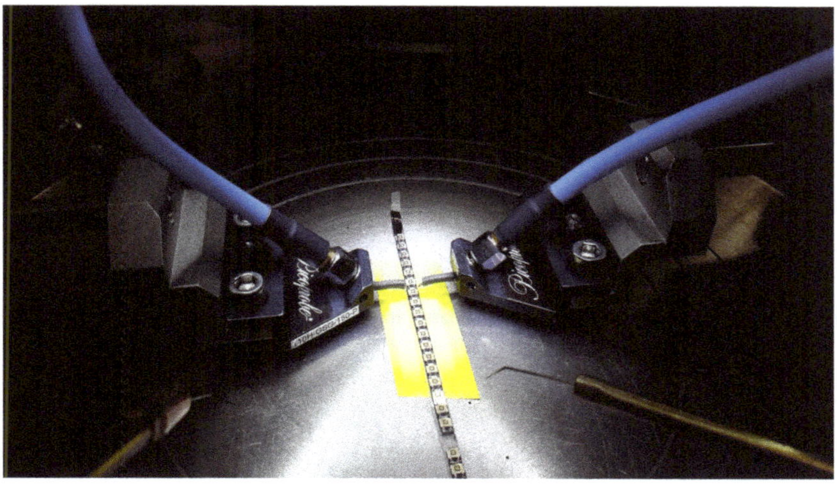

Figure 4.14. Photograph of RF-MEMS series ohmic switch physical samples (see figure 4.12 and figure 4.13) accommodated on the probe station for S-parameters' measurements.

(High Frequency Range, HFR). A photograph of a few RF-MEMS physical samples placed on the probe station for RF measurements is shown in figure 4.14. GSG probes for S-parameters' characterisation, as well as DC probes to provide biasing for the actuation of the elevated MEMS membranes are visible.

(a)

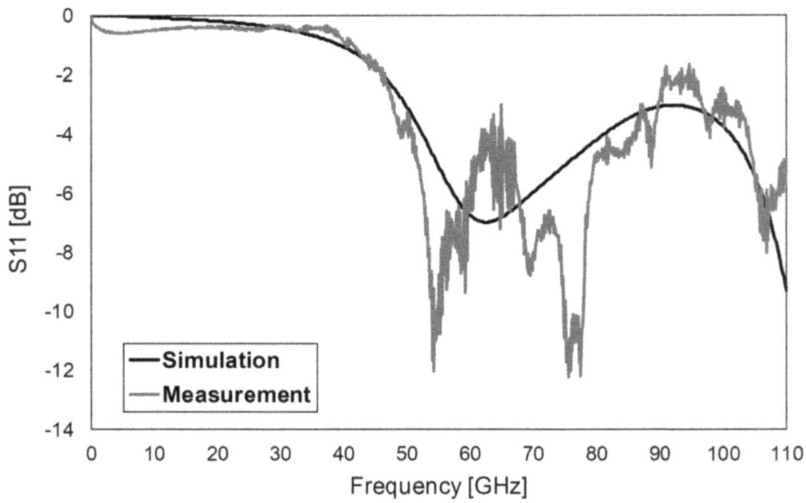

(b)

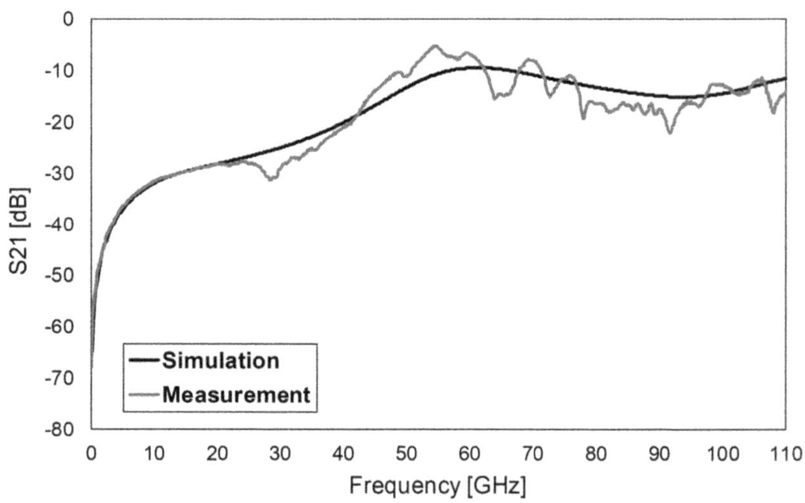

Figure 4.15. Simulated versus measured S-parameters' characteristics of the RF-MEMS device in figure 4.12 from 10 MHz up to 110 GHz when the DUT is OFF (switch OPEN; MEMS in the rest position). (a) Reflection (S11 parameter). (b) Isolation (S21 parameter).

Comparisons of HFSS simulations against S-parameters' measurements are reported in figure 4.15 and figure 4.16 for the RF-MEMS switch in OPEN (OFF state) and CLOSED (ON state) configurations, respectively. Concerning the latter, the MEMS elevated membrane was pulled-in by applying a DC bias level of 50 V. In more detail, figure 4.15(a) shows the reflection parameter (S11) and figure 4.15(b) the input/output isolation (S21) of the OPEN micro-relay. Figure 4.16(a) reports the reflection parameter (S11) and figure 4.16(b) the loss (S21) of the CLOSE switch, respectively.

Recalling the validation purposes discussed above, the results just mentioned highlight a very good accuracy of the simulated traces against experimental

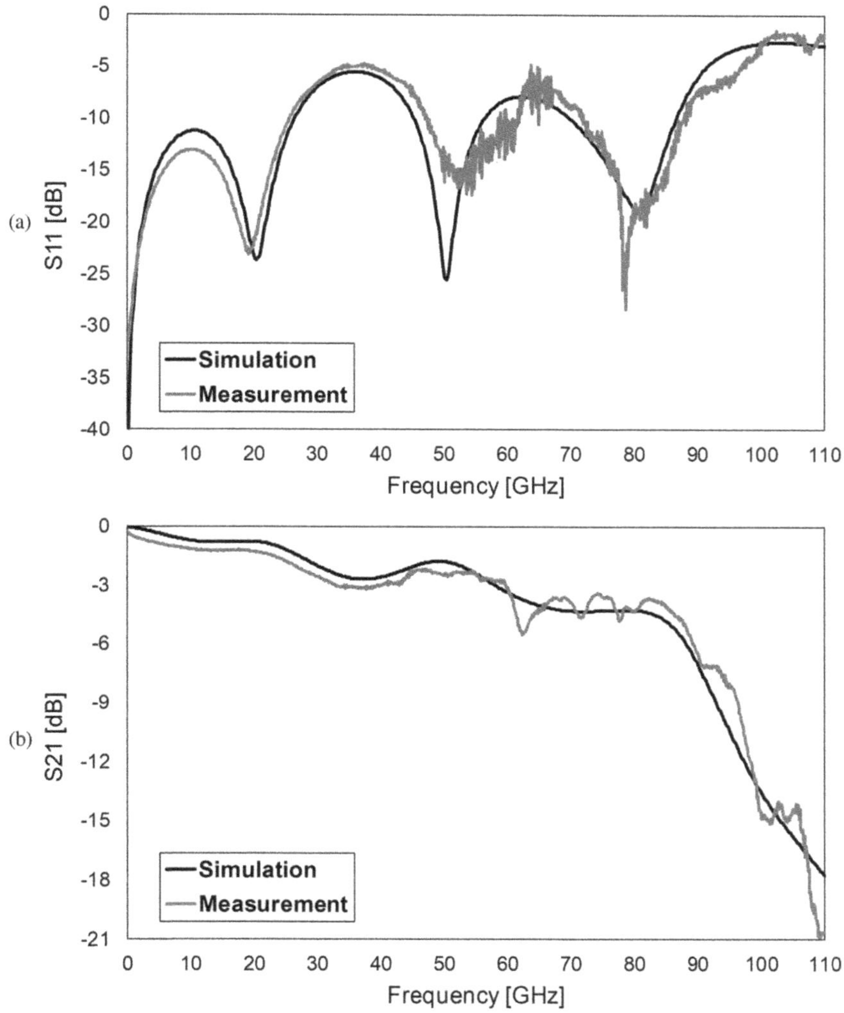

Figure 4.16. Simulated versus measured S-parameters' characteristics of the RF-MEMS device in figure 4.12 from 10 MHz up to 110 GHz when the DUT is ON (switch CLOSE; pulled-in MEMS). (a) Reflection (S11 parameter). (b) Loss (S21 parameter).

measurements, thus proving that the 3D model-building procedure, the assigned effective material properties, as well as the simulator settings are properly chosen and defined. The methodology's robustness is also proven by the fact that the accuracy of simulations does not get worse when the state of the DUT is commuted from the isolating (OFF state) to conducting (ON state) configuration. In addition, it should also be borne in mind that the analysed frequency range is very wide, as it spans from 10 MHz up to 110 GHz. These considerations further support the broad usability of the discussed simulation approach, both concerning RF-MEMS structures with different geometries/configurations, as well as diverse frequency ranges of interest.

Transcending the accomplished validation task, the results reported in figure 4.15 and figure 4.16 offer an effective insight in order to start a discussion on the suitability of RF-MEMS devices for 5G applications. To this purpose, it is useful to refer to the set of target specifications discussed at the end of section 4.1, with particular reference to isolation and loss.

Looking at the plot in figure 4.15(b), isolation (S21 when the switch is OFF) is better than −30 dB up to 14 GHz, and better than −20 dB up to 40 GHz. Nonetheless, between 40 GHz and 110 GHz isolation is never better than −20 dB, and the worst value it scores (in the simulated trace) is −10 dB around 62 GHz. However, in agreement with the target specification discussed in section 4.1, isolation should be better than −30/−40 dB, up to the highest frequency possible. According to that, the RF-MEMS switch design reported in figure 4.12 is able to satisfy the requirement on isolation just up to 14 GHz. Therefore, effort must be invested with the aim of improving the switching unit design.

Similarly, figure 4.16(b) provides meaningful insight concerning the RF-MEMS micro-relay loss while conducting (S21 when the switch is ON). In the discussion developed in section 4.1, loss should be ideally better than −1 dB, up to the highest frequency possible. According to the simulated trace in figure 4.16(b), loss is better than −1 dB up to 24 GHz, better than −2.6 dB up to 55 GHz, and better than −4.5 dB up to 80 GHz. Above the latter frequency, however, the S21 worsens quite steeply, reaching a poorest value of about −18 dB at 110 GHz.

In conclusion, although the RF-MEMS switch topology reported in figure 4.12 is not meant for 5G applications, but rather reported here for validation purposes, it exhibits promising characteristics in terms of S-parameters. Therefore, it is a valuable case study to be employed as a starting point for further design optimisations.

4.4.1.2 Design variations and performance optimisation aimed at 5G requirements

In this section, a few examples of RF-MEMS switch/switching unit design variations are going to be proposed, in order to investigate their impact on the S-parameters' characteristics, in light of the target specification discussed at the beginning of this chapter.

At first, the switching unit clamped–clamped topology in figure 4.12 is replaced by a cantilever-type geometry, such as the one reported in figure 4.17.

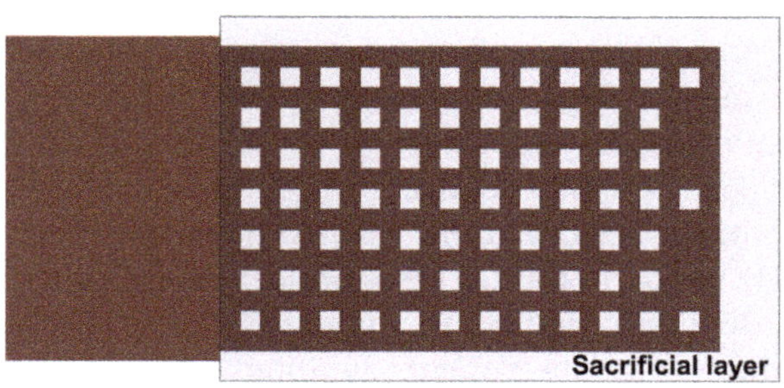

Figure 4.17. Layout of the RF-MEMS series ohmic cantilever-type (i.e. single-hinged) switch.

The suspended membrane is hinged only on one side, while the other end of the elevated gold layer is free. Underneath it, metal contacts (not shown in figure 4.17) are placed, thus ensuring input/output metal-to-metal continuity, when the MEMS membrane is pulled-in. Beside the substantial difference between the clamped–clamped and cantilever geometry just mentioned, other aspects must be highlighted. Concerning the micro-relay placement with respect to the CPW framing it, clamped–clamped geometries are typically deployed transversally with respect to the RF signal line. On the other hand, cantilever geometries can easily be accommodated in-line along the signal path, as in the case shown in figure 4.17. This leads to more compact arrangements of the switching device, without the need for recesses in the RF ground planes to allocate the MEMS membrane, as in the case of clamped–clamped configurations. Such characteristics suggest smaller influence of cantilever versus clamped–clamped switches on loss and mismatch of the micro-relay when ON (i.e. CLOSE switch) [33–35]. Concerning the electromechanical properties of cantilever-type micro-relays, having the membrane hinged only on one end leads intrinsically to lower pull-in voltage levels, if compared to clamped–clamped geometries. Nonetheless, on the other side, the limited restoring force exposes cantilevered MEMS micro-relays to malfunctioning due to charge entrapment and micro-welding, making them, in general, less robust than clamped–clamped switches versus power handling and cycling [36–38].

The simulated S21 characteristics of the switch topologies in figure 4.12 and in figure 4.17 (isolation when OFF; loss when ON) are reported in figure 4.18. In order to have a consistent comparison of the two geometries, in the HFSS 3D model the intrinsic switch was modified exclusively, while the surrounding CPW structure was unaltered.

Probing in more detail, the plot in figure 4.18(a) compares the isolation (S21) of the two geometries when both micro-switches are OFF. In general terms, the original clamped–clamped design solution in figure 4.12 performs better. The two characteristics are quite similar in the frequency range from 51 GHz to 65 GHz. On the other hand, from 200 MHz up to 51 GHz, isolation of the clamped–clamped micro-relay is always better, with a difference compared to the cantilever that reaches a maximum of about 10 dB around 20 GHz. From 65 GHz up to 110 GHz the isolation divide between the two geometries is less pronounced. In any case, the geometry in figure 4.12 scores values better than a few dB with respect to the design solution in figure 4.17.

The fact that the clamped–clamped configuration outperforms the cantilever switch in terms of isolation can be explained in quite a straightforward fashion. The double-hinged switch has two contact areas above the RF line. When ON, such areas establish metal-to-metal contact ensuring the RF signal flows across the device. However, when OFF, overlapped contact pads realise two small parasitic capacitances placed in a series configuration, which couple the input and output branches. On the other hand, the cantilever membrane has just one contact area, which means a unique parasitic capacitance when OPEN. Since the contacts' area was kept the same in the two geometries, the clamped–clamped switch is characterised by a parasitic OFF state capacitance that is half of the one realised by the

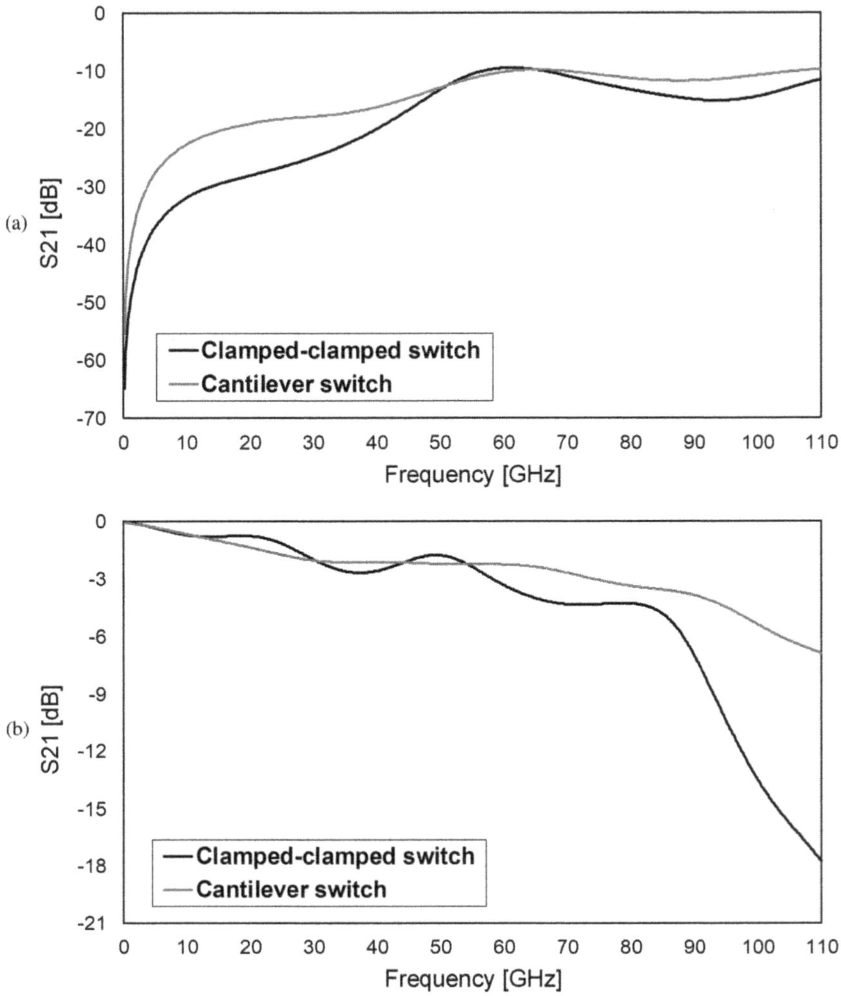

Figure 4.18. Simulated S-parameters' characteristics of the clamped–clamped and cantilever series switches, depicted in figure 4.12 and in figure 4.17, respectively, from 200 MHz up to 110 GHz. (a) Isolation (S21 parameter) when micro-relays are OFF (i.e. OPEN switches). (b) Loss (S21 parameter) when micro-relays are ON (i.e. CLOSE switches).

cantilever (two equal capacitors in series), thus ensuring better isolation. This issue can be mitigated by reducing the contact area in the cantilever design, albeit influencing the switch loss when ON.

In continuing, figure 4.18(b) compares the loss introduced by the clamped–clamped and cantilever switches, when conducting (ON state), indeed overturning the situation described above. Considering the observed frequency range as a whole, the loss introduced by the cantilever is better than that of the clamped–clamped micro-relay. In more detail, the two curves are nearly identical up to 12 GHz. In between 12 and 31 GHz, the cantilever performance is slightly worse than in the case of the double-hinged device, being in any case better than −2 dB. Nonetheless, between 31 and 64 GHz the cantilever loss characteristic exhibits pronounced flatness, worsening from −2 to −2.3 dB from the lower to the higher frequency previously indicated. In the same range, the clamped–clamped loss oscillates between −2.7 and −1.8 dB up to 50 GHz, and then degrades to −4 dB at 64 GHz. Again, between 64 and 82 GHz the cantilever loss is always better than −3.5 dB, while the clamped–clamped loss is better than −4.4 dB. Finally, from 82 GHz up to 110 GHz both characteristics become more significantly lossy. However, while the cantilever S21 trace is always better than −7 dB, the clamped–clamped loss reaches −18 dB at 110 GHz.

This solution has the unique advantage of modifying the switching unit intrinsic topology in order to improve device performance. Such a strategy leads unavoidably to dead ends in terms of conflicting characteristics optimisation. In this regard, the example of the clamped–clamped versus cantilever switch geometry, comprehensively reported above, is quite significant. Having just one contact area (cantilever switch) minimises the series ON state resistance, which leads to reduced loss when compared against devices with double contact areas (double-hinged switch). Nonetheless, a single contact area means doubling the OFF state capacitance that, in turn, brings worse isolation of the cantilever against the clamped–clamped geometry. Given the background just depicted, the only trade-off possible is sacrificing isolation in the name of smaller loss, or vice versa.

On a completely different reference plane lie plenty of opportunities in terms of performance optimisation, if one pursues the strategy of making the design more complex. Looking in this direction, a valuable case study is reported in the following. When an RF-MEMS switching unit topology is optimised against the desired performance figure in terms of loss when CLOSED, the isolation (when OPEN) can be increased by adding redundant switching units working as isolation boosters. A reference layout of this concept is shown in figure 4.19.

The actual series switching unit is preceded and followed by two micro-relays with the same geometry as the suspended MEMS membrane, albeit configured as shunt ohmic micro-relays. Shunt switches have a dual behaviour with respect to series ones, as they are CLOSED when the MEMS is OFF, and OPEN when the MEMS is ON. In the latter case, they divert (short) the signal to RF ground. Therefore, when the central micro-relay is OFF (OPEN switch), the two isolation boosters can be actuated (pulled-in) in order to short to ground the RF signal flowing across the intrinsic switch because of the parasitic capacitive coupling. This helps improve the isolation of the overall switching device in figure 4.19, as reported in figure 4.20.

When the isolation boosting units are not activated (OFF state), the only series RF-MEMS switch does not exhibit outstanding isolation (S21 parameter). Diversely, when the shunt ohmic micro-relays are actuated (ON state), the isolation shifts evidently towards improved performance. In detail, the S21 trace of the OPEN switch with isolation boosters ON in figure 4.20 scores enhanced values of 10 dB up to 25 dB when compared to isolation with boosters OFF, from 200 MHz up to around 84 GHz. Above the latter frequency, and up to 110 GHz, the two curves are

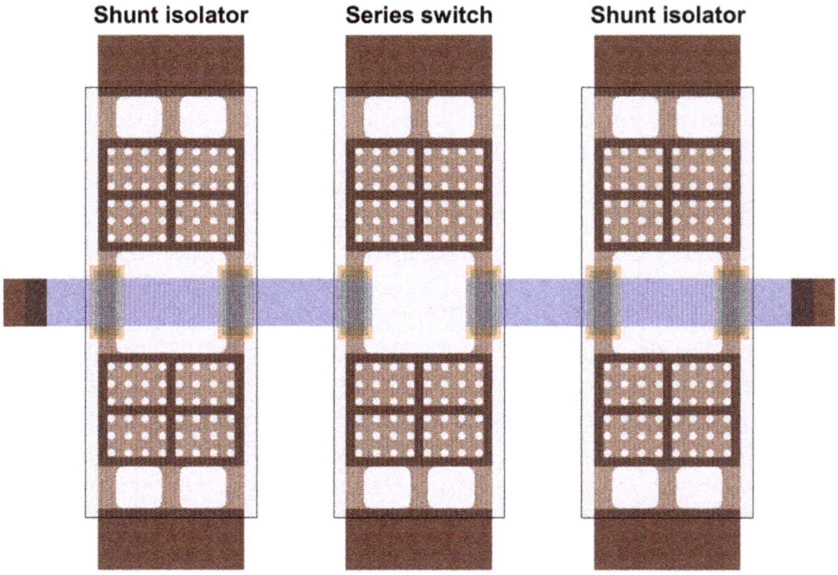

Figure 4.19. Layout of an RF-MEMS ohmic series switch with two isolation boosting units, placed before and after the intrinsic micro-relay, based upon shunt ohmic switches. The buried RF underpass is visible. It is continuous underneath the shunt units and interrupted underneath the series switch.

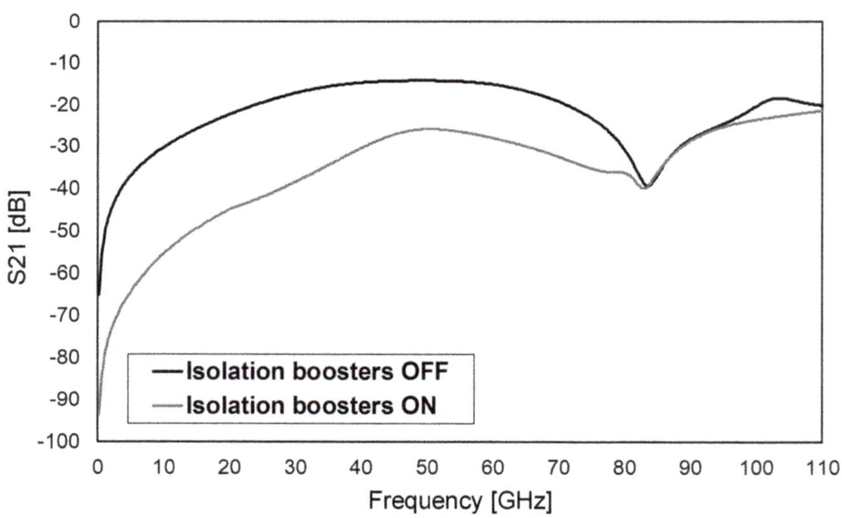

Figure 4.20. Simulated isolation (S21) parameter of the switching device shown in figure 4.19 when the isolation boosting units are OFF and ON, from 200 MHz up to 110 GHz.

quite close to each other. Recalling the 5G reference specifications listed in section 4.1, the isolation of the design concept in figure 4.19 with isolation boosters activated reaches the target of being better than −30 dB from 200 MHz up to around 90 GHz.

Of course, the inclusion of isolation boosting units affects the loss performance of the switch when ON, as additional parasitic capacitance couples the shunt ohmic switches to RF ground when they are OFF. Therefore, increasing the geometry complexity does not rule out, by itself, the rising of conflicting characteristics. On the other hand, one should bear in mind that adding redundancy to the RF-MEMS design concept under development introduces significant sets of DoFs the developer can play with. For example, still referring to the case in figure 4.19, one can choose the intrinsic switching unit tailoring its loss (when ON) characteristics to the best performance possible, without paying too much attention to its conflicting figure in terms of isolation, as the latter can be enhanced by introducing additional switching stages.

4.4.2 Modelling based on equivalent lumped-element networks

The target of this concluding section is to provide an agile outlook of RF-MEMS modelling techniques based on equivalent lumped-element networks. With similarity to the discussion previously developed around electromechanical modelling of MEMS, prediction of RF/electromagnetic behaviour can be performed following diverse approaches and by means of different tools. From a conceptual point of view, the approach to RF equivalent networks is very close to that of device behavioural description based on compact models, developed in section 4.3.3. As a matter of fact, the aim is to attain accurate prediction of RF-MEMS electromagnetic behaviour by keeping a limited computation load. Thereof, multiple DoFs design optimisation runs can be performed quickly, while subsystem/system-level simulations are also enabled. To this purpose, compact models describing electromechanical behaviour can be enriched by equivalent lumped-element networks, thus allowing the prediction of the whole mixed-domain RF-MEMS behaviour in quite a straightforward manner.

Equivalent lumped-element networks are widely used in RF/microwave modelling of electronic devices and circuits. Substantially, starting from experimental datasets (e.g. S-parameters), accurate simulated data (e.g. from already validated FEM models), as well as being based upon mathematical models, a network of capacitors, inductors, resistors and other components is derived. Such an architecture is able to describe quite accurately the electromagnetic behaviour of the original device, within an established range of frequency. It should be highlighted that the three sources of data for the network extraction, i.e. measurements, simulations and analytical models, can be exploited in synergy, as well as exclusively. This means that, depending on the circumstances, a lumped-element network of a certain device can be based, for instance, just on measured datasets, or on a part of experiments and of simulations, and so on. This suggests the pronounced flexibility of the discussed approach, which is also confirmed when probing into the scientific literature. In this regard, the variety of electronic components modelled by means of such a methodology is significantly wide. Several techniques and examples of lumped-element networks describing semiconductor devices, transistors and circuits are available [39–41]. Furthermore, valuable contributions discussing equivalent networks of passives such as waveguides, inductors, transformers, and so on, are also available [42–47]. Following the same path, modelling based on equivalent lumped-element networks was also investigated in the field of RF-MEMS devices [11, 48–52].

4.4.2.1 Selected case study and methodology

The example that is going to be discussed in the following is based upon elementary 1-bit (ON/OFF) RF-MEMS attenuator designs, previously discussed in [53, 54]. Microphotographs of the devices for which the lumped-element networks will be extracted are reported in figure 4.21.

The RF-MEMS attenuator modules are available in two configurations, namely series and shunt, reported in figure 4.21(a) and figure 4.21(b), respectively. The RF-MEMS switch and surrounding CPW frame are unaltered in both variants. What changes is how the resistive load is deployed. In the series device, the polycrystalline silicon resistor is in-line on the RF central path, while in the shunt device two polycrystalline silicon resistors are placed in shunt-to-RF ground configuration. Therefore, in the series device in figure 4.21(a) the RF signal is attenuated when the

(a)

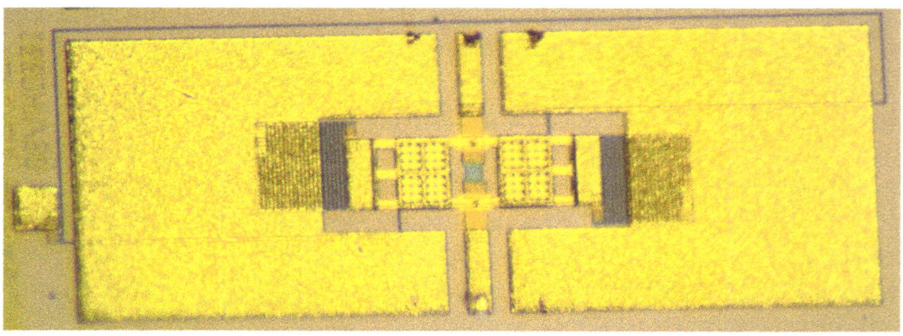

(b)

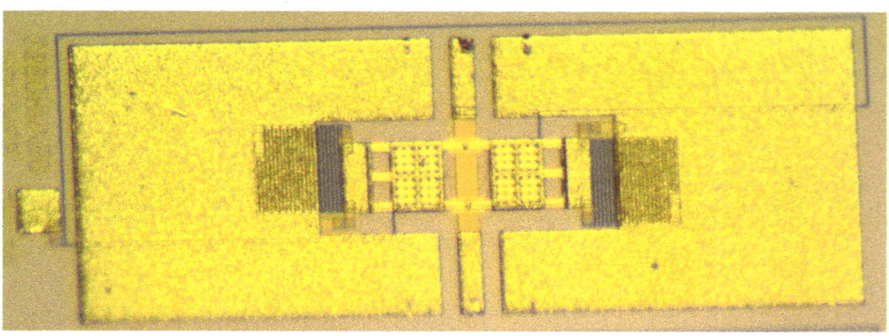

Figure 4.21. Microphotographs of the RF-MEMS 1-bit (ON/OFF) power attenuators discussed in [53, 54]. (a) Series design, featuring one resistive load placed in-line on the RF path. (b) Shunt design, featuring two resistive loads placed in shunt-to-RF ground configuration. © 2016 IEEE. Reprinted, with permission, from [54].

MEMS switch is OFF, while the load resistor is shorted when the MEMS switch is ON. However, in the shunt device in figure 4.21(b) the RF signal is not attenuated when the MEMS switch is OFF, while the two shunt loading resistors are inserted (in parallel to one another) when the MEMS switch is ON.

The circumstance of design variations, in which, for the most part, the RF-MEMS device layout is kept fixed while just limited portions are modified, such as in the case discussed here, is particularly favourable for extracting equivalent lumped-element networks. To this purpose, a critical aspect must be carefully evaluated. Lumped networks able to mimic a measured/simulated curve, or the response of a mathematical model in a limited range of frequencies, can be arbitrarily defined. Nonetheless, this would not bring any added value to the analysis of RF-MEMS electromagnetic behaviour. However, what really makes sense is constructing lumped-element network topologies that capture the physical characteristics of the device under analysis, thus enabling a coherent description of it.

At this stage, a practical example can help one to understand the aforementioned critical aspect. Let us recall the basic equivalent network of a CPW, already discussed in chapter 1 and reported in figure 1.4(c). The architecture makes sense from a physical point of view. The series resistance and inductance are those of the metal RF line, while the shunt capacitance and resistance model the capacitive coupling and losses to ground, respectively. If the CPW geometry was modified, it is quite straightforward to envisage how this would reflect on the topology in figure 1.4(c). For instance, if the gap (between the RF line and the ground planes) is reduced, the capacitance to ground increases. If the substrate is replaced with a more lossy material, the resistance to ground will decrease (i.e. loss will increase). Similarly, if the CPW length is increased, both resistive and inductive series contributions will rise.

Following a different approach, various topologies, other than the one in figure 1.4(c), can be put arbitrarily together to effectively reproduce the S-parameters' characteristics of a CPW, also via automated computer-based extraction routines. Nonetheless, when a robust conceptual link between each lumped element and the physical feature(s) it describes is missing, impairing limitations start to emerge. First, the network architecture does not make sense (partially or totally) from a physical point of view. Therefore, it is useless to develop a better understanding around the behaviour of the RF-MEMS under analysis. Then, the extracted network architecture is valid exclusively in the context within which it was derived. This means that if the geometry DoFs of the device are altered, as well as if the topology is modified (such as when stepping from the device in figure 4.21(a) to that in figure 4.21(b)), the lumped network can be neither parameterised nor partially reused/ported. On the other hand, it should be once again extracted from scratch.

This discussion clarifies the importance of building the equivalent lumped-element network architecture, always sticking to sound physical considerations, and increasing its complexity in a smooth fashion, i.e. adding reactive/resistive elements always linked to real effects, and not just because they improve the match to target measured datasets.

For the same reason, once the starting network architecture is defined, the tentative value of each lumped element should be also carefully set. Such numbers should always be distilled starting from physical considerations and not following random approaches. To this purpose, possessing sound knowledge both of how the device is structured and of the basic lumped-elements' analytical models gives a significant helping hand. For instance, if the developer, besides the 2D layout, knows also the thickness of the metal and insulating layers, he/she can determine quite confidently the value of the parasitic OFF state capacitance of an RF-MEMS. Similarly, if the series inductance contribution due to a portion of the RF line has to be determined, the developer is prompted to calculate the inductance per unit length, on the basis of the line dimensions and material properties. Therefore, it will be easy to derive a value making physical sense.

Once the initial values of each lumped element are determined following the approach just discussed, they can be tuned within a limited range of variability around the starting setting, observing how the network S-parameters are affected. In this regard, the following aspect should always be borne in mind. Given a certain network topology that is kept fixed, there is no unique set of lumped-elements' values yielding a good match between the network response and the target datasets in a certain frequency range. Nonetheless, just one among those sets of values sticks better to the physical features of the analysed devices and, therefore, it is useful to build behavioural considerations bearing effective significance.

After this articulated discussion, it is worth recalling in brief a statement reported above. Having target devices that are very similar to each other, such as in figure 4.21(a) and figure 4.21(b), is particularly favourable to verify the soundness of the equivalent lumped-element network topologies. As a matter of fact, the more the architecture is based upon solid physical considerations, the less the network configuration and the lumped-elements values need to be modified in order to maintain accuracy. Ideally, still referring to the examples in figure 4.21, the network topology should exclusively require the modification of the contact resistances to reproduce the measured curves of the ON and OFF state, as well as just the reconfiguration of the load resistor, to pass from the series to the shunt attenuator device (figure 4.21(a) and figure 4.21(b), respectively).

4.4.2.2 Extracted networks and discussion of their behaviour

In order to develop equivalent lumped-elements' networks, a proper simulation environment is necessary as the first item. As a matter of fact, any tool featuring symbolic composition of analogue circuits, libraries of passive lumped-elements based on mathematical models and allowing S-parameters analysis is suitable to pursue the target discussed here. Just to mention a few commercial software tools, the Advanced Design System (ADS) by Keysight Technologies [55], the Spectre Circuit Simulator within Cadence [56] and the HyperLynx Analog within Mentor [57] are examples of exploitable simulation environments. Some of these tools also offer the possibility to run automated optimisation algorithms, thus speeding up the identification of appropriate values for the network's lumped elements.

In the examples discussed below, Quite Universal Circuit Simulation (QUCS) [58] is selected as a software tool for equivalent lumped-element networks extraction and analysis. The QUCS is an Integrated Circuits (ICs) simulator featuring a Graphical User Interface (GUI) for networks assembly at a schematic level. A fair variety of component libraries is available, spanning from lumped passives, such as reactive elements and resistances, to waveguide components, again to active semiconductor devices. QUCS enables various types of simulation analyses, such as large-signal, small-signal and noise behaviour. Finally, an easy-to-use GUI allows a fast arrangement of results, according to the needs of the developer. Also, importantly, QUCS is released under a General Public License (GNU-GPL). Therefore, it can be freely used, shared and modified by the end users.

Having provided such a context, a QUCS equivalent lumped-element network schematic corresponding to the series RF-MEMS attenuator module (see figure 4.21(a)), with the micro-relay in the OFF state (rest position), is shown in figure 4.22. The input/output CPW portions, visible in figure 4.21(a), are modelled by the CPW symbols, available in the QUCS waveguide components library. The polycrystalline silicon series load resistor is modelled by the R_{lse} component, while direct input/output coupling, taking place as frequency rises, is modelled by the parallel of C_{air} and R_{air}. The lower branch, in parallel to the attenuation load, represents the path across the MEMS suspended membrane. Since the micro-relay is OFF, parasitic capacitance builds across the underlying contact areas and the suspended gold fingers [13]. The latter is modelled by the C_{off} elements, while L_{se} is the series inductance the RF signal encounters when travelling from the input to the output terminations. The network is stimulated with two RF power sources matched to 50 Ω, i.e. P_1 and P_2, respectively. S-parameters analysis is set, and the material properties of the CPW elements, such as the substrate resistivity, dielectric constant and thickness of metal, are defined according to the values previously listed in table 4.1.

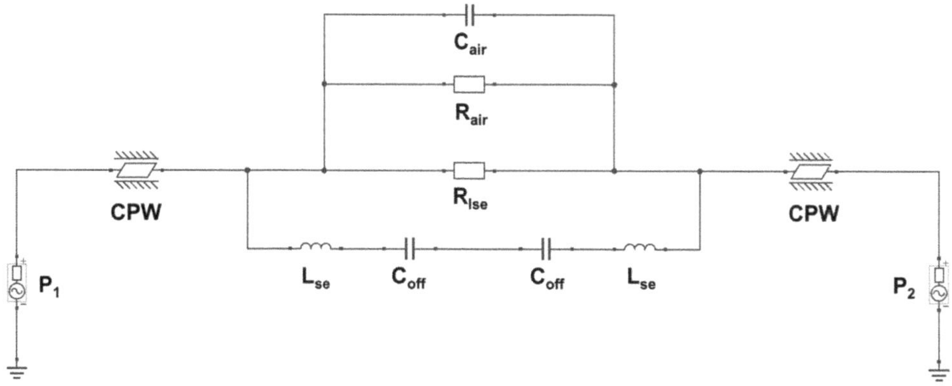

Figure 4.22. Equivalent lumped-element network of the series RF-MEMS attenuator module shown in figure 4.21 (a) when the micro-relay is OFF (rest position).

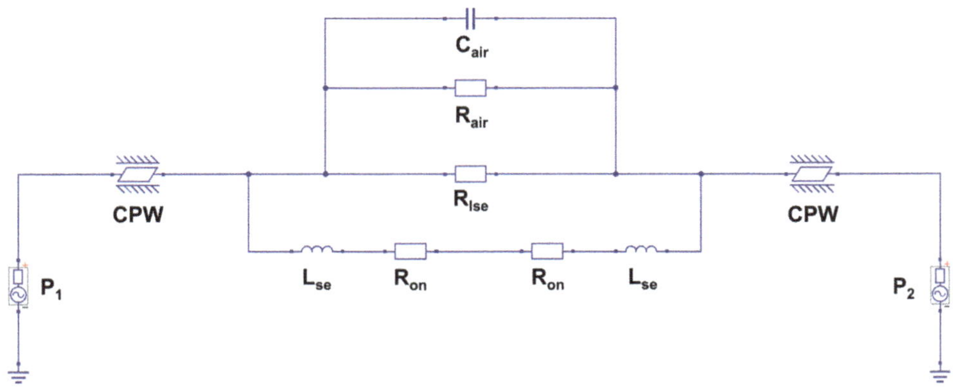

Figure 4.23. Equivalent lumped-element network of the series RF-MEMS attenuator module shown in figure 4.21(a) when the micro-relay is ON (pulled-in).

A QUCS equivalent lumped-element network schematic corresponding to the series RF-MEMS attenuator module (see figure 4.21(a)), with the micro-relay in the ON state (pulled-in), is shown in figure 4.23. In this case, the polycrystalline silicon resistor is shorted by the pulled-in gold MEMS membrane. Therefore, the attenuation is zeroed.

After comparing the network architecture in figure 4.22 (micro-relay OFF) with that in figure 4.23 (micro-relay ON), the only difference is that OFF state parasitic capacitances C_{off} are replaced by ON state series contact resistors R_{on}, as is straightforward to figure out, when thinking about how ohmic RF-MEMS switches behave [59]. The extracted optimal values of all the lumped elements in the networks reported in figure 4.22 and in figure 4.23, as well as those of the the networks in figure 4.26 and in figure 4.27, discussed in the following, are listed in table 4.2.

In the following, a comparison between the target experimental S-parameters' traces and the simulated output of extracted lumped-elements networks will be discussed. Plots are arranged always according to the same fashion. First, a comparison of the measured and simulated reflection (S11) parameter is reported. Then, the absolute difference (in dB) between the measurement and simulation is plotted for the reflection parameter at both ports (S11 and S22). Furthermore, the same comparisons are reported for the transmission parameter. To this purpose, the measured and simulated loss/attenuation (S21) is reported, together with the absolute difference of measurements and simulations related to both transmission parameters (S12 and S21). In light of the explanation just developed, the results related to the series RF-MEMS attenuator module in the OFF state—device in figure 4.21(a); lumped network in figure 4.22—are plotted in figure 4.24.

From a qualitative point of view, the equivalent lumped-element network reproduces quite satisfactorily both the measured reflection (S11) and attenuation (S21) characteristics, as shown in figure 4.24(a) and figure 4.24(c), respectively. The upward step, taking place in the measured reflection between 50 GHz and 60 GHz, corresponds to a downward step in the attenuation curve. Such a characteristic is not present in the simulated curves. However, it must be observed that the absolute error

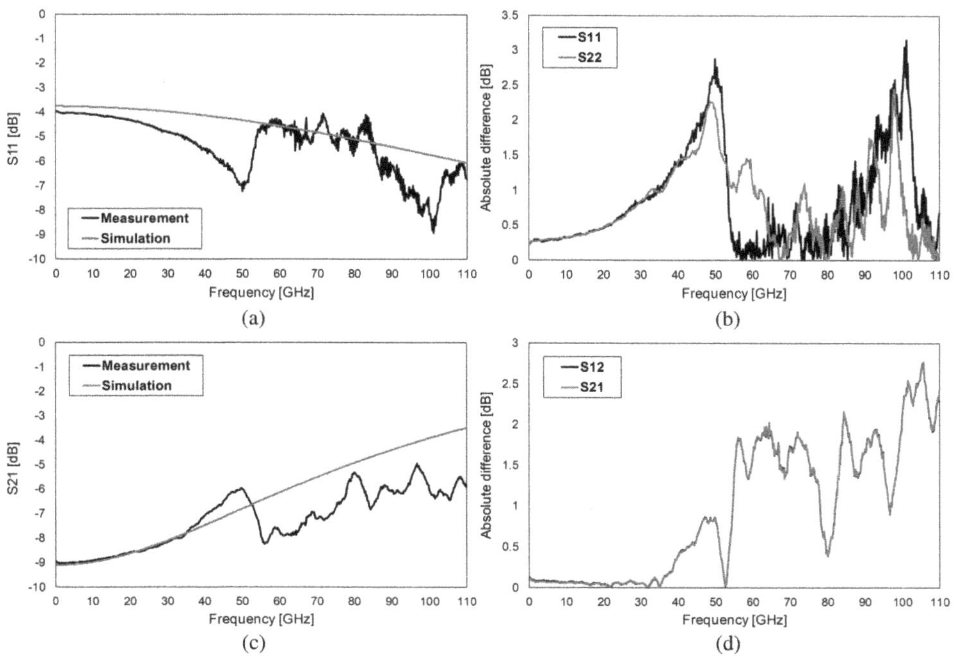

Figure 4.24. Measured S-parameters' characteristics of the series RF-MEMS attenuator module in figure 4.21(a) when the micro-relay is OFF and response of the equivalent circuit in figure 4.22, from 200 MHz up to 110 GHz. (a) Measured versus simulated reflection (S11) characteristic. (b) Absolute difference (in dB) between the measured and simulated reflection characteristics at both ports (S11 and S22). (c) Measured versus simulated attenuation (S21) characteristic. (d) Absolute difference (in dB) between the measured and simulated attenuation characteristics at both ports (S12 and S21).

of simulations versus measurements is never too large. Looking at the plot in figure 4.24(b), the difference between the experimental and simulated reflection (both concerning S11 and S22) is below 1 dB in most of the frequency range, while it is below 3 dB in two subranges centred around 50 GHz and 100 GHz. However, the difference of attenuation (both concerning S12 and S21) in figure 4.24(d) is below 1 dB up to 55 GHz, below 2 dB up to 100 GHz and below 3 dB up to 110 GHz.

The results related to the series RF-MEMS attenuator module in the ON state—device in figure 4.21(a); lumped network in figure 4.23—are plotted in figure 4.25.

Also in this circumstance, the qualitative prediction operated by the equivalent lumped-element network is quite satisfactory, for both the reflection (S11) and loss (S21) characteristics, reported in figure 4.25(a) and in figure 4.25(c), respectively. In particular, the error of the simulated loss (S12 and S21) is quite low, especially up to around 100 GHz, as shown in figure 4.25(d). In contrast, the difference between the simulated and measured reflection (S11 and S22) is more pronounced, as visible in figure 4.25(b). Nonetheless, when the RF-MEMS device is conducting, the reflection scores larger values, in the order of a few tens of dB. Therefore, larger error levels can be expected and tolerated. Moreover, since equivalent networks enable approximate behavioural description of physical devices, it is quite difficult to

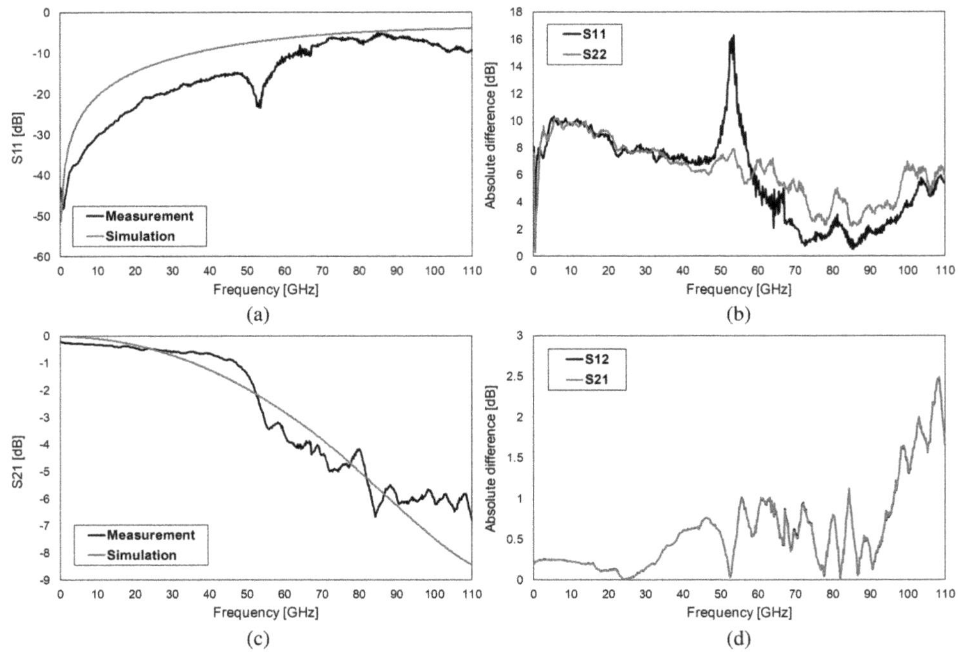

Figure 4.25. Measured S-parameters' characteristics of the series RF-MEMS attenuator module in figure 4.21(a) when the micro-relay is ON and response of the equivalent circuit in figure 4.23, from 200 MHz up to 110 GHz. (a) Measured versus simulated reflection (S11) characteristic. (b) Absolute difference (in dB) between the measured and simulated reflection characteristics at both ports (S11 and S22). (c) Measured versus simulated loss (S21) characteristic. (d) Absolute difference (in dB) between the measured and simulated loss characteristics at both ports (S12 and S21).

obtain an accurate prediction of all the S-parameters, especially on a very wide frequency range, such as the one considered here, spanning from 200 MHz up to 110 GHz. Still referring to figure 4.25, it is possible to tailor the lumped-elements' values in order to achieve a better match of reflection. However, this would bring a larger error of the transmission parameters. In this case, the accuracy of the S12 and S21 parameters was considered to be more critical than that of S11 and S22.

Stepping forth, a QUCS equivalent lumped-element network schematic corresponding to the shunt RF-MEMS attenuator module (see figure 4.21(b)), with the micro-relay in the OFF state (rest position), is shown in figure 4.26. In this configuration, the resistive load is not selected and the device does not realise any RF signal attenuation.

A comparison between the equivalent networks of the series and shunt RF-MEMS attenuation modules, reported in figure 4.22 and in figure 4.26, respectively, highlights their similarities and differences. On a general reference plane, the architectures are very similar. The series attenuation resistive load is replaced in figure 4.26 by the resistive (R_{up}) and inductive (L_{up}) series contributions of the underpass connecting the input/output terminations, visible in figure 4.21(b). Moreover, a branch to RF ground accounts for the shunt attenuating resistor.

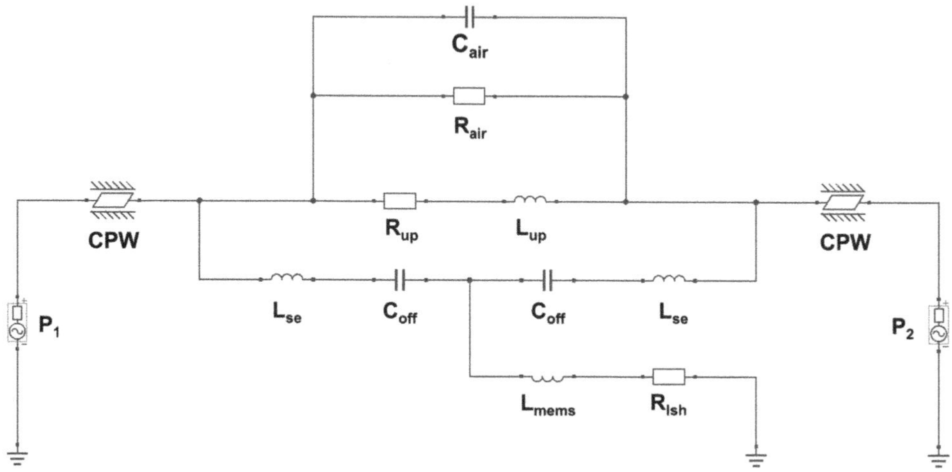

Figure 4.26. Equivalent lumped-element network of the shunt RF-MEMS attenuator module shown in figure 4.21(b) when the micro-relay is OFF (rest position).

It features the inductance of the path to ground (L_{mems}) and, of course, the polycrystalline silicon shunt resistive load (R_{lsh}). In addition, a QUCS equivalent lumped-element network schematic corresponding to the shunt RF-MEMS attenuator module (see figure 4.21(b)), with the micro-relay in the ON state (pulled-in), is shown in figure 4.27.

For the case of the series RF-MEMS device, when referring to the shunt device the micro-relay OFF and ON states are modelled simply by replacing parasitic series capacitances (C_{off} in figure 4.26) with series contact resistances (R_{on} in figure 4.27). Of course, when C_{off} capacitors are inserted, the shunt branch featuring the resistive load does not influence the S-parameters' behaviour of the network. However, when the R_{on} resistances are included in the architecture, part of the RF signal is shorted to ground through R_{lsh}. The results related to the shunt RF-MEMS attenuator module in the OFF state—device in figure 4.21(b); lumped network in figure 4.26—are plotted in figure 4.28.

The match between the measured and simulated reflection (S11) characteristic is very good up to around 60 GHz, but gets worse above that frequency, as shown in the S11 plot in figure 4.28(a) as well as in the error plot of (S11 and S22) reported in figure 4.28(b). In detail, with reference to the latter graph, the difference is below 2 dB up to 60 GHz, while in the upper part of the frequency range it increases up to around 16 dB. When looking at the loss (S21) characteristic, a satisfactory match between simulations and measurements is noticeable in figure 4.28(c). The differences between the transmission parameters (S12 and S21) are quite limited up to 80 GHz, while from 80 to 110 GHz they increase to the worst value of about 8 dB.

Finally, the results related to the shunt RF-MEMS attenuator module in the ON state—device in figure 4.21(b); lumped network in figure 4.27—are plotted in figure 4.29. In this circumstance, from a general point of view, the match between the measurements and simulations is not as good as in the cases discussed previously.

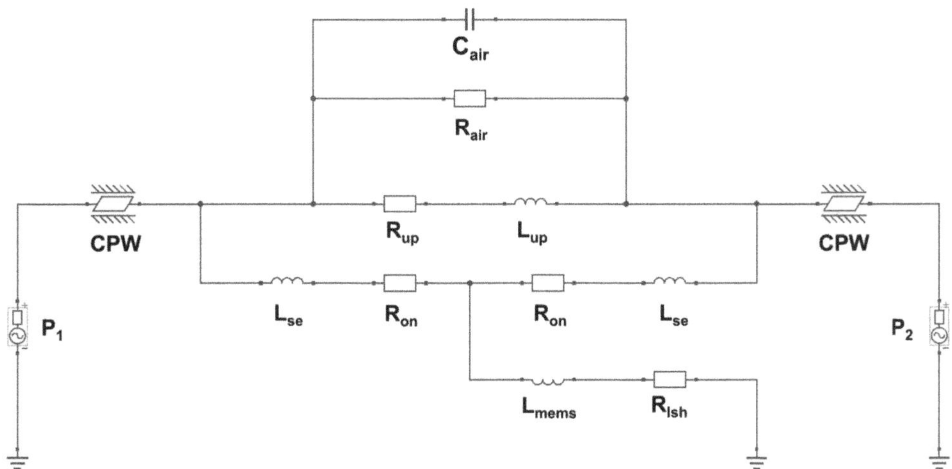

Figure 4.27. Equivalent lumped-element network of the shunt RF-MEMS attenuator module shown in figure 4.21(b) when the micro-relay is ON (pulled-in).

Table 4.2. Summary of the optimal values extracted for all the lumped elements included in the equivalent networks reported in figure 4.22, figure 4.23, figure 4.26 and figure 4.27.

Element name	Description	Value
C_{air}	Direct input/output capacitive coupling through air	12 fF
R_{air}	Direct input/output resistive coupling (loss) through air	10 GΩ
L_{se}	Series inductance of the RF signal line	70 pH
R_{on}	Metal-to-metal resistance between the pulled-in MEMS and the contact area	100 mΩ
C_{off}	Capacitive coupling between the MEMS (rest position) and the contact area	7 fF
R_{lse}	Polycrystalline silicon series resistor	185 Ω
R_{up}	Series resistance of the RF signal underpass	500 mΩ
L_{up}	Series inductance of the RF signal underpass	150 pH
L_{mems}	Inductance of the MEMS micro-switch	40 pH
R_{lsh}	Polycrystalline silicon shunt resistor	42 Ω
W_{CPW}	Width of the input/output CPW signal line	56 μm
G_{CPW}	Gap of the input/output CPW signal line	50 μm
L_{CPW}	Length of the input/output CPW signal line	200 μm

The qualitative match between the measured and simulated reflection (S11) characteristics, reported in figure 4.29(a), is not fully satisfactory. While the measurement trace is substantially flat, the simulated one bends upward as the frequency increases. On the other hand, looking at the attenuation (S21) characteristic in figure 4.29(c), a good match is visible only up to 60 GHz. Observing the plots related to the difference between the measurements and simulations, both the ones concerning reflection (S11 and S22) and attenuation (S12 and S21) characteristics—figure 4.29(b) and figure 4.29(d), respectively—exhibit a reasonably limited value up to around 60 GHz only.

In conclusion, the methodology just discussed for the definition and extraction of equivalent lumped-element networks of RF-MEMS devices provides a significant helping hand to the developer/designer. As reported in the previous pages, it is possible to extract network architectures with validity on very wide frequency ranges. In this scope, the observed range from 200 MHz up to 110 GHz is extremely wide, and it was employed mainly for explanatory purposes, with the aim of challenging quite harshly the robustness of the discussed approach. In real cases, what the RF-MEMS developer needs are lumped networks exhibiting good accuracy in a range from 5–10 GHz up to 20–30 GHz at most. In light of the plots previously commented upon and reported in figure 4.24, figure 4.25, figure 4.28 and figure 4.29, it clearly emerges that restricting the analysed frequency span from 110 GHz down to a few tens of GHz would lead to significantly improved levels of accuracy.

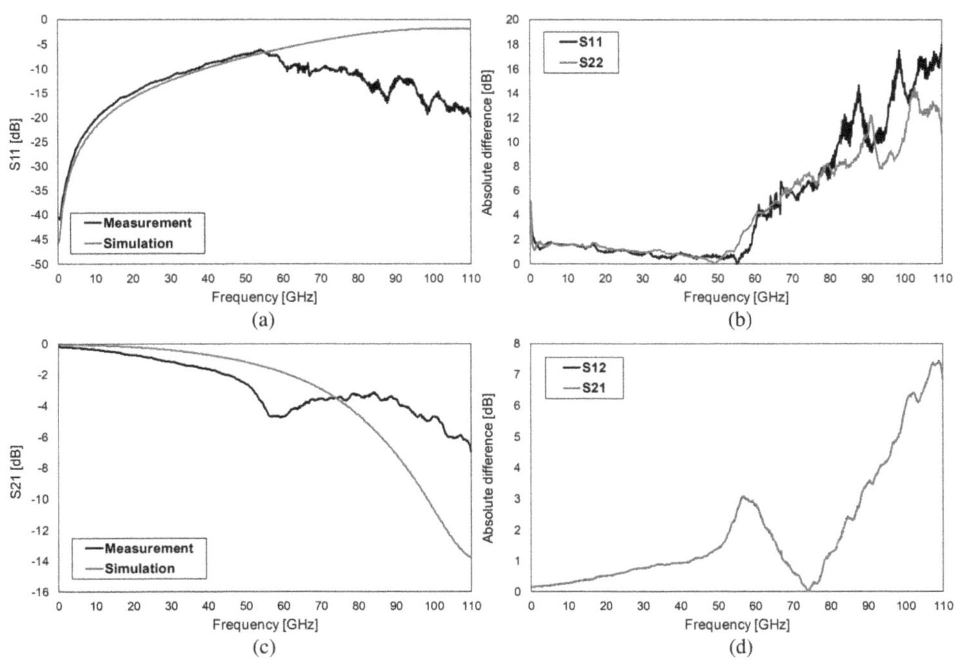

Figure 4.28. Measured S-parameters' characteristics of the shunt RF-MEMS attenuator module in figure 4.21(b) when the micro-relay is OFF, and response of the equivalent circuit in figure 4.26, from 200 MHz up to 110 GHz. (a) Measured versus simulated reflection (S11) characteristic. (b) Absolute difference (in dB) between the measured and simulated reflection characteristics at both ports (S11 and S22). (c) Measured versus simulated loss (S21) characteristic. (d) Absolute difference (in dB) between the measured and simulated loss characteristics at both ports (S12 and S21).

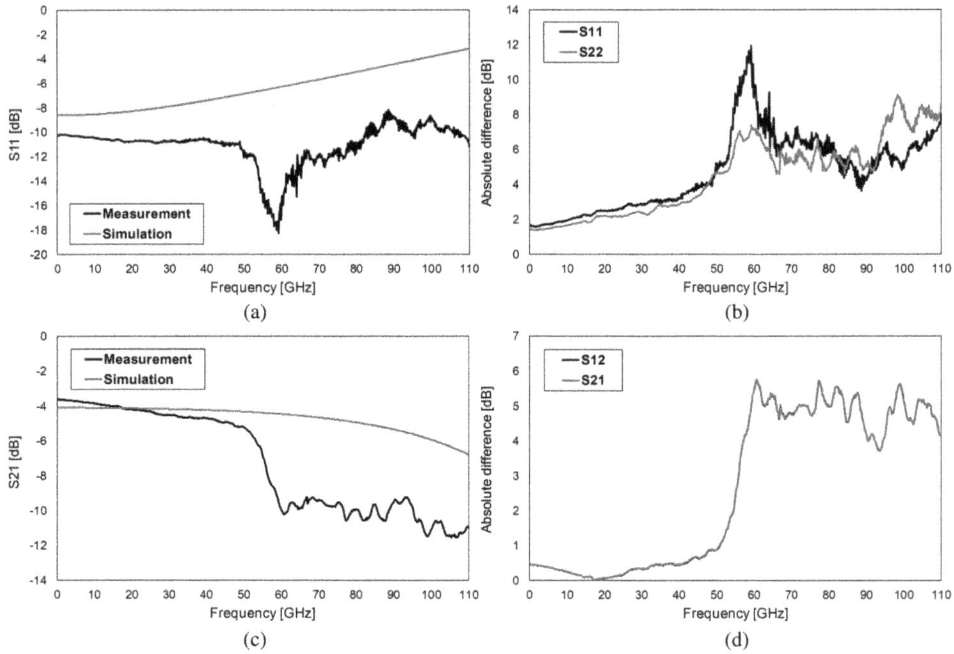

Figure 4.29. Measured S-parameters' characteristics of the shunt RF-MEMS attenuator module in figure 4.21(b), when the micro-relay is ON, and response of the equivalent circuit in figure 4.27, from 200 MHz up to 110 GHz. (a) Measured versus simulated reflection (S11) characteristic. (b) Absolute difference (in dB) between the measured and simulated reflection characteristics at both ports (S11 and S22). (c) Measured versus simulated attenuation (S21) characteristic. (d) Absolute difference (in dB) between the measured and simulated attenuation characteristics at both ports (S12 and S21).

4.5 Conclusion

This chapter builds on the previously developed discussion on future 5G applications, presenting the analysis of RF-MEMS passive components by means of more practical and hands-on approaches. Nonetheless, before stepping into detail, a comprehensive analysis of target performances was developed. As a matter of fact, the preceding chapter 3 on the 5G scenario listed system-level characteristics that mobile networks and infrastructures will have to satisfy. Therefore, the initial part of this chapter was dedicated to deriving a plausible set of specifications at the device level that RF passive components should be able to reach in order to be suitable for 5G applications. Providing these numbers, case studies and examples of RF-MEMS design optimisation were reviewed.

First, the focus was concentrated on the coupled electromechanical characteristics of MEMS devices. In this regard, a few switching unit topologies were placed under the spotlight, capitalising on FEM simulations aimed at investigating their characteristics and performance. In particular, the influence of the flexible suspensions geometry on the pull-in and pull-out voltages was analysed in static conditions. Therefore, the dynamic behaviour of RF-MEMS switches was also studied, showing how the mechanical stiffness of suspensions can influence the opening/closing switching time in transient conditions. In particular, it was highlighted how the RF-MEMS designer very often has to deal with conflicting specifications, and how proper trade-offs must be sought, trying to achieve satisfactory characteristics without impairment of other device features, at the same time. In particular, it was demonstrated that lowering the pull-in voltage leads to MEMS switches with slower actuated/release dynamic response.

Then, discussion around the electromagnetic modelling of RF-MEMS was developed. At first, the selected FEM tool underwent a comprehensive validation phase, taking experimental S-parameters' datasets as reference targets. The validation was necessary to ensure accurate prediction of the physical device's behaviour, through tailoring the 3D model assembly and the simulator settings on the actual characteristics of a real selected technology platform. Furthermore, a validated electromagnetic simulation method was employed to discuss a few RF-MEMS switching unit design concepts. In particular, the influences of the suspended membrane and of the contact geometry on critical parameters like isolation (when OPEN) and loss (when CLOSED) were discussed. As previously done concerning the electromechanical behaviour, the presence of conflicting characteristics in the electromagnetic domain was reported. In particular, the trade-off between isolation and loss was analysed, also introducing an increase of redundancy as a possible strategy to improve the RF-MEMS device characteristics as a whole.

Finally, the discussion on RF modelling was completed by an extensive overview of a methodology for extracting RF-MEMS equivalent lumped-element networks. Such architectures are very useful both to comprehend in-depth the physical causes lying behind the RF behaviour of the micro-devices, as well as to easily parameterise and/or co-simulate the MEMS part together with the rest of the system/subsystem. Eventually, all the reported characteristics and performance of the selected RF-MEMS case studies were commented on against the set of 5G target specifications mentioned in the beginning of the chapter, thus closing the whole discussion developed throughout this work.

References

[1] Giacomozzi F, Mulloni V, Colpo S, Iannacci J, Margesin B and Faes A 2011 A flexible fabrication process for RF MEMS devices *Rom. J. Inf. Sci. Technol* **14** 259–68 https://imt.ro/romjist/Volum14/Number14_3/pdf/07-Giacomozzi.pdf

[2] Iannacci J 2013 *Practical Guide to RF-MEMS* 1st edn (Weinheim: Wiley) p 372

[3] Iannacci J and Gaddi R 2010 *Mixed-Domain Simulation and Wafer-Level Packaging of RF-MEMS Devices* 1st edn (Saarbrücken: Lambert Academic) p 282

[4] Senturia S D 2001 *Microsystem Design* 1st edn (New York: Springer) p 689

[5] Mentor *Tanner* L-Edit IC *Layout* https://eda-solutions.com/app/uploads/2020/06/tanner-mems-design-ds.pdf

[6] Culpepper M L *Creating DXF Files For The Waterjet* https://ocw.mit.edu/courses/mechanical-engineering/2-000-how-and-why-machines-work-spring-2002/tools/2_000_DXF_Tutorial.pdf

[7] *FreeCAD* www.freecadweb.org/

[8] *The FreeCAD Manual* www.freecadweb.org/wiki/Import_Export

[9] ANSYS *Workbench Platform* www.ansys.com/products/platform

[10] Harrar D II *MEMS Multiphysics Simulation in ANSYS Workbench* www.ozeninc.com/downloads/PRESENTATION-Mulitiphysics_Simulation_for_MEMS_Using_Workbench.pdf

[11] Iannacci J, Gaddi R and Gnudi A 2007 Non-linear electromechanical RF model of a MEMS varactor based on VerilogA© and lumped-element parasitic network *Proc. 37th Europ. Microw. Conf. EuMC (Munich, Oct. 2007)* pp 544–47

[12] Pacheco S P, Katehi L P B and Nguyen C T C 2000 Design of low actuation voltage RF MEMS switch *Proc. IEEE MTT-S Int. Microw. Symp. (Boston, MA, June 2000)* **1** 165–68

[13] Iannacci J, Gaddi R and Gnudi A 2010 Experimental validation of mixed electromechanical and electromagnetic modeling of RF-MEMS devices within a standard IC simulation environment *IEEE J. Microelectromechan. Syst.* **19** 526–37

[14] *ANSYS—Cantilever Beam Modal Analysis* https://confluence.cornell.edu/display/SIMULATION/ANSYS+-+Cantilever+Beam+Modal+Analysis

[15] Sun Y, Fang D and Soh A K 2006 Thermoelastic damping in micro-beam resonators *Int. J. Solids Struct.* **43** 3213–29

[16] Künzig T, Niessner M, Wachutka G, Schrag G and Hammer H 2010 The effect of thermoelastic damping on the total Q-factor of state-of-the-art MEMS gyroscopes with complex beam-like suspensions *Procedia Eng.* **5** 1296–9

[17] De S K and Aluru N R 2006 Theory of thermoelastic damping in electrostatically actuated microstructures *Phys. Rev.* B **74** 144305

[18] Iannacci J, Del Tin L, Gaddi R, Gnudi A and Rangra K J 2005 Compact modeling of a MEMS toggle-switch based on modified nodal analysis *Proc. Symp. on Design, Test, Integration and Packaging of MEMS/MOEMS DTIP 2005 (Montreux, June 2005)* pp 411–6

[19] Iannacci J and Gaddi R 2005 Nodal modelling of uneven electrostatic transduction in MEMS ed C Di Natale, A D'Amico, G Martinelli, M C Carotta and V Guidi *Proc. 9th Italian Conf. on Sensors and Microsystems* (Singapore: World Scientific) pp 385–9

[20] Niessner M, Schrag G, Iannacci J and Wachutka G 2011 Macromodel-based simulation and measurement of the dynamic pull-in of viscously damped RF-MEMS switches *Sensors Actuators* A **172** 269–79

[21] Fruehling A, Yang W and Peroulis D 2012 Cyclic evolution of bouncing for contacts in commercial RF MEMS switches *Proc. IEEE 25th Int. Conf. on Micro Electro Mechanical Syst. MEMS (Paris, Jan. 2012)* pp 688–91

[22] Peschot A, Poulain C, Valadares C, Reig B, Bonifaci N and Lesaint O 2014 Evolution of contact bounces in MEMS switches under cycling *Proc. IEEE 60th Holm Conf. on Electrical Contacts (New Orleans, LA, Oct. 2014)* pp 1–6

[23] Tavassolian N, Koutsoureli M, Papaioannou G, Lacroix B and Papapolymerou J 2010 Dielectric charging in capacitive RF MEMS switches: the effect of electric stress *Proc. Asia-Pacific Microw. Conf. APMC (Yokohama, Dec. 2010)* pp 1833–6

[24] Zaghloul U, Bhushan B, Pons P, Papaioannou G, Coccetti F and Plana R 2011 Different stiction mechanisms in electrostatic MEMS devices: nanoscale characterization based on adhesion and friction measurements *Proc. Int. Solid-State Sens., Actuators and Microsystems Conf. TRANSDUCERS (Beijing, June 2011)* pp 2478–81

[25] Tazzoli A, Iannacci J and Meneghesso G 2011 A positive exploitation of ESD events: micro-welding induction on ohmic MEMS contacts *Proc. Electrical Overstress/Electrostatic Discharge Symp. EOS/ESD (Anaheim, CA, Sept. 2011)* pp 1–8

[26] Iannacci J, Faes A, Repchankova A, Tazzoli A and Meneghesso G 2011 An active heat-based restoring mechanism for improving the reliability of RF-MEMS switches *Microelectron. Reliab.* **51** 1869–73
[27] Ansys H F S S www.ansys.com/products/electronics/ansys-hfss
[28] Casini F, Farinelli P, Mannocchi G, DiNardo S, Margesin B, De Angelis G, Marcelli R, Vendier O and Vietzorreck L 2010 High performance RF-MEMS SP4T switches in CPW technology for space applications *Proc. European Microw. Conf. EuMW (Paris, Sep.–Oct. 2010)* pp 89–92
[29] Diaferia F *et al* 2014 Compact 12×12 switch matrix integrating RF MEMS switches in LTCC hermetic packages *Proc. European Microw. Conf. EuMW (Rome, Oct. 2014)* pp 199–202
[30] Giacomozzi F, Mulloni V, Colpo S, Faes A, Sordo G and Girardi S 2013 RF-MEMS devices packaging by using quartz caps and epoxy polymer sealing rings *Proc. Symp. on Design, Test, Integration and Packaging of MEMS/MOEMS DTIP (Barcelona, April 2013)* pp 1–6
[31] Iannacci J 2013 Simulation techniques (commercial tools) *Practical Guide to RF-MEMS* 1st edn (Weinheim: Wiley) ch 4 pp 121–31
[32] Iannacci J and Tschoban C 2017 RF-MEMS for future mobile applications: experimental verification of a reconfigurable 8-bit power attenuator up to 110 GHz *J. Micromech. Microeng.* **27** 044003
[33] Huang J T, Hsu Y K, Lo Y C, Lee K Y, Chen C K and Tsai T C 2010 Design and fabrication of low-insertion loss and high-isolation CMOS-MEMS switch for microwave applications *Proc. Int. Microsystems Packaging Assembly and Circuits Technol. Conf (Taipei, Oct. 2010)* pp 1–3
[34] Fomani A A and Mansour R R 2009 Miniature RF MEMS switch matrices *Proc. IEEE MTT-S Int. Microw. Symp. (Boston, MA, June 2009)* pp 1221–4
[35] Shen H, Gong S and Barker N S 2008 DC-contact RF MEMS switches using thin-film cantilevers *Proc. Europ. Microw. Integrated Circuit Conf. EuMIC (Amsterdam, Oct. 2008)* pp 382–5
[36] Kanthamani S, Raju S and Abhai Kumar V 2008 Design of low actuation voltage RF MEMS cantilever switch *Proc. Int. Conf. on Recent Adv. in Microw. Theory and Applications (Jaipur, Nov. 2008)* pp 584–6
[37] Muley C A and Naveed S A 2013 Modelling of cantilever based MEMS RF switch *Proc. Int. Conf. on Computing, Commun. and Networking Technol. ICCCNT (Tiruchengode, July 2013)* pp 1–5
[38] Shalaby M M, Wang Z, Chow L L-W, Jensen B D, Volakis J L, Kurabayashi K and Saitou K 2009 Robust design of RF-MEMS cantilever switches using contact physics modeling *IEEE Trans. Ind. Electron.* **56** 1012–21
[39] Homayouni S M, Schreurs D and Nauwelaers B 2009 Wide-band multi-bias equivalent circuit extraction for FinFET transistors accommodating high-frequency kink-behaviours *Proc. Europ. Microw. Integrated Circuits Conf. EuMIC (Rome, Sept. 2009)* pp 885–8
[40] Wada T, Nakajima R, Obiya H, Ogami T, Koshino M, Kawashima M and Nakajima N 2014 A miniaturized broadband lumped element circulator for reconfigurable front-end system *Proc. IEEE MTT-S Int. Microw. Symp. IMS2014 (Tampa, FL, June 2014)* pp 1–3
[41] Essaadali R, Jarndal A, Kouki A B and Ghannouchi F M 2016 A new GaN HEMT equivalent circuit modeling technique based on x-parameters *IEEE Trans. Microw. Theory Tech.* **64** 2758–77

[42] Arcioni P, Castello R, Perregrini L, Sacchi E and Svelto F 1998 An improved lumped-element equivalent circuit for on silicon integrated inductors *Proc. IEEE Radio and Wireless Conf. RAWCON (Colorado Springs, CO, Aug. 1998)* pp 301–4

[43] Kang J, Sun L, Wen J and Zhao M 2009 An equivalent lumped-circuit model for on-chip symmetric intertwined transformer *Proc. IEEE Int. Conf. on ASIC (Changsha, Hunan, Oct. 2009)* pp 674–7

[44] Horng T S, Jau J K, Tsai Y S and Huang C S A decomposition and reconstruction scheme for broadband modeling of on-chip passive components using the modified T-equivalent circuit topology *Proc. IEEE Radio Frequency integrated Circuits (RFIC) Symp* (Long Beach, CA, June 2005) pp 299–302

[45] Hwang J, Jung W and Kim S 2015 Coupling analysis and equivalent circuit model of the IC stripline method *Proc. Asia-Pacific Symp. on Electromag. Compatibility APEMC (Taipei, May 2015)* pp 650–3

[46] Zhou Y and Chen Y 2008 Lumped-element equivalent circuit models for distributed microwave directional couplers *Proc. Int. Conf. on Microw. and Millimeter Wave Technol. (Nanjing, April 2008)* pp 131–4

[47] Yang K, Yin W Y and Mao J F 2007 Wideband lumped element model for on-chip (a) symmetrical coupled interconnects on lossy silicon substrate *Proc. Asia-Pacific Microw. Conf. (Bangkok, Dec. 2007)* pp 1–4

[48] Chen K L, Salvia J, Potter R, Howe R T and Kenny T W 2009 Performance evaluation and equivalent model of silicon interconnects for fully-encapsulated RF MEMS devices *IEEE Trans. Adv. Packag.* **32** 402–9

[49] Grinde C, Nygaard K H, Due-Hansen J and Fjeldly T A 2009 Lumped modeling of a novel RF MEMS double-disk resonator system *Proc. Symp. on Design, Test, Integration & Packag. of MEMS/MOEMS DTIP (Rome, April 2009)* pp 14–8

[50] Unlu M, Topalli K, Demir S, Civi O A, Koc S and Akin T 2004 A parametric modeling study on distributed MEMS transmission lines *Proc. Europ. Microw. Conf. EuMW (Amsterdam, Oct. 2004)* pp 1157–60

[51] Klymyshyn D M, Martin Borner M, Haluzan D T, Santosa E G, Melissa Schaffer M, Achenbach S and Mohr J 2010 Vertical hgh-Q RF-MEMS devices for reactive lumped-element circuits *IEEE Trans. Microw. Theory Techn.* **58** 2976–86

[52] Iannacci J 2014 RF-MEMS: a development flow driving innovative device concepts to high performance components and networks for wireless applications *Proc. Microelectron. Syst. Symp. MESS (Vienna, May 2014)* pp 1–6

[53] Iannacci J, Huhn M, Tschoban C and Pötter H 2016 RF-MEMS technology for 5G: series and shunt attenuator modules demonstrated up to 110 GHz *IEEE Electron Device Lett.* **37** 1336–9

[54] Iannacci J, Reyes J, Maaß U, Huhn M, Ndip I and Pötter H 2016 RF-MEMS for 5G mobile communications: a basic attenuator module demonstrated up to 50 GHz *Proc. IEEE SENSORS (Orlando, FL, Oct.–Nov. 2016)* pp 1–3

[55] Keysight Technologies *Advanced Design System (ADS)* https://keysight.com/it/en/products/software/pathwave-design-software/pathwave-advanced-design-system.html

[56] Cadence *Spectre Simulation Platform—Block-level, Chip-level, and Mixed-signal Simulation* https://cadence.com/en_US/home/tools/custom-ic-analog-rf-design/circuit-simulation/spectre-simulation-platform.html

[57] Mentor *HyperLynx Analog* https://cadlog.it/prototipazione-virtuale/hyperlynx-analog/hyperlynx-analog-4/

[58] *Quite Universal Circuit Simulator* http://qucs.sourceforge.net/

[59] Iannacci J 2010 Mixed-domain fast simulation of RF and microwave MEMS-based complex networks within standard IC development frameworks *Advanced Microwave Circuits and Systems* ed V Zhurbenko 1st edn (Rijeka: INTECH) pp 313–38

Chapter 5

A comprehensive approach to the compact multi-physical modelling and simulation of RF-MEMS

5.1 Introductive considerations on the compact modelling approach

As emerged in the discussion previously developed in this book, and in particular through the last chapter, the behaviour of Radio Frequency Micro-ElectroMechanical Systems (RF-MEMS) devices is rather complex. In fact, the general characteristic of Micro-ElectroMechanical Systems (MEMS) is that of coupling at least two different physical domains, one of which is always the mechanical/structural one. In the particular case, here under examination, of electrostatically controlled MEMS for RF applications, the involved domains are structural mechanics, electrostatics and electromagnetism, with the mechanical part coupled to electrostatics (i.e. electromechanical behaviour), and to the Radio Frequency (RF) domain. Having said that, a first attempt to derive a multi-physical model, describing the overall behaviour of an RF-MEMS device, would rely on putting together the theory of mechanics and elasticity [1], of electrostatics [2], as well as of microwaves and electromagnetics [3], leading to a formal problem too complex to be agilely addressed, as indicated in figure 5.1.

Then, it is straightforward that some simplifications must be applied to the purely analytical approach just mentioned. Starting the discussion from a rather general perspective, the possible strategies to build up more agile models, commencing from the full analytical description of the problem, are basically two. On one hand, numerical approaches can be followed, while, on the other, compact and simplified approaches are also viable, as schematically reported in figure 5.2.

When numerical approaches are pursued, the solution of the mathematical problem is simplified and approximated, in such a way that its resolution can

doi:10.1088/978-0-7503-4199-8ch5

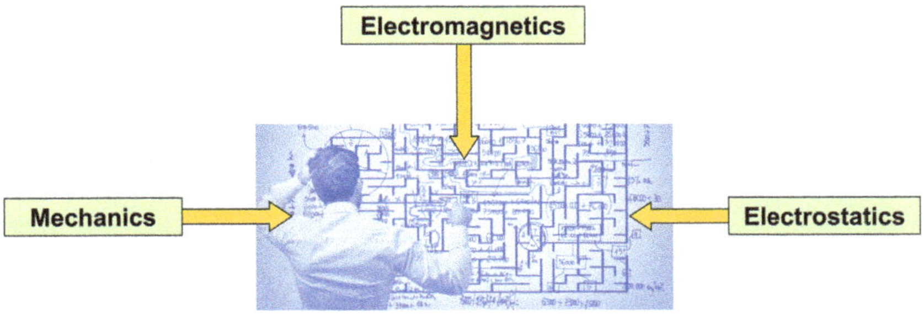

Figure 5.1. A combination of the full theories of mechanics, electrostatics and electromagnetics to model RF-MEMS leads to the definition of problems that are too complex to be easily solved.

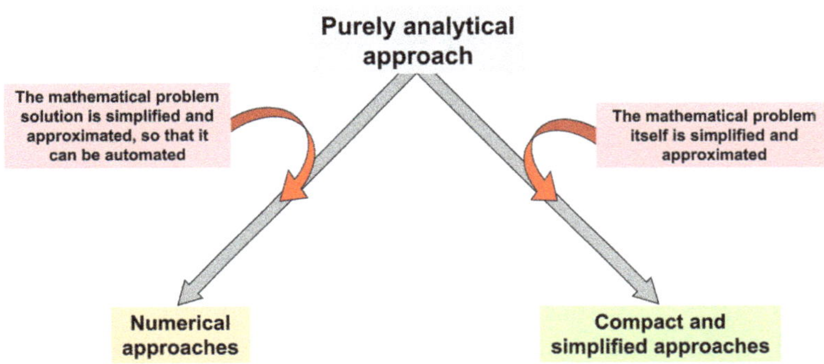

Figure 5.2. Fundamental simplification approaches to derive agile descriptive models starting from the full analytical description of the problem to be solved.

effectively be automated. For instance, this is the case with the software tools based on the Finite Element Method (FEM) approach, which will be mentioned later. Alternatively, when the compact and simplified strategy is deployed, it is the mathematical formulation itself to be reduced in complexity, in order to make viable and affordable its solution in a closed form. Of course, the two mentioned strategies are not exclusive, as proper combinations of both can be an efficient strategy, depending on the complexity of the initial problem to be solved, as well as on the desired accuracy of the results and on the expected range of validity of the model.

Bearing in mind the mentioned strategies, a plethora of different approaches to the modelling and simulation of RF-MEMS and, more in general, of whatever type of problem, unfolds against those interested in predicting the behaviour of a novel design concept, to be optimised with respect of certain goals and specifications. To this end, the scheme depicted in figure 5.3 summarises some of the most relevant modelling and simulation techniques available, when following numerical or compact and simplified solution approaches.

Starting from the numerical approaches, the FEM, the Boundary Element Method (BEM) and the Finite Difference Method (FDM) are mentioned. Within such methodologies, FEM-based tools are certainly the most common and diffused, especially when dealing with structural mechanics, thermodynamics and the

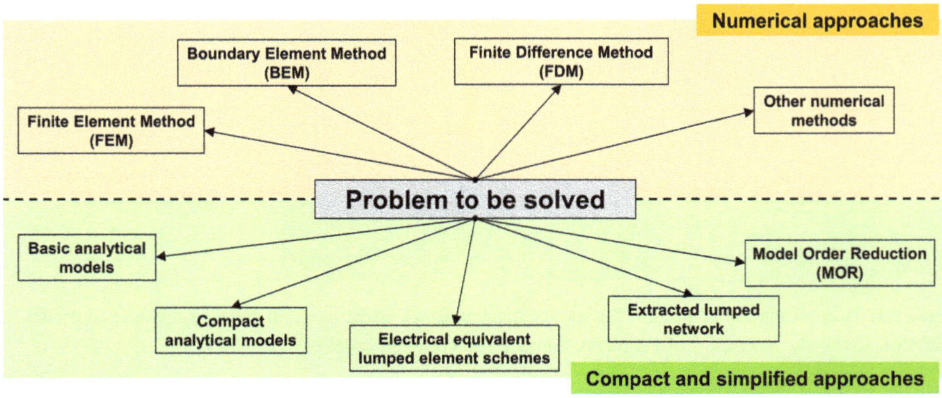

Figure 5.3. Constellation of significant modelling and simulation techniques, available when following numerical or compact and simplified approaches.

dynamics of fluids. In brief, FEM is a discretisation method, in which the mathematical model is subdivided into smaller and non-overlapping components of simple geometry, called elements [4, 5]. Each element response is expressed in terms of a finite number of Degrees of Freedom (DoFs), which are characterised as the value of a function (or functions) at a set of nodal points. The response of the mathematical model is approximated to that of the discrete model obtained by connecting the elements. In contrast, the BEM approach is a technique to approximate and simplify the resolution of equations without, in fact, performing any volumetric discretisation, limiting it to the sole boundaries of elements [6], while the FDM, likewise FEM, divides the volume into elements and solves differential equations by approximating derivatives with finite differences [7]. The FEM, BEM and FDM approaches exhibit advantages and disadvantages, not discussed here for the sake of brevity, leading to the selection of the most appropriate method to depend on several factors, like the physical domain(s) to be analysed, the desired accuracy of results, as well as the computational complexity required. Other numerical approaches are not reported in detail here, despite being viable and accurate, similar to the others discussed above.

Stepping now into the domain of compact and simplified modelling and simulation approaches (lower half in figure 5.3), there are several items worth being mentioned and briefly reviewed as to their main details. The first and simplest solution is that of basic analytical models. In this case, the mathematical formulation of the problem is rather simple, and it is sufficient to describe with reasonable accuracy a well-circumstanced behavioural characteristic of the studied system. Basic analytical models often rely on drastic simplifications of the system geometry, therefore limiting their range of validity to a narrow set of cases. A typical example of this approach is the modelling of the pull-in/pull-out characteristic of an electrostatically controlled MEMS actuator [8], in which the transducer is described as an ideal parallel plate capacitor, while the elastic behaviour of the membrane under deformation is approximated to a 1 DoF ideal spring, i.e. deformable just along the vertical axis. Differently, the compact analytical modelling approach, likewise basic models, introduces a certain degree of simplification with respect to the full mathematical formulation, yet enabling an enhanced spatial description of the physical effects to be studied. In practical terms, the behaviour of the studied system is described according to a certain number of input/output nodes, and the set of relationships between such nodes is defined by the chosen mathematical model [9–12]. Therefore, compact models are by their nature significantly flexible, as one can add DoFs to the model, therefore increasing the number of geometrical and physical features taken into account. On the other hand, the analytical model, defining the link between the nodes, can be increased or lowered in complexity, depending on the desired trade-offs in terms of accuracy and ease of computation.

A rather different modelling approach is that of electrical equivalent lumped-element schemes and networks, based on the concept of conjugate power variables. In practical terms, the exchange of power between lumped elements can always be expressed as the combination of two magnitudes, named effort and flow, that refer to a certain physical domain [13]. These considerations pose an analogy between different physical domains, like electrical, mechanical, magnetic, and so on.

A typical case in which electrical-equivalent-network-based modelling can be effectively exploited is in the description of the mechanical behaviour of MEMS. In fact, the typical mass-spring-damper mechanical lumped network describing the mechanical device characteristics can be transformed into an electrical equivalent resistor, inductor, capacitor (RLC) network, where the following analogies are satisfied [14–16]:

- Spring elastic energy → capacitor electrostatic energy.
- Mass potential energy → inductor magnetic energy.
- Mechanical energy dissipated by the damper (air damping) → electrical energy dissipated by the resistor.

Moving on in reviewing the modelling approaches inserted in figure 5.3, the next item is that of extracted lumped networks. This method is particularly suitable in the description of electrical and electronic devices and circuits, especially if working in the Alternating Current (AC) and RF regimes [17–28]. Starting from a basic yet solid knowledge of the physics and working principles of a certain device, e.g. a transistor, a micro-switch, etc, a network architecture, featuring in the first instance resistors, capacitors and inductors, is assembled, in order to capture the main electrical, electronic and/or electromagnetic behaviour of the studied device. Subsequently, suitable first attempt values are assigned to the network components, basing them on physical and mathematical considerations. Next, such values are optimised, in order to reach satisfactory fitting against the experimentally measured characteristics of the analysed device, and/or data obtained by means of accurate and validated simulation approaches (e.g. FEM simulations). The mentioned optimisation can be done by hand if the network architecture is simple, as well as by means of automated optimisation software tools. Since both the circuit architecture definition and the values assigned to its components are carried out on the basis of physical considerations (i.e. are not randomly fixed just to fit measurements), the obtained models are flexible to a certain extent, in terms of ranges of application and validity. This means that a unique architecture can accurately predict the changed behaviour of the device, when some of its geometry and/or material properties are varied. In addition, the extracted network can be expanded, in order to include the modelling of second-order behavioural characteristics, like, for instance, parasitic effects and unwanted couplings in RF devices. Examples of how the extracted lumped-element network approach can be applied to RF-MEMS devices, with different extent of inclusion of second-order effects and non-idealities, are reported in [29–32].

Finally, to complete the discussion of the constellation reported in figure 5.3, Model Order Reduction (MOR) is a branch of systems and control theory which studies the properties of dynamical systems in application for reducing their complexity, while preserving (to the greatest possible extent) their input-output behaviour [33]. MOR is of particular interest when dealing with MEMS, as the geometry, working principles and characteristics of microsystems are wider, in terms of physical domains and their coupling, than transistor-like devices. MOR algorithms are also suitable to be fully automated, indeed relieving the MEMS designer/scientist from the task of particularising the chosen modelling and descriptive methodologies every time the device design changes. Starting from a complete set of FEM simulations, the MOR identifies a subset of critical nodes, capturing the salient features of the system behaviour while ignoring all the other terminations. This leads to a reduced-order system that can be treated as a lumped-element network. The MOR also enables and eases simulation at the system level [34–39].

Now that the main modelling and simulation approaches have been reviewed, the following graph in figure 5.4 provides a comparison between the complexity and the flexibility of usage and adaptation of each method.

In fact, as it is straightforward to realise, the more a modelling approach targets simplicity and ease of computation, the more it is tailored to the specific device

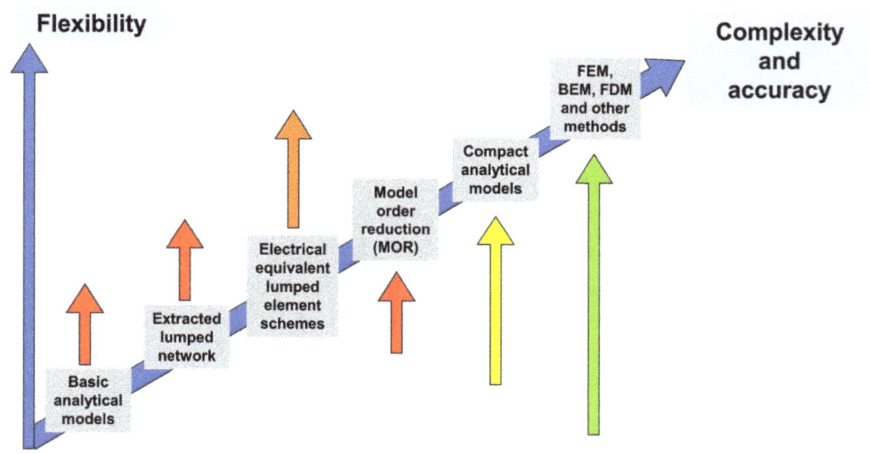

Figure 5.4. Visual graph building a comparison of the flexibility of usage against complexity and accuracy, of the modelling and simulation techniques previously discussed and listed in figure 5.3.

under analysis, therefore the less it is flexible and portable. In the following pages, a specific modelling and simulation approach to the description of RF-MEMS is going to be reported, discussed and validated.

5.2 Developed models and software library implementation

Capitalising on the previous discussion, a multi-physical modelling and simulation approach, purposely developed for RF-MEMS, is now going to be introduced and discussed. In particular, the simulation kit is a software library of basic components that can be suitably assembled together at schematic level, in order to compose actual electrostatically driven RF-MEMS devices like switches and varactors, as well as complex networks (e.g. multi-state phase shifters, attenuators, etc). The mentioned simple components are building blocks of typical MEMS geometries, like straight flexible suspensions, and parallel plate electromechanical transducers (i.e. air gaps), each provided with a schematic symbol. The library also features Boundary Conditions (BCs) and stimuli elements, like mechanical anchoring points and forces/torques. Therefore, the symbols just have to be instanced and properly wired to each other in order to build a schematic of the RF-MEMS component to be simulated. According to the compact analytical modelling approach, each element of the library has a certain number of input/output nodes, and their relationships are defined by suitable analytical models, as will be reported in the following pages. Such models are expressed and appropriately simplified, in order to be in a closed form, and therefore easily implemented in a programming language. In this specific case, the chosen code is VerilogA [40–43], that is, one of the various types of Hardware Description Languages (HDLs) [44–47], and the library is developed and utilised within the Cadence Integrated Circuits (ICs) development software framework (www.cadence.com). Nonetheless, it has to be highlighted that the library can be ported to other HDL-based languages, as well as migrated to different simulation environments for electronic circuits and devices.

The following subsections will delve into the technical details of the analytical models behind the most relevant items included in the software library. As the mentioned models mainly describe the mechanical (static and dynamic), electrical and coupled electromechanical behaviour of RF-MEMS, a proper description of the RF characteristics capitalises on the exploitation of extracted lumped networks, built around electromechanical compact models. The latter aspect will be discussed in more detail in the next section, reporting on the validation and practical exploitation in real cases of the behavioural software library.

5.2.1 Rigid plate electromechanical transducer

The aim of this subsection is that of reporting on the mechanical formulation concerned with the first fundamental element, implemented in the software library, necessary to compose schematics of RF-MEMS devices: that is, the rigid plate electromechanical model. The following content is divided into modules, according to the physical domain/domains involved (e.g., mechanical model, electromechanical transduction, and so on).

5.2.1.1 Four Degrees of Freedom (4 DoFs) mechanical model

The first step is that of describing the rigid plate according to a suitable reference system. The reference frame origin is defined in the center, and the local reference system (*xyz*) is assumed to coincide with the global one (*XYZ*). The main geometrical features of the plate are the width (W), length (L) and the thickness (T), along the x, y and z axes, respectively, as reported by the schematic in figure 5.5.

Since in the majority of practical cases the plate thickness is much smaller than the width and length ($T \ll W, L$), the specific point along T chosen as the reference system origin is irrelevant. It is assumed here that such a point corresponds to the plate center of mass, under the hypothesis that the material is homogeneous. In the 4 DoFs implementation reported here, the rigid plate allows three linear displacements, along the three coordinated axes (*xyz*), plus rotation around the vertical axis (z). This means that, according to the model, the rigid plate always remains parallel with respect to the underlying electrodes and contacts. Having said that, the main physical effect involved is the body inertia, expressed by the well-known equation (5.1) as

$$F = mA \tag{5.1}$$

where F is the force applied to the body, m is the mass and A is its acceleration. The main issue to be solved is how to compose the set of external forces and torques applied to the plate, in order to define their resultant contributions. Considering the typical operation conditions and topologies of MEMS and RF-MEMS structures, it is reasonable to assume that the only points at which external stimuli can be applied to the plate are its four vertexes (nodes). Starting from this statement, the nodes in this model are addressed as North-East (*NE*), South-East (*SE*), South-West (*SW*) and North-West (*NW*), as indicated in the schematic reported in figure 5.6.

The external applied forces and torques are referred to the Center of Mass (*CM*) that is placed at the mid-point of W, L and T. Thereafter, the forces applied to each node have to be decomposed in their contributions along the X and Y directions. Then, the arms of each node with respect to the center on the XY plane must be derived, in order to transform the imposed forces into the corresponding torques applied to the *CM*. It is worth highlighting that the forces' components and arms are

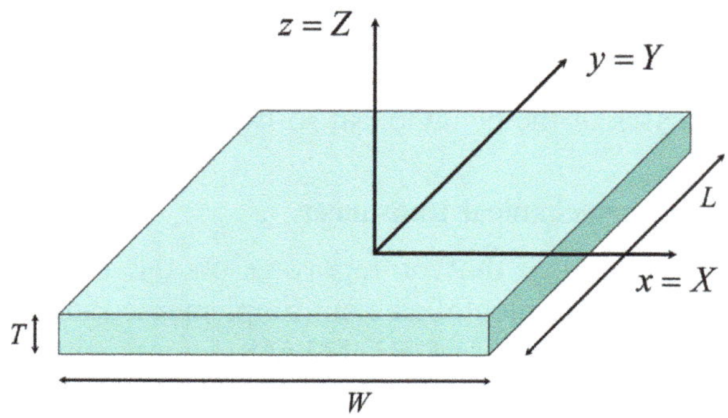

Figure 5.5. Schematic of the rigid plate, with the reference systems and its dimensions indicated. [51] John Wiley & Sons. © 2013 Wiley-VCH Verlag GmbH & Co. KGaA.

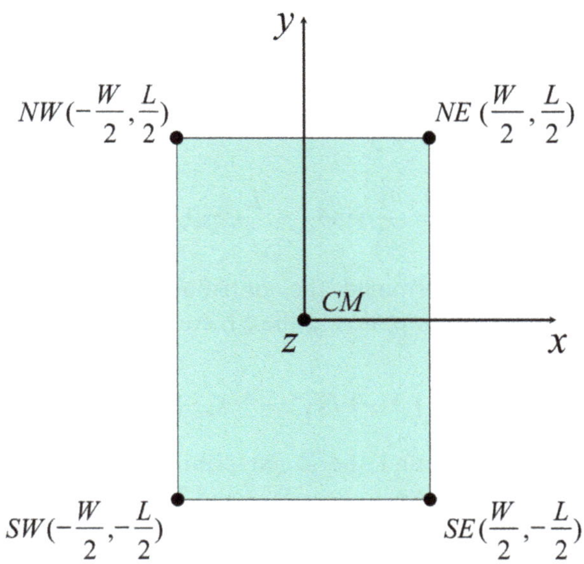

Figure 5.6. Schematic of the rigid plate with its corners coordinates in the local reference system. [51] John Wiley & Sons. © 2013 Wiley-VCH Verlag GmbH & Co. KGaA.

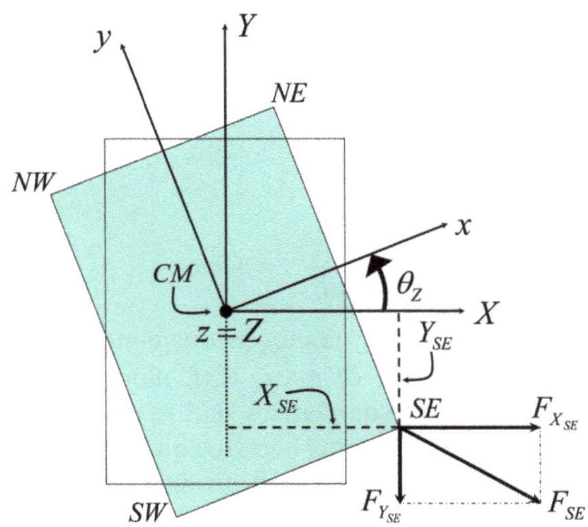

Figure 5.7. Schematic of the plate rotation around the vertical axis when an external force is applied to the *SE* corner. [51] John Wiley & Sons. © 2013 Wiley-VCH Verlag GmbH & Co. KGaA.

referred to the global system (*XYZ*). In fact, when describing different components joined together, the consistency between them has to be maintained by choosing a common and unique reference system. An example of plate displacement due to the application of an external force to the *SE* corner is reported in figure 5.7.

The segments labelled as X_{SE} and Y_{SE} are the projections of the *SE* corner in the global reference system (*XYZ*), and depend on the coordinates of the *SE* point in the local frame (*xyz*) and on the angle θ_Z. Such segments can be easily calculated by exploiting a rotation matrix [29], yielding the expressions in equation (5.2a) and equation (5.2b):

$$X_{SE} = \frac{W}{2}\cos\theta_Z + \frac{L}{2}\sin\theta_Z \tag{5.2a}$$

$$Y_{SE} = \frac{W}{2}\sin\theta_Z - \frac{L}{2}\cos\theta_Z. \tag{5.2b}$$

Once X_{SE} and Y_{SE} are determined, the contributions to the Z axis torque of the CM, due to the external forces applied to the SE corner, can be calculated according to equation (5.3):

$$T_{\theta_{ZCM}} = |\, Y_{SE} \,|\, F_{\Delta X_{SE}} - |\, X_{SE} \,|\, F_{\Delta Y_{SE}} \tag{5.3}$$

where $T_{\theta_{ZCM}}$ is the torque around the Z axis, and where $F_{\Delta X_{SE}}$ and $F_{\Delta Y_{SE}}$ are the effective forces applied to the SE node, along the X and Y directions, respectively, expressed as follows:

$$F_{\Delta X_{SE}} = F_{X_{SE}} - F_{X_{CM}} \tag{5.4a}$$

$$F_{\Delta Y_{SE}} = F_{Y_{SE}} - F_{Y_{CM}}. \tag{5.4b}$$

In the previous equation (5.4a) and equation (5.4b), inertia is taken into account and applied to the CM, by means of the terms $F_{X_{CM}}$ and $F_{Y_{CM}}$, expressed as follows:

$$F_{X_{CM}} = -m\frac{d^2 p_{X_{CM}}}{dt^2} = -m a_{X_{CM}} \tag{5.5a}$$

$$F_{Y_{CM}} = -m\frac{d^2 p_{Y_{CM}}}{dt^2} = -m a_{Y_{CM}} \tag{5.5b}$$

where m is the plate mass (assuming homogeneous material), while $p_{X_{CM}}$ and $p_{Y_{CM}}$ describe the position of the center of mass, CM, along the X and Y directions, respectively, and their double time derivatives lead to the acceleration components, expressed by $a_{X_{CM}}$ and $a_{Y_{CM}}$. Now, coming back to equation (5.3), the inclusion of the momentum of inertia of the whole plate leads to the following reformulation:

$$T_{\theta_{ZCM}} = -I_Z\frac{d^2\theta_Z}{dt^2} \,|\, Y_{SE} \,|\, F_{\Delta X_{SE}} - |\, X_{SE} \,|\, F_{\Delta Y_{SE}} \tag{5.6}$$

where I_Z is the momentum of inertia of the plate with respect to the Z axis, expressed as follows [48, 49]:

$$I_Z = \frac{1}{12}m(L^2 + W^2). \tag{5.7}$$

Now, with the aim of describing all the external allowed stimuli accounted for by the 4 DoFs rigid plate mechanical model, the schematic in figure 5.8 is reported.

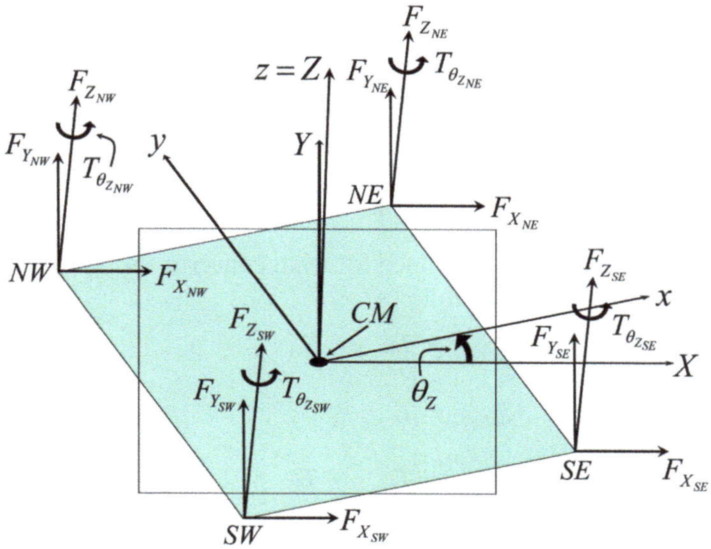

Figure 5.8. Schematic of the 4 DoFs rigid plate model with all the possible stimuli highlighted. [51] John Wiley & Sons. © 2013 Wiley-VCH Verlag GmbH & Co. KGaA.

With the visual support provided in figure 5.8, the forces previously reported in equation (5.4a) and equation (5.4b) are now expanded for the three linear DoFs at each plate node, as follows:

$$F_{\Delta X_{NE}} = F_{X_{NE}} - F_{X_{CM}} \tag{5.8a}$$

$$F_{\Delta Y_{NE}} = F_{Y_{NE}} - F_{Y_{CM}} \tag{5.8b}$$

$$F_{\Delta Z_{NE}} = F_{Z_{NE}} - F_{Z_{CM}} \tag{5.8c}$$

$$F_{\Delta X_{SE}} = F_{X_{SE}} - F_{X_{CM}} \tag{5.8d}$$

$$F_{\Delta Y_{SE}} = F_{Y_{SE}} - F_{Y_{CM}} \tag{5.8e}$$

$$F_{\Delta Z_{SE}} = F_{Z_{SE}} - F_{Z_{CM}} \tag{5.8f}$$

$$F_{\Delta X_{SW}} = F_{X_{SW}} - F_{X_{CM}} \tag{5.8g}$$

$$F_{\Delta Y_{SW}} = F_{Y_{SW}} - F_{Y_{CM}} \tag{5.8h}$$

$$F_{\Delta Z_{SW}} = F_{Z_{SW}} - F_{Z_{CM}} \tag{5.8i}$$

$$F_{\Delta X_{NW}} = F_{X_{NW}} - F_{X_{CM}} \tag{5.8j}$$

$$F_{\Delta Y_{NW}} = F_{Y_{NW}} - F_{Y_{CM}} \tag{5.8k}$$

$$F_{\Delta Z_{NW}} = F_{Z_{NW}} - F_{Z_{CM}} \tag{5.8l}$$

where $F_{X_{CM}}$ and $F_{Y_{CM}}$ are reported in previous equation (5.5a) and equation (5.5b), while $F_{Z_{CM}}$ is expressed as

$$F_{Z_{CM}} = -m\frac{d^2P_{Z_{CM}}}{dt^2} = -mA_{Z_{CM}}. \tag{5.9}$$

A set of similar equations is also defined for what concerns the momentum of inertia around the Z axis, expressed as follows:

$$T_{\Delta\theta_{Z_{NE}}} = T_{\theta_{Z_{NE}}} - T_{\theta_{Z_{CM}}} \tag{5.10a}$$

$$T_{\Delta\theta_{Z_{SE}}} = T_{\theta_{Z_{SE}}} - T_{\theta_{Z_{CM}}} \tag{5.10b}$$

$$T_{\Delta\theta_{Z_{SW}}} = T_{\theta_{Z_{SW}}} - T_{\theta_{Z_{CM}}} \tag{5.10c}$$

$$T_{\Delta\theta_{Z_{NW}}} = T_{\theta_{Z_{NW}}} - T_{\theta_{Z_{CM}}} \tag{5.10d}$$

where $T_{\Delta\theta_{Z_{NE}}}$, $T_{\Delta\theta_{Z_{SE}}}$, $T_{\Delta\theta_{Z_{SW}}}$, and $T_{\Delta\theta_{Z_{NW}}}$ are the external torques applied to the NE, SE, SW and NW nodes, respectively. Whereas, $T_{\theta_{Z_{CM}}}$ corresponds to equation (5.6), and is here expanded as follows:

$$\begin{aligned} T_{\theta_{Z_{CM}}} = -I_Z\frac{d^2\theta_Z}{dt^2} &- sgn(Y_{NE})|\ Y_{NE}\ |\ F_{\Delta X_{NE}} + sgn(X_{NE})|\ X_{NE}\ |\ F_{\Delta Y_{NE}} \\ &- sgn(Y_{SE})|\ Y_{SE}\ |\ F_{\Delta X_{SE}} + sgn(X_{SE})|\ X_{SE}\ |\ F_{\Delta Y_{SE}} - sgn(Y_{SW}) \\ &|\ Y_{SW}\ |\ F_{\Delta X_{SW}} + sgn(X_{SW})|\ X_{SW}\ |\ F_{\Delta Y_{SW}} - sgn(Y_{NW})|\ Y_{NW}\ |\ F_{\Delta X_{NW}} \\ &+ sgn(X_{NW})|\ X_{NW}\ |\ F_{\Delta Y_{NW}}. \end{aligned} \tag{5.11}$$

In order to complete the mechanical model of the 4 DoFs rigid plate, the relative displacement of each node with respect to the CM in the global frame has to be defined. To this end, the relative displacement of a generic point P in an arbitrary instant is defined as the difference of the instantaneous P_{INST} and initial P_{INIT} position:

$$\Delta P = P_{INST} - P_{INIT}. \tag{5.12}$$

Eventually, since the simplified 4 DoFs model does not allow any rotation of the plate around the X and Y axes, relative displacements of each node with respect to the CM must never occur. Such a condition is ensured by imposing proper constraints, both for linear displacements and rotations. In particular, the linear constraints are based on equation (5.12) and are reported in the following set of equations (5.13):

$$\Delta Z_{NE} = 0 \tag{5.13a}$$

$$\Delta Z_{SE} = 0 \tag{5.13b}$$

$$\Delta Z_{SW} = 0 \tag{5.13c}$$

$$\Delta Z_{NW} = 0. \tag{5.13d}$$

Finally, the set of angular constraints is expressed in the following set of equations (5.14):

$$\theta_{Z_{NE}} - \theta_{Z_{CM}} = 0 \tag{5.14a}$$

$$\theta_{Z_{SE}} - \theta_{Z_{CM}} = 0 \tag{5.14b}$$

$$\theta_{Z_{SW}} - \theta_{Z_{CM}} = 0 \tag{5.14c}$$

$$\theta_{Z_{NW}} - \theta_{Z_{CM}} = 0. \tag{5.14d}$$

5.2.1.2 Six Degrees of Freedom (6 DoFs) mechanical model

The extension of the rigid plate mechanical model from 4 DoFs, previously discussed, to the full 6 DoFs requires accounting also for the rotation around the *X* and *Y* axis at each node. Therefore, the set of relationships previously reported in equation (5.10) must be extended to the *X* and *Y* axes, as well. With that established, a schematic in figure 5.9 shows the plate with all the possible linear and angular stimuli allowed by having 6 DoFs at each node.

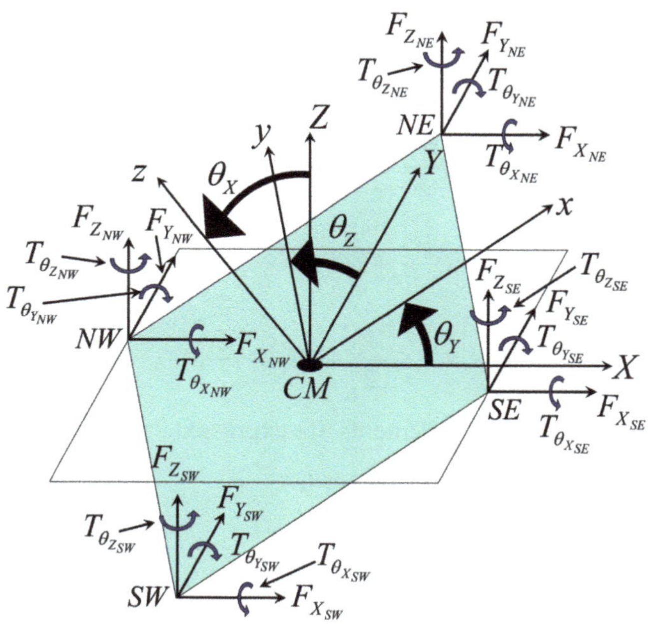

Figure 5.9. Schematic of the rigid plate highlighting all the 6 DoFs at each node. [51] John Wiley & Sons. © 2013 Wiley-VCH Verlag GmbH & Co. KGaA.

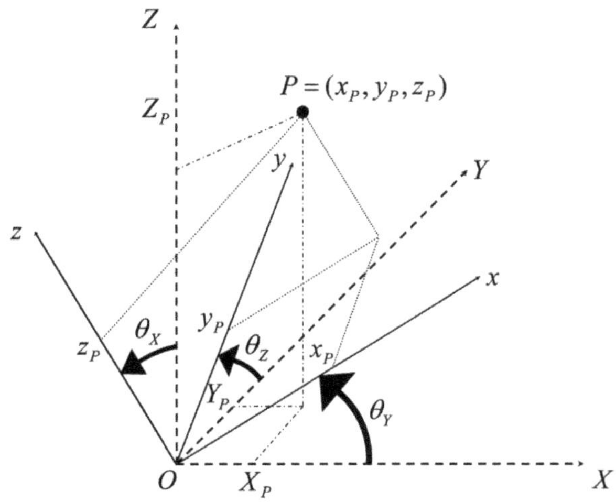

Figure 5.10. Scheme of the local reference system rotation around the three axes of the global fixed system, defined by the θ_X, θ_Y and θ_Z angles. [51] John Wiley & Sons. © 2013 Wiley-VCH Verlag GmbH & Co. KGaA.

Given this case, out-of-plane displacements are now possible, and each generic point, P, on the plate surface is identified by three vectors in the fixed reference plane, as shown by the scheme in figure 5.10.

Focusing, as an example, on the NE corner, its coordinates expressed in the local reference systems are as follows:

$$NE(x, y, z) = \left(\frac{W}{2}, \frac{L}{2}, 0\right). \tag{5.15}$$

Now, in order to derive the three arms, X_{NE}, Y_{NE} and Z_{NE}, a rotation matrix [29, 48] has to be exploited once again, yielding the following expressions:

$$X_{NE} = (C_Y C_Z)\frac{W}{2} + (-C_X S_Z + S_X S_Y C_Z)\frac{L}{2} \tag{5.16a}$$

$$Y_{NE} = (C_Y S_Z)\frac{W}{2} + (C_X C_Z + S_X S_Y S_Z)\frac{L}{2} \tag{5.16b}$$

$$Z_{NE} = -S_Y\frac{W}{2} + (S_X C_Y)\frac{L}{2}. \tag{5.16c}$$

In which the rotation matrix elements are expressed as reported in the following:

$$C_X = \cos\theta_X \tag{5.17a}$$

$$S_X = \sin\theta_X \tag{5.17b}$$

$$C_Y = \cos\theta_Y \tag{5.17c}$$

$$S_Y = \sin\theta_Y \tag{5.17d}$$

$$C_Z = \cos\theta_Z \tag{5.17e}$$

$$S_Z = \sin\theta_Z. \tag{5.17f}$$

Following a similar approach, it is straightforward to derive the projections of all the other nodes. After such a step is completed, the equations describing the torque of the *CM* around the *X* and *Y* axis can be formulated, while equation (5.11), referring to the torque around the *Z* axis, still holds validity. The formerly mentioned torques are then expressed as reported in the following equation (5.18) and equation (5.19), respectively:

$$\begin{aligned} T_{\theta_{X_{CM}}} = {} & -I_X\frac{d^2\theta_X}{dt^2} - sgn(Z_{NE})|\, Z_{NE} \,|\, F_{\Delta Y_{NE}} + sgn(Y_{NE})|\, Y_{NE} \,|\, F_{\Delta Z_{NE}} \\ & - sgn(Z_{SE})|\, Z_{SE} \,|\, F_{\Delta Y_{SE}} + sgn(Y_{SE})|\, Y_{SE} \,|\, F_{\Delta Z_{SE}} - sgn(Z_{SW}) \\ & |\, Z_{SW} \,|\, F_{\Delta Y_{SW}} + sgn(Y_{SW})|\, Y_{SW} \,|\, F_{\Delta Z_{SW}} - sgn(Z_{NW})|\, Z_{NW} \,|\, F_{\Delta Y_{NW}} \\ & + sgn(Y_{NW})|\, Y_{NW} \,|\, F_{\Delta Z_{NW}} \end{aligned} \tag{5.18}$$

$$\begin{aligned} T_{\theta_{Y_{CM}}} = {} & -I_Y\frac{d^2\theta_Y}{dt^2} + sgn(Z_{NE})|\, Z_{NE} \,|\, F_{\Delta X_{NE}} - sgn(X_{NE})|\, X_{NE} \,|\, F_{\Delta Z_{NE}} \\ & + sgn(Z_{SE})|\, Z_{SE} \,|\, F_{\Delta X_{SE}} - sgn(X_{SE})|\, X_{SE} \,|\, F_{\Delta Z_{SE}} + sgn(Z_{SW}) \\ & |\, Z_{SW} \,|\, F_{\Delta X_{SW}} - sgn(X_{SW})|\, X_{SW} \,|\, F_{\Delta Z_{SW}} + sgn(Z_{NW})|\, Z_{NW} \,|\, F_{\Delta X_{NW}} \\ & - sgn(X_{NW})|\, X_{NW} \,|\, F_{\Delta Z_{NW}} \end{aligned} \tag{5.19}$$

where I_X and I_Y is the momentum of inertia around the *X* axis and *Y* axis, respectively, expressed as

$$I_X = \frac{1}{12}m(L^2 + T^2) \tag{5.20}$$

$$I_Y = \frac{1}{12}m(W^2 + T^2). \tag{5.21}$$

Once all the magnitudes have been calculated, each relative displacement of angular rotation must keep a link with the initial position and with the fixed reference frame. To do so, relationships like the previous equation (5.12) need to be defined, thus considering an initial and instantaneous contribution to each magnitude. Eventually, the 6 DoFs plate, likewise the 4 DoFs one, has to behave as a rigid body. Such a condition is ensured by imposing the following relationships between the rotation angles of the four edges, with respect to the *CM*:

$$\theta_{X_{NE}} - \theta_{X_{CM}} = 0 \tag{5.22a}$$

$$\theta_{Y_{NE}} - \theta_{Y_{CM}} = 0 \tag{5.22b}$$

$$\theta_{Z_{NE}} - \theta_{Z_{CM}} = 0 \tag{5.22c}$$

$$\theta_{X_{SE}} - \theta_{X_{CM}} = 0 \tag{5.22d}$$

$$\theta_{Y_{SE}} - \theta_{Y_{CM}} = 0 \tag{5.22e}$$

$$\theta_{Z_{SE}} - \theta_{Z_{CM}} = 0 \tag{5.22f}$$

$$\theta_{X_{SW}} - \theta_{X_{CM}} = 0 \tag{5.22g}$$

$$\theta_{Y_{SW}} - \theta_{Y_{CM}} = 0 \tag{5.22h}$$

$$\theta_{Z_{SW}} - \theta_{Z_{CM}} = 0 \tag{5.22i}$$

$$\theta_{X_{NW}} - \theta_{X_{CM}} = 0 \tag{5.22j}$$

$$\theta_{Y_{NW}} - \theta_{Y_{CM}} = 0 \tag{5.22k}$$

$$\theta_{Z_{NW}} - \theta_{Z_{CM}} = 0. \tag{5.22l}$$

5.2.1.3 Contact model for the 4 DoFs and 6 DoFs rigid plate cases

Now that the mechanical behaviour of the rigid plate has been modelled, a crucial extension is that of including the physical contact of the plate itself with an underlying surface. This is the typical situation of a MEMS/RF-MEMS transducer that is brought to the pull-in condition. The contact is modelled by accounting for a suitable mechanical constraint, as soon as a certain condition takes place for what concerns the vertical displacement of the plate. Starting from the simple case of the 4 DoFs model, where the plate always remains parallel to the substrate, an initial air gap distance is defined, and any instantaneous upward/downward displacement of the plate is referred to such a starting condition. When the air gap reaches zero (i.e. pull-in), a large reaction force is applied to the *CM*, obtained by defining a fictitious contact material with a very large elastic constant. In the balance of forces, this condition does not allow the plate to further displace downwards. In formal terms, the contact force is expressed as $F_{Z_{TOUCH}}$ and is added to equation (5.1) as follows:

$$F_{Z_{CM}} = -mA_{Z_{CM}} + F_{Z_{TOUCH}}. \tag{5.23}$$

In order to discriminate the cases of collision and no collision, a binary flag, called TC_{FLAG} (Touch Flag), is defined as follows:

$$TC_{FLAG} = \begin{cases} 1 \text{ if } (Z_{CM_{INST}} < 0) \text{ and } (| Z_{CM_{INST}} | \geqslant | Z_{GAP} |) \\ 0 \qquad\qquad\qquad \text{otherwise} \end{cases} \tag{5.24}$$

where $Z_{CM_{INST}}$ is the instantaneous displacement of the *CM* along *Z*, and Z_{GAP} is the initial air gap, i.e. the distance between the suspended plate and the underlying electrode or substrate. Now that suitable variables are defined, it is possible to impose the values to $F_{Z_{TOUCH}}$ depending on TC_{FLAG}, as defined in the following relationships:

$$F_{Z_{TOUCH}} = \begin{cases} K_{TOUCH}(Z_{GAP} + Z_{CM_{INST}}) & \text{if } TC_{FLAG} = 1 \\ 0 & \text{if } TC_{FLAG} = 0 \end{cases} \tag{5.25}$$

where K_{TOUCH} is a fictitious elastic constant with a very large value (e.g. in the range of 10^9 N/m).

The extension of the contact model to the 6 DoFs rigid plate case requires accounting for a collision condition taking place at the single node, rather than directly in the *CM*. As a matter of fact, in the full DoFs plate, collision can occur at each single node, or at two contiguous nodes, or, again, to all four nodes, in the final parallel approach and landing of the plate on the underlying surface. For explanation purposes, only the case of a collision at the *SE* corner is going to be detailed. To this end, the schematic in figure 5.11 depicts the circumstance here at stake.

The first step to be addressed is the definition of a suitable flag variable, similar to equation (5.24), for the *SE* corner, as follows:

$$TC_{SE_{FLAG}} = \begin{cases} 1 \text{ if } (Z_{SE_{INST}} < 0) \text{ and } (\mid Z_{SE_{INST}} \mid \geqslant \mid Z_{GAP} + Z_{SE_{INIT}} \mid) \\ 0 \quad \text{otherwise} \end{cases} \tag{5.26}$$

where $Z_{SE_{INIT}}$ and $Z_{SE_{INST}}$ are the initial and instantaneous positions of the *SE* corner along the *Z* axis, calculated by means of relationships similar to those reported in the previous equation (5.16). Subsequently, the contact force $F_{Z_{SE_{TOUCH}}}$, similar to that in equation (5.25), is determined as follows:

$$F_{Z_{SE_{TOUCH}}} = \begin{cases} K_{TOUCH}(Z_{GAP} + Z_{SE_{INST}} + Z_{SE_{INIT}}) & \text{if } TC_{SE_{FLAG}} = 1 \\ 0 & \text{if } TC_{SE_{FLAG}} = 0. \end{cases} \tag{5.27}$$

Eventually, the contribution of $F_{Z_{SE_{TOUCH}}}$ due to the collision of the SE corner acts on the balance of torques referring to the *CM*. In detail, $F_{Z_{SE_{TOUCH}}}$ contributes to the

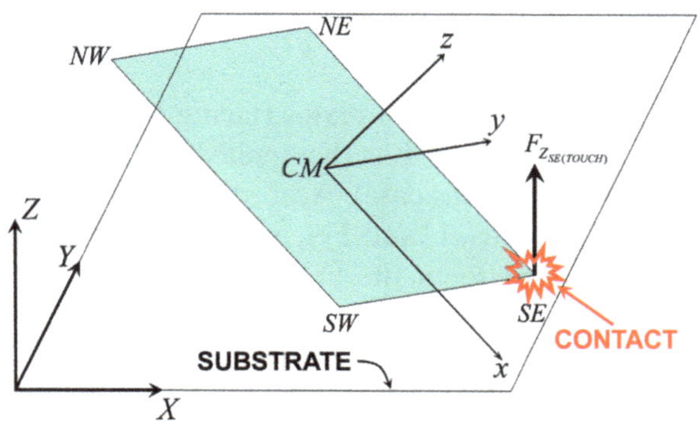

Figure 5.11. Schematic of the 6 DoFs rigid plate, when collision of the *SE* conner with the substrate occurs. [51] John Wiley & Sons. © 2013 Wiley-VCH Verlag GmbH & Co. KGaA.

torques around the X and Y axes, and therefore equation (5.18) and equation (5.19) have to be expanded as follows:

$$\begin{aligned} T_{\theta_{XCM}} = & -I_X\frac{d^2\theta_X}{dt^2} - sgn(Z_{NE})|\ Z_{NE}\ |\ F_{\Delta Y_{NE}} + sgn(Y_{NE})|\ Y_{NE}\ |\ F_{\Delta Z_{NE}} \\ & - sgn(Z_{SE})|\ Z_{SE}\ |\ F_{\Delta Y_{SE}} + sgn(Y_{SE})|\ Y_{SE}\ |(F_{\Delta Z_{SE}} + F_{Z_{SETOUCH}}) \\ & - sgn(Z_{SW})|\ Z_{SW}\ |\ F_{\Delta Y_{SW}} + sgn(Y_{SW})|\ Y_{SW}\ |\ F_{\Delta Z_{SW}} - sgn(Z_{NW}) \\ & |\ Z_{NW}\ |\ F_{\Delta Y_{NW}} + sgn(Y_{NW})|\ Y_{NW}\ |\ F_{\Delta Z_{NW}} \end{aligned} \tag{5.28}$$

$$\begin{aligned} T_{\theta_{YCM}} = & -I_Y\frac{d^2\theta_Y}{dt^2} + sgn(Z_{NE})|\ Z_{NE}\ |\ F_{\Delta X_{NE}} - sgn(X_{NE})|\ X_{NE}\ |\ F_{\Delta Z_{NE}} \\ & + sgn(Z_{SE})|\ Z_{SE}\ |\ F_{\Delta X_{SE}} - sgn(X_{SE})|\ X_{SE}\ |(F_{\Delta Z_{SE}} + F_{Z_{SETOUCH}}) \\ & + sgn(Z_{SW})|\ Z_{SW}\ |\ F_{\Delta X_{SW}} - sgn(X_{SW})|\ X_{SW}\ |\ F_{\Delta Z_{SW}} + sgn(Z_{NW}) \\ & |\ Z_{NW}\ |\ F_{\Delta X_{NW}} - sgn(X_{NW})|\ X_{NW}\ |\ F_{\Delta Z_{NW}}. \end{aligned} \tag{5.29}$$

The full DoFs contact model has to be completed by deriving similar contributions for the collision occurrence at all the other nodes, not reported here for the sake of brevity.

5.2.1.4 Electrostatic transduction model of the rigid plate

This subsection introduces the modelling of the actual transduction between the electrical and mechanical domains, by reporting on the electrostatic model implemented for the 4 DoFs and 6 DoFs rigid plate. In the case of the 4 DoFs plate, the model is rather simple, as the calculation of the parallel plates capacitance C and of the electrostatic attraction force F_{EL} is based on well-known formulas [13], shown in the following equation (5.30) and equation (5.31), respectively:

$$C = \varepsilon\frac{WL}{g} \tag{5.30}$$

$$F_{EL} = \frac{1}{2}\varepsilon V^2\frac{WL}{g^2} \tag{5.31}$$

where WL is the surface of the parallel plates (facing each other), g is the vertical distance between them (i.e. air gap), V is the applied voltage, and ε is the dielectric constant of the insulator (air, in this case). According to the approach and notation adopted in the previous formulas, C and F_{EL} must be expressed accounting for the instantaneous air gap referred to in the fixed reference system. To this end, the previous equation (5.30) and equation (5.31) are rewritten as reported in the following equation (5.32) and equation (5.33), respectively:

$$C = \frac{WL}{(Z_{GAP} + Z_{CM_{INST}})} \tag{5.32}$$

$$F_{Z_{EL}} = \frac{1}{2}\varepsilon V^2 \frac{WL}{(Z_{GAP} + Z_{CM_{INST}})^2}. \tag{5.33}$$

Thereafter, the balance of forces in the CM, previously reported in equation (5.23), is expanded as shown in equation (5.34), thus completing the 4 DoFs rigid plate electrostatic model:

$$F_{Z_{CM}} = -mA_{Z_{CM}} + F_{Z_{TOUCH}} - F_{Z_{EL}}. \tag{5.34}$$

The extension of the electrostatic model to the 6 DoFs rigid plate has to rely on a point-based, rather than surface-based, calculation of the capacitance and attraction force, as the tilted plate, in the most general case, will have a different and unique distance of each point on its surface from the underlying fixed electrode. Despite that the complexity of the analytical formulation unavoidably increases, it is still manageable in a closed form, as will be discussed in the following. To this end, the notation to identify the distance components of the generic point P on the tilted plate surface and the substrate is described by the schematic in figure 5.12.

Starting from such a description, it is possible to rewrite the previous equation (5.32) and equation (5.33) in the most generalised case, by a double integral formulation on the plate surface, i.e. along L and W, as reported in the following:

$$C = \varepsilon \int_{-\frac{W}{2}}^{\frac{W}{2}} \int_{-\frac{L}{2}}^{\frac{L}{2}} \frac{dxdy}{(Z_{GAP} + Z_{CM_{INST}} + Z_{P_{INST}})} \tag{5.35}$$

$$F_{Z_{EL}} = \frac{1}{2}\varepsilon V^2 \int_{-\frac{W}{2}}^{\frac{W}{2}} \int_{-\frac{L}{2}}^{\frac{L}{2}} \frac{dxdy}{(Z_{GAP} + Z_{CM_{INST}} + Z_{P_{INST}})^2}. \tag{5.36}$$

Thereafter, following the same approach discussed above, $F_{Z_{EL}}$, which is expressed in the plate local reference system, is split in its components in the global fixed frame by

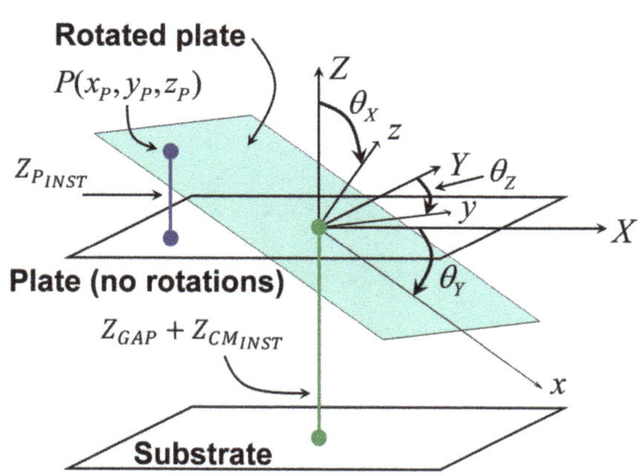

Figure 5.12. Scheme of the vertical distance between a generic point P of the tilted plate and the substrate. [51] John Wiley & Sons. © 2013 Wiley-VCH Verlag GmbH & Co. KGaA.

means of a rotation matrix. This yields $F_{X_{EL}}$, $F_{Y_{EL}}$ and $F_{Z_{EL}}$, which contribute to the balance of forces in the *CM* being expressed in the following:

$$F_{X_{CM}} = -mA_{X_{CM}} - F_{X_{EL}} \tag{5.37}$$

$$F_{Y_{CM}} = -mA_{Y_{CM}} + F_{Y_{EL}} \tag{5.38}$$

$$F_{Z_{CM}} = -mA_{Z_{CM}} + F_{Z_{TOUCH}} - F_{Z_{EL}}. \tag{5.39}$$

The subsequent step consists in defining the contribution that the electrostatic attraction force builds, in terms of torque, when the plate is tilted. Following the same approach as before, the torque contribution can be calculated in a double integral closed form on the plate surface. As the electrostatic transduction, calculated in the local reference frame, gives a contribution when the plate is tilted around the *X* and/or the *Y* axis, while no contribution is given for the rotation around the vertical axis *Z*, electrostatic torques are calculated as equation (5.40) and equation (5.41) report:

$$T_{\theta x_{EL}} = \frac{1}{2}\varepsilon V^2 \int_{-\frac{W}{2}}^{\frac{W}{2}} \int_{-\frac{L}{2}}^{\frac{L}{2}} \frac{y_P}{(Z_{GAP} + Z_{CM_{INST}} + Z_{P_{INST}})^2} dxdy \tag{5.40}$$

$$T_{\theta y_{EL}} = \frac{1}{2}\varepsilon V^2 \int_{-\frac{W}{2}}^{\frac{W}{2}} \int_{-\frac{L}{2}}^{\frac{L}{2}} \frac{x_P}{(Z_{GAP} + Z_{CM_{INST}} + Z_{P_{INST}})^2} dxdy. \tag{5.41}$$

As before, starting from $T_{\theta x_{EL}}$ and $T_{\theta y_{EL}}$, computed in the local frame, their components in the global fixed frame, $T_{\theta X_{EL}}$, $T_{\theta Y_{EL}}$ and $T_{\theta Z_{EL}}$, are obtained exploiting a rotation matrix. Once derived, the latter contributions are added to the total balance of torques, as follows:

$$\begin{aligned} T_{\theta_{xCM}} = -I_X\frac{d^2\theta_X}{dt^2} &- sgn(Z_{NE})|\, Z_{NE} \,|\, F_{\Delta Y_{NE}} + sgn(Y_{NE})|\, Y_{NE} \,|(F_{\Delta Z_{NE}} + F_{Z_{NETOUCH}}) \\ &- sgn(Z_{SE})|\, Z_{SE} \,|\, F_{\Delta Y_{SE}} + sgn(Y_{SE})|\, Y_{SE} \,|(F_{\Delta Z_{SE}} + F_{Z_{SETOUCH}}) \\ &- sgn(Z_{SW})|\, Z_{SW} \,|\, F_{\Delta Y_{SW}} + sgn(Y_{SW})|\, Y_{SW} \,|(F_{\Delta Z_{SW}} + F_{Z_{SWTOUCH}}) \\ &- sgn(Z_{NW})|\, Z_{NW} \,|\, F_{\Delta Y_{NW}} + sgn(Y_{NW})|\, Y_{NW} \,|(F_{\Delta Z_{NW}} + F_{Z_{NWTOUCH}}) - T_{\theta X_{EL}} \end{aligned} \tag{5.42}$$

$$\begin{aligned} T_{\theta_{yCM}} = -I_Y\frac{d^2\theta_Y}{dt^2} &+ sgn(Z_{NE})|\, Z_{NE} \,|\, F_{\Delta X_{NE}} - sgn(X_{NE})|\, X_{NE} \,|(F_{\Delta Z_{NE}} + F_{Z_{NETOUCH}}) \\ &+ sgn(Z_{SE})|\, Z_{SE} \,|\, F_{\Delta X_{SE}} - sgn(X_{SE})|\, X_{SE} \,|(F_{\Delta Z_{SE}} + F_{Z_{SETOUCH}}) \\ &+ sgn(Z_{SW})|\, Z_{SW} \,|\, F_{\Delta X_{SW}} - sgn(X_{SW})|\, X_{SW} \,|(F_{\Delta Z_{SW}} + F_{Z_{SWTOUCH}}) \\ &+ sgn(Z_{NW})|\, Z_{NW} \,|\, F_{\Delta X_{NW}} - sgn(X_{NW})|\, X_{NW} \,|(F_{\Delta Z_{NW}} + F_{Z_{NWTOUCH}}) - T_{\theta Y_{EL}} \end{aligned} \tag{5.43}$$

$$T_{\theta_{zCM}} = -I_Z\frac{d^2\theta_Z}{dt^2} - sgn(Y_{NE})|\ Y_{NE}\ |\ F_{\Delta X_{NE}} + sgn(X_{NE})|\ X_{NE}\ |\ F_{\Delta Y_{NE}} - sgn(Y_{SE})|\ Y_{SE}\ |\ F_{\Delta X_{SE}} + sgn(X_{SE})|\ X_{SE}\ |\ F_{\Delta Y_{SE}} - sgn(Y_{SW}) |\ Y_{SW}\ |\ F_{\Delta X_{SW}} + sgn(X_{SW})|\ X_{SW}\ |\ F_{\Delta Y_{SW}} - sgn(Y_{NW})|\ Y_{NW}\ |\ F_{\Delta X_{NW}} + sgn(X_{NW})|\ X_{NW}\ |\ F_{\Delta Y_{NW}} - T_{\theta Z_{EL}}. \tag{5.44}$$

The last feature of the rigid plate electrostatic model to be discussed here is the inclusion of openings on the surface of the plate itself, necessary in order to enable proper removal of the sacrificial layer and release of the MEMS suspended membrane [50]. The presence of holes exerts significant effects on the mechanical behaviour of the rigid plate in dynamic conditions, as well as on the electrostatic transduction. Concerning the former, in fact, the mass of the plate is lower when holes are etched in it. In addition, more relevantly, the presence or absence of holes modifies significantly the impact of the viscous damping effect, not discussed here for the sake of brevity. With reference to the latter, instead, holes reduce the surface of the movable electrode, therefore changing the capacitance and electrostatic attraction forces with respect to the case of a plate without openings. This is the focus of the current subsection.

A schematic representation of the plate surface with a casual distribution (on the XY plane) of rectangular holes on the surface, along with all the significant geometrical features and spacings, is reported in figure 5.13.

Starting from this sketch, the surface covered by the holes, labelled as WL_{Holes}, can be calculated as follows:

$$WL_{Holes} = W_H L_H mn \tag{5.45}$$

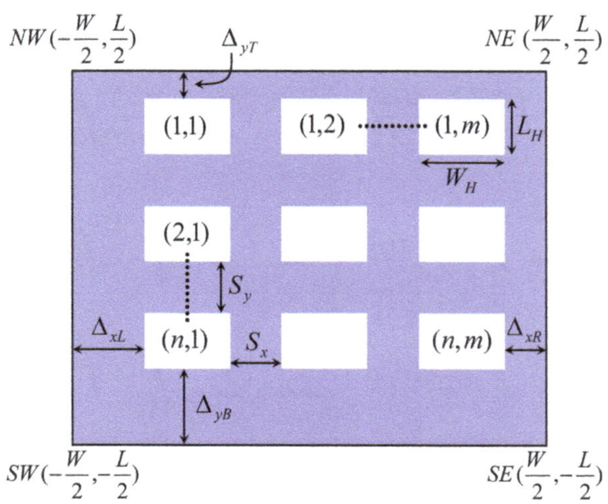

Figure 5.13. Schematic of the rigid plate surface with a casual distribution of rectangular holes, where all the relevant dimensions and distances are indicated. [51] John Wiley & Sons. © 2013 Wiley-VCH Verlag GmbH & Co. KGaA.

where W_H and L_H are the holes' width and length, respectively, along the X and Y axis, while m and n are the number of holes along the X and Y axes, respectively. Since it is straightforward to calculate, the effective mass on the plate with holes, labelled as m_{EFF}, is reported here, despite not being directly used in the electrostatic model:

$$m_{EFF} = \rho(WLT - W_H L_H Tmn) \tag{5.46}$$

where ρ is the density of the material of the rigid plate. Now, the formulation of the capacitance and electrostatic attraction force in the case of the 4 DoFs plate can be rewritten accounting for the holes' presence, thanks to the linear dependence of both magnitudes with respect to the surface of the plate. In light of that, a *negative* contribution of capacitance and force, calculated over the (void) area of the holes, can be subtracted from the previous equation (5.32) and equation (5.33), yielding the following relationships:

$$C = \varepsilon\left[\frac{WL}{Z_{GAP} + Z_{CM_{INST}}} - \frac{WL_{Holes}}{Z_{GAP} + Z_{CM_{INST}}}\right] \tag{5.47}$$

$$F_{Z_{EL}} = \frac{1}{2}\varepsilon V^2\left[\frac{WL}{(Z_{GAP} + Z_{CM_{INST}})^2} - \frac{WL_{Holes}}{(Z_{GAP} + Z_{CM_{INST}})^2}\right]. \tag{5.48}$$

The same approach is exploited to extend the validity of the model to the case of a 6 DoFs rigid plate. To this end, the capacitance, electrostatic linear force and electrostatic torque around the local x and y axes, accounting for holes, are formulated as in the following equation (5.49), equation (5.50), equation (5.51) and equation (5.52), respectively:

$$C = \varepsilon\left[\int_{-\frac{W}{2}}^{\frac{W}{2}}\int_{-\frac{L}{2}}^{\frac{L}{2}}\frac{dxdy}{Z_P} - \sum_{i=0}^{m-1}\sum_{j=0}^{n-1}\int_{a(i)}^{b(i)}\int_{c(j)}^{d(j)}\frac{dxdy}{Z_P}\right] \tag{5.49}$$

$$F_{Z_{EL}} = \frac{1}{2}\varepsilon V^2\left[\int_{-\frac{W}{2}}^{\frac{W}{2}}\int_{-\frac{L}{2}}^{\frac{L}{2}}\frac{dxdy}{Z_P^2} - \sum_{i=0}^{m-1}\sum_{j=0}^{n-1}\int_{a(i)}^{b(i)}\int_{c(j)}^{d(j)}\frac{dxdy}{Z_P^2}\right] \tag{5.50}$$

$$T_{\theta x_{EL}} = \frac{1}{2}\varepsilon V^2\left[\int_{-\frac{W}{2}}^{\frac{W}{2}}\int_{-\frac{L}{2}}^{\frac{L}{2}}\frac{y_P}{Z_P^2}dxdy - \sum_{i=0}^{m-1}\sum_{j=0}^{n-1}\int_{a(i)}^{b(i)}\int_{c(j)}^{d(j)}\frac{y_P}{Z_P^2}dxdy\right] \tag{5.51}$$

$$T_{\theta y_{EL}} = \frac{1}{2}\varepsilon V^2\left[\int_{-\frac{W}{2}}^{\frac{W}{2}}\int_{-\frac{L}{2}}^{\frac{L}{2}}\frac{x_P}{Z_P^2}dxdy - \sum_{i=0}^{m-1}\sum_{j=0}^{n-1}\int_{a(i)}^{b(i)}\int_{c(j)}^{d(j)}\frac{x_P}{Z_P^2}dxdy\right]. \tag{5.52}$$

In the previous equations, the integrals limits $a(i)$, $b(i)$, $c(j)$ and $d(j)$, are expressed as follows:

$$a(i) = -\frac{W}{2} + \Delta_{xL} + i(W_H + S_x) \tag{5.53}$$

$$b(i) = -\frac{W}{2} + \Delta_{xL} + i(W_H + S_x) + W_H \tag{5.54}$$

$$c(j) = \frac{L}{2} - \Delta_{yT} - L_H - j(L_H + S_y) \tag{5.55}$$

$$d(j) = \frac{L}{2} - \Delta_{yT} - j(L_H + S_y) \tag{5.56}$$

where S_x and S_y is the spacing between holes along the x and y axes, respectively, accounting for the most general case of uneven placement of the openings on the plate surface, expressed as reported in the following equation (5.57a) and equation (5.57b):

$$S_x = \frac{W - \Delta_{xL} - \Delta_{xR} - W_H m}{m - 1} \tag{5.57a}$$

$$S_y = \frac{L - \Delta_{yT} - \Delta_{yB} - L_H n}{n - 1}. \tag{5.57b}$$

In previous equations, the magnitudes defined as Δ_{xL}, Δ_{xR}, Δ_{yT} and Δ_{yB} are the distances of the holes from the four plate edges (left, right, top, bottom), as visible in figure 5.13. Finally, as done before, the results of equation (5.50), equation (5.51) and equation (5.52) must be decomposed in their contributions in the global reference frame, by means of a rotation matrix.

The rigid plate model is completed by a fringing field formulation, addressing the electrostatic behaviour, and by a viscous damping model, not reported here for the sake of brevity, but available for any in-depth study in [51]. Eventually, figure 5.14 depicts the symbol corresponding to a rigid plate model, to be instantiated and wired within a Cadence schematic to compose the RF-MEMS structure to be simulated.

5.2.2 Elastic straight (and slender) beam suspension

The aim of this section is that of describing the other fundamental component included in the VerilogA software library, that, along with the rigid plate just discussed, enables one to build schematics corresponding to most parts of electrostatically controlled RF-MEMS devices discussed in the literature and available on the market. The model at stake is the elastic suspension, realising the spring element of the system that generates a mechanical restoring force, when plate transducers displace from their rest position due to electrostatic actuation force (or when subjected to an external acceleration). In particular, the suspension is configured as a straight slender beam, therefore with a length much larger than its width and thickness (i.e. section), in order to fall into the so-called Euler beam approximation [52]. A schematic of a generic straight and slender beam is reported in figure 5.15, where the length (L), width (W) and thickness (T), along with the ends A and B, are highlighted.

Despite a full mathematical description of the elastic beam was developed, in the following pages the most relevant equations are going to be reported, for the sake of

Figure 5.14. Rigid plate symbol in Cadence.

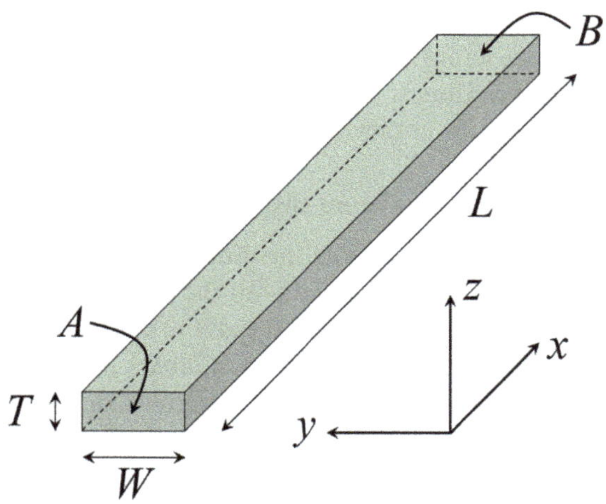

Figure 5.15. Schematic of a generic elastic straight beam. [51] John Wiley & Sons. © 2013 Wiley-VCH Verlag GmbH & Co. KGaA.

brevity. Moreover, some of the relevant steps in deriving the model are similar to those already described for the rigid plate. In any case, the full theoretical basis of structural mechanics, along with more in-depth details of the models' implementation, are available in [48] and [51], respectively.

5.2.2.1 Simplified 2 DoFs mechanical model

For the purposes of explanation, in this section a very simplified mechanical model of a flexible beam is briefly reported. This is because the full DoFs model builds upon the same approach, and therefore its understanding is much simpler if observed in a basic circumstance. To this end, a schematic of a slender beam subjected to a longitudinal force applied along the local *X* axis, first to the *A* end (i.e. F_{xA}, causing the u_{xA} deformation) and then to the *B* end (i.e. F_{xB}, causing the u_{xB} deformation), is shown in figure 5.16.

At this point, a few considerations are necessary. The case reported in the previous schematic is basically that of a one geometrical DoF deformable beam, as the only displacement possible is along its length. However, differently from the rigid plate, the slender beam is deformable. Therefore, each end has to be considered geometrically independent from each other. This is the reason why the same DoF, in this case the displacement along the longitudinal axis, has to be accounted for twice, yielding, in fact, 2 DoFs. Consequently, the full model implementation, discussed below, will feature 12 DoFs, six at each end.

Another important consideration emerges from the analysis of the schematic in figure 5.16. The theoretical basis of the mechanical model leverages the principle of superposition [48]. Therefore, whenever a certain stimulus is applied to one end of the beam (in this case only elongation/compression is possible), all the DoFs at the other end are supposed to be constrained. This is the reason why the *B* end is anchored when the *A* end is stimulated, and vice versa.

After these considerations, the constitutive equation of the flexible beam, neglecting for simplicity the damping effect, is formulated as follows:

$$\boldsymbol{F} = k\boldsymbol{u} + m\ddot{\boldsymbol{u}}. \tag{5.58}$$

The use of bold characters in equation (5.58) indicates that it is expressed in a matrix form, in this case of a 2 by 2 rank, as it describes a 2 DoFs model. In particular, $\boldsymbol{F}$, $\boldsymbol{u}$ and $\ddot{\boldsymbol{u}}$, are vertical vectors indicating the force, the displacement and the acceleration, expressed as the double derivative of displacement with respect to time, at

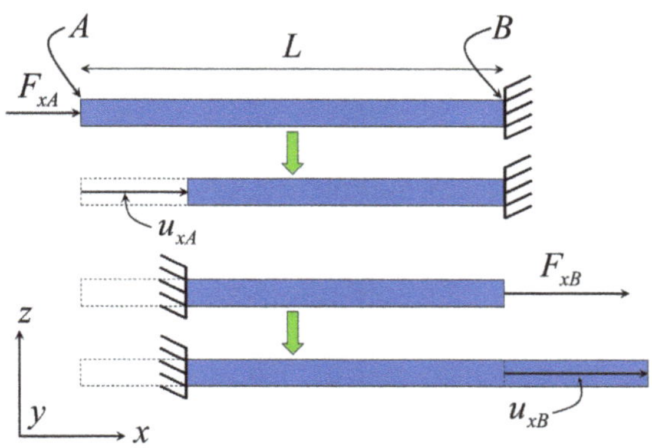

Figure 5.16. Schematic of a 2 DoFs beam with stimuli applied along the longitudinal direction at the two ends. [51] John Wiley & Sons. © 2013 Wiley-VCH Verlag GmbH & Co. KGaA.

the nodes A and B, respectively, as reported in the following equation (5.59a), equation (5.59b) and equation (5.59c):

$$\boldsymbol{F} = \begin{bmatrix} F_{xA} \\ F_{xB} \end{bmatrix} \tag{5.59a}$$

$$\boldsymbol{u} = \begin{bmatrix} u_{xA} \\ u_{xB} \end{bmatrix} \tag{5.59b}$$

$$\boldsymbol{a} = \ddot{\boldsymbol{u}} = \begin{bmatrix} \ddot{u}_{xA} \\ \ddot{u}_{xB} \end{bmatrix}. \tag{5.59c}$$

Coming back to equation (5.58), it covers the elastic (deformable) behaviour and the inertia of the flexible beam. The elastic characteristics are modelled by the stiffness matrix $\boldsymbol{k}$, that in this case has a 2 by 2 dimension, and is formulated as follows [1, 48]:

$$\boldsymbol{k} = \frac{WTE}{L}\begin{bmatrix} 1 & -1 \\ -1 & 1 \end{bmatrix} \tag{5.60}$$

where E is the Young's modulus of the material. With a similar approach, the inertial behaviour of the beam is described in equation (5.58) by the 2 by 2 mass matrix $\boldsymbol{m}$, formulated as reported in the following:

$$\boldsymbol{m} = \rho WTL\begin{bmatrix} \frac{1}{3} & \frac{1}{6} \\ \frac{1}{6} & \frac{1}{3} \end{bmatrix} \tag{5.61}$$

where ρ is the density of the material. In conclusion, the full matrix formulation of equation (5.58) is expressed as follows:

$$\begin{bmatrix} F_{xA} \\ F_{xB} \end{bmatrix} = \frac{WTE}{L}\begin{bmatrix} 1 & -1 \\ -1 & 1 \end{bmatrix}\begin{bmatrix} u_{xA} \\ u_{xB} \end{bmatrix} + \rho WTL\begin{bmatrix} \frac{1}{3} & \frac{1}{6} \\ \frac{1}{6} & \frac{1}{3} \end{bmatrix}\begin{bmatrix} \ddot{u}_{xA} \\ \ddot{u}_{xB} \end{bmatrix}. \tag{5.62}$$

More details on the mechanical model are available in [1, 48], while additional information on the model implementation, e.g. also concerning the damping effect term to be included in equation (5.58), is reported in [51].

5.2.2.2 Extended 12 DoFs mechanical model

This brief section discusses the extension of the previous 2 DoFs mechanical model to 12 DoFs, i.e. six at beam end A and another independent six at the beam end B. The schematic in figure 5.17 reports the beam with all the DoFs labelled and indicated.

As mentioned above, the full formulation of the model is not reported here for the sake of brevity, as the theory is available in [1, 48], while the model implementation

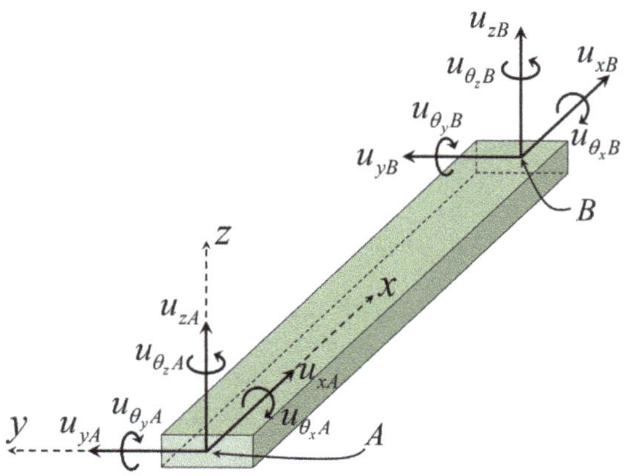

Figure 5.17. Schematic of the flexible straight beam with 12 DoFs, i.e., six at each end *A* and *B*. [51] John Wiley & Sons. © 2013 Wiley-VCH Verlag GmbH & Co. KGaA.

is discussed in [51]. What is worth mentioning is that in the case of 12 DoFs, equation (5.58) features 12 element vectors and a 12 by 12 formulation of the stiffness (***k***) and mass (***m***) matrix. Concerning the latter one, it is useful to mention that its mathematical formulation is based on the so-called Hermitian Shape Functions (HSFs) [48]. As is known, the whole description of the mechanical behaviour of the beam is led back to the two ends *A* and *B*, which are the unique points through which the beam model is linked to the external world. Having said that, HSFs are necessary to determine, point by point along the beam length, the extent of deformation, given an external stimulus imposed to one of the two beam ends. The main conditions of validity for HSFs is that each applied stimulus, regardless if it is a linear force or a torque around an axis, is applied one at a time, with all the DoFs at the other end of the beam constrained. This means that there is a specific HFS per each DoF stimulated, and when in real conditions one or two ends will be deformed by two or more stimuli at the same time, the deformed shape of the beam, point by point along its length, is the superposition of all the involved HSFs. As an example, the schematics in figure 5.18(a) and figure 5.18(b) depict the beam configuration when a linear deformation u_{zA} along the Z axis of the A end, and when a torque $u_{\theta yA}$ around the Y axis of the A end are imposed, respectively.

5.2.2.3 Full electrostatic model

The previous discussion on HSFs is motivated by what is going to be discussed in this section. In brief, the theory behind the beam mechanical model uses such a set of functions, which are subsequently incorporated into the model. Nonetheless, the electrostatic model developed in the VerilogA library makes use of HSFs, as well. In particular, the beam surface is discretised in a suitable (and variable by the user) number n of small rigid plates, each of width W and length L/n, as shown in the scheme reported in figure 5.19.

By doing so, the tilting and placement of each rigid plate small element is calculated in the global frame and kept coherent with all the other elements, based

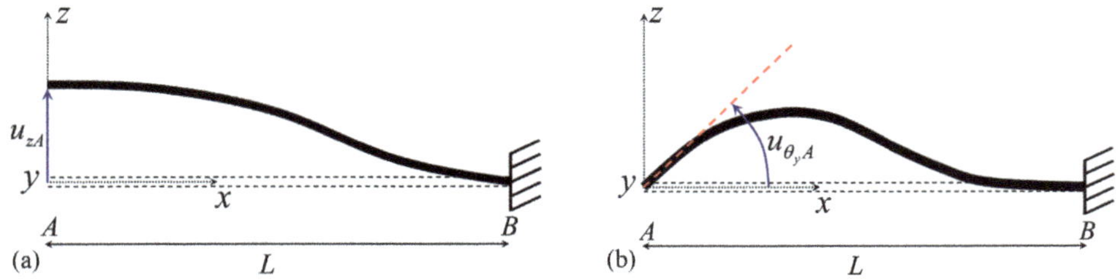

Figure 5.18. Deformed beam shapes, according to the HSFs description, when (a) a linear deformation u_{zA} is applied along the Z axis of the A end, and when (b) a torque $u_{\theta yA}$ is imposed around the Y axis of the A end. [51] John Wiley & Sons. © 2013 Wiley-VCH Verlag GmbH & Co. KGaA.

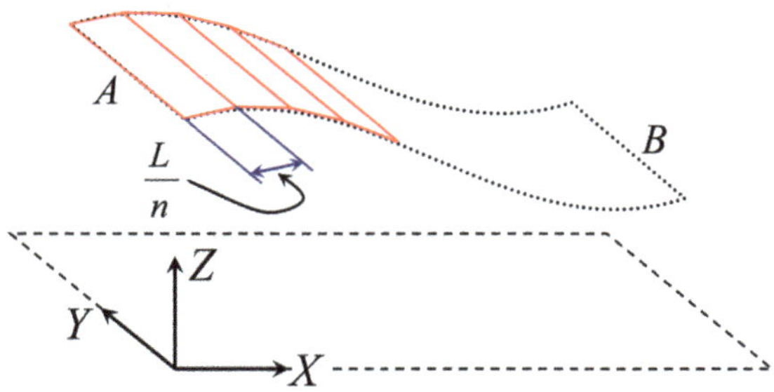

Figure 5.19. Scheme of the flexible beam discretised in a series of small rigid plates. [51] John Wiley & Sons. © 2013 Wiley-VCH Verlag GmbH & Co. KGaA.

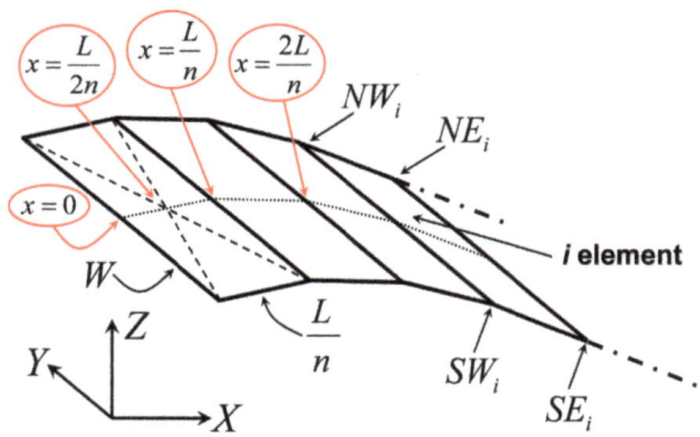

Figure 5.20. Scheme indicating the most relevant coordinates of the ith small rigid plate element, along the beam length. [51] John Wiley & Sons. © 2013 Wiley-VCH Verlag GmbH & Co. KGaA.

on the superposition of the involved HSFs and on the displacements and rotations of each beam end. As visible in figure 5.20, each small rigid plate element has its own North-East (NE), South-East (SE), South-West (SW) and North-West (NW) corners, as previously discussed for the rigid plate model.

Capitalising on the previous discussion, the capacitance, electrostatic attraction force and torques can be calculated for the deformable beam, exploiting the same model developed before for the rigid plate. In particular, the capacitance of the generically deformed beam is expressed as follows:

$$C = L\varepsilon \sum_{i=0}^{n-1} \int_{\frac{i}{n}}^{\frac{i+1}{n}} \int_{-\frac{W}{2}}^{\frac{W}{2}} \frac{d\xi dy}{Z_L(\xi, y, \theta_{X_i}, \theta_{Y_i})} \tag{5.63}$$

where ξ is a non-dimensional variable, indicating the generic point x along the length of the beam, normalised with respect of the length L itself, as utilised in the HSFs. Moreover, Z_L is a local coordinate that accounts for the local tilting angles of the ith small rigid plate element [51]. Concerning electrostatic forces and torques, differently from the rigid plate, in the case of the flexible beam the contributions are not applied to the CM, but suitably weighted and applied back to the A and B ends. For the same reason, forces and torques must be computed twice, i.e. once per each end. Thereafter, the electrostatic attraction force at the A and B ends, indicated as $F_{zA_{EL}}$ and $F_{zB_{EL}}$, respectively, are expressed in the following equation (5.64) and equation (5.65):

$$F_{zA_{EL}} = \frac{L}{2}\varepsilon \sum_{i=0}^{n-1}\left(\psi_{zA}\left(\frac{1}{n}\left(i + \frac{1}{2}\right)\right) \int_{\frac{i}{n}}^{\frac{i+1}{n}} \int_{-\frac{W}{2}}^{\frac{W}{2}} \frac{V^2 d\xi dy}{Z_L(\xi, y, \theta_{X_i}, \theta_{Y_i})^2}\right) \tag{5.64}$$

$$F_{zB_{EL}} = \frac{L}{2}\varepsilon \sum_{i=0}^{n-1}\left(\psi_{zB}\left(\frac{1}{n}\left(i + \frac{1}{2}\right)\right) \int_{\frac{i}{n}}^{\frac{i+1}{n}} \int_{-\frac{W}{2}}^{\frac{W}{2}} \frac{V^2 d\xi dy}{Z_L(\xi, y, \theta_{X_i}, \theta_{Y_i})^2}\right) \tag{5.65}$$

where ψ_{zA} and ψ_{zB} are the suitable HSFs necessary to describe the configuration of each element [51]. With a similar approach, the electrostatic torques around the X and Y axes of the A and B ends, labelled as $T_{\theta xA_{EL}}$, $T_{\theta xB_{EL}}$, $T_{\theta yA_{EL}}$ and $T_{\theta yB_{EL}}$, are expressed in the following equation (5.66), equation (5.67), equation (5.68) and equation (5.69), respectively:

$$T_{\theta xA_{EL}} = \frac{L}{2}\varepsilon \sum_{i=0}^{n-1}\left(\left(1 - \frac{1}{n}\left(i + \frac{1}{2}\right)\right) \int_{\frac{i}{n}}^{\frac{i+1}{n}} \int_{-\frac{W}{2}}^{\frac{W}{2}} \frac{V^2 y d\xi dy}{Z_L(\xi, y, \theta_{X_i}, \theta_{Y_i})^2}\right) \tag{5.66}$$

$$T_{\theta xB_{EL}} = \frac{L}{2}\varepsilon \sum_{i=0}^{n-1}\left(\frac{1}{n}\left(i + \frac{1}{2}\right) \int_{\frac{i}{n}}^{\frac{i+1}{n}} \int_{-\frac{W}{2}}^{\frac{W}{2}} \frac{V^2 y d\xi dy}{Z_L(\xi, y, \theta_{X_i}, \theta_{Y_i})^2}\right) \tag{5.67}$$

$$T_{\theta yA_{EL}} = \frac{L}{2}\varepsilon \sum_{i=0}^{n-1}\left(\psi_{\theta_y A}\left(\frac{1}{n}\left(i + \frac{1}{2}\right)\right) \int_{\frac{i}{n}}^{\frac{i+1}{n}} \int_{-\frac{W}{2}}^{\frac{W}{2}} \frac{V^2 \xi d\xi dy}{Z_L(\xi, y, \theta_{X_i}, \theta_{Y_i})^2}\right) \tag{5.68}$$

$$T_{\theta yB_{EL}} = \frac{L}{2}\varepsilon \sum_{i=0}^{n-1}\left(\psi_{\theta_y B}\left(\frac{1}{n}\left(i + \frac{1}{2}\right)\right) \int_{\frac{i}{n}}^{\frac{i+1}{n}} \int_{-\frac{W}{2}}^{\frac{W}{2}} \frac{V^2 \xi d\xi dy}{Z_L(\xi, y, \theta_{X_i}, \theta_{Y_i})^2}\right) \tag{5.69}$$

where $\psi_{\theta_y A}$ and $\psi_{\theta_y B}$ are the suitable HSFs necessary to describe the configuration of each element [51]. Accordingly, the symbol of a beam model within the VerilogA library in the Cadence environment is shown in figure 5.21.

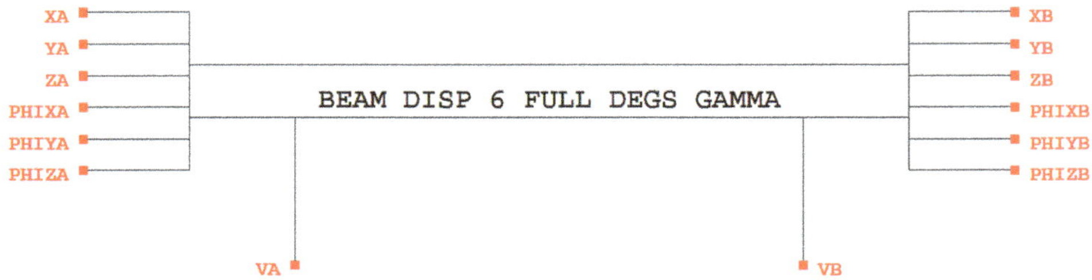

Figure 5.21. Symbol of a deformable beam as defined in the VerilogA library within Cadence.

The beam model is completed by the description of the fringing effect of the electric field at the beam edges, as well as by the viscous damping module, not reported here for the sake of brevity, and available in [51].

5.2.3 Mechanical constraints and force/displacement sources

This last brief subsection lists the main additional elements, developed and included in the VerilogA library, that complement the previously discussed rigid plate and elastic beam models, making it possible to compose schematics of entire RF-MEMS devices, ready for simulation. The objects mentioned in the following are all in the 6 DoFs configuration, despite the library including elements with fewer DoFs, as well. The first element is the anchoring point, shown in figure 5.22, necessary to impose mechanical constraints to the end of a flexible beam, as well as the flexible suspensions of physical MEMS membranes which are anchored to the substrate.

When a fixed mechanical boundary condition has to be imposed to the nodes of a rigid plate, a different element is necessary, i.e. the anchoring point with a minimum resistance shown in figure 5.23.

In fact, according to how the *through* and *across* magnitudes related to each node are managed in [51], connecting together the nodes of a rigid body with a fixed displacement value is like shorting two voltage sources in the electrical domain. This is the reason why a fictitious resistance has to be added in between, as is done by the element in figure 5.23. The same problem arises when two rigid bodies are to be connected together. For this reason, the minimum resistance interposing element, reported in the following figure 5.24, is also available in the software library.

Furthermore, in the composition of a schematic it might happen that some nodes remain floating and unconnected. This is not a problem, as the simulation can be equally performed, despite any warning messages generated. In order to avoid this circumstance, the unconnected pins element, reported in the following figure 5.25, is included in the library. It has to be wired to the pins supposed to remain unconnected.

Eventually, the last two elements to be reviewed in this subsection are necessary to provide mechanical stimuli to the MEMS device assembled in the schematic. In particular, the item reported in the following figure 5.26 applies linear mechanical forces, as well as torques, around one or more axes. The force and torque can be freely defined by the user, along with the rotation angles, that split the vector in its components along and around the three axes.

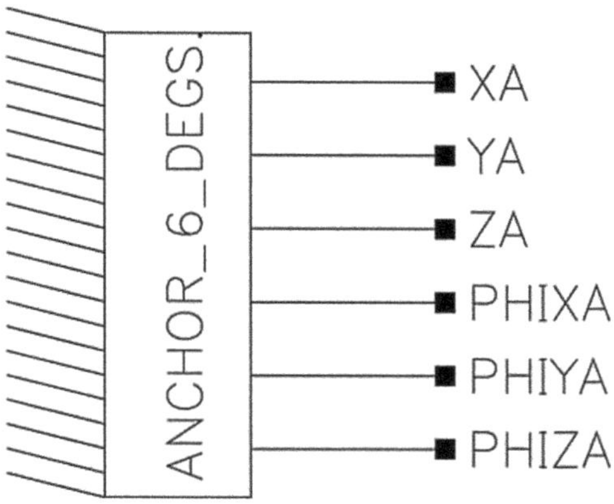

Figure 5.22. Symbol of the 6 DoFs anchoring point.

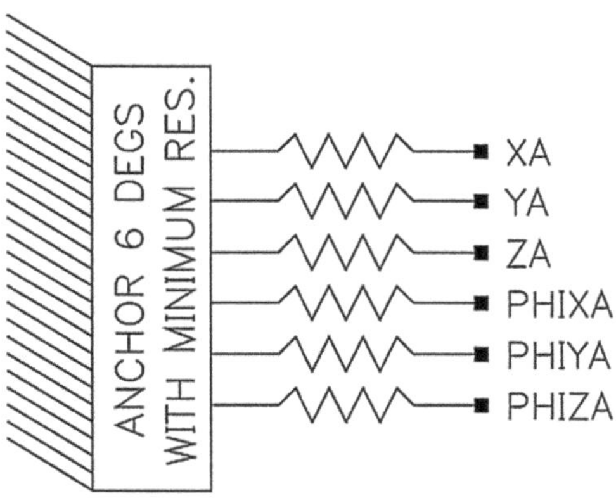

Figure 5.23. Symbol of the 6 DoFs anchoring point with minimum resistance.

Figure 5.24. Symbol of the 6 DoFs minimum resistance interconnect.

Following the same approach, the element reported in figure 5.27 provides linear displacements and angular rotations to the pins to which it is connected.

This subsection concludes the discussion of the behavioural models developed and included in the VerilogA library. In the following pages, the tool will be exploited for validation purposes, as well as to report significant examples of its exploitation in the multi-physical simulation of RF-MEMS devices and networks.

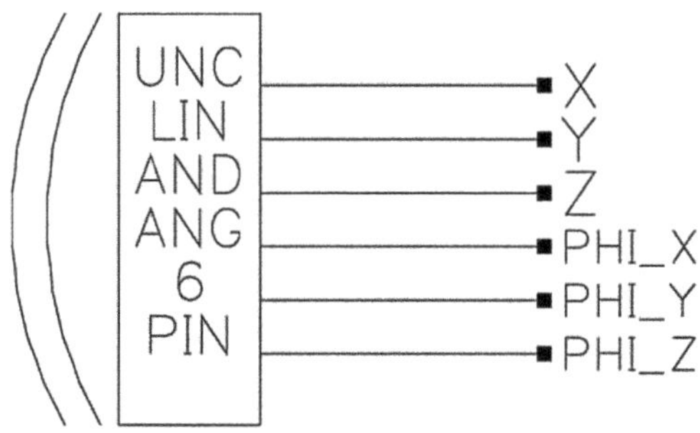

Figure 5.25. Symbol of the 6 DoFs unconnected pins termination.

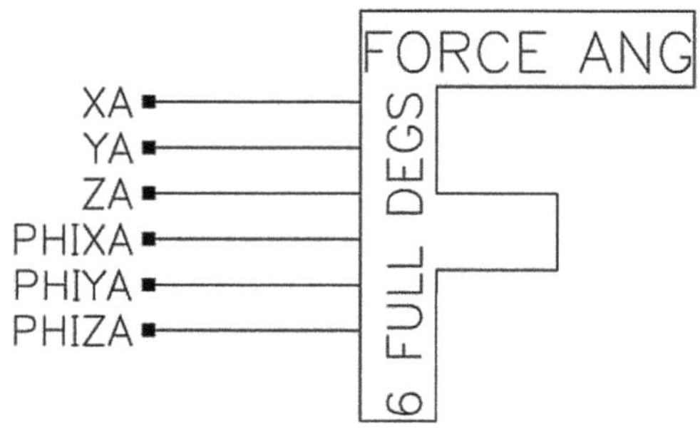

Figure 5.26. Symbol of the 6 DoFs source of linear forces and torques.

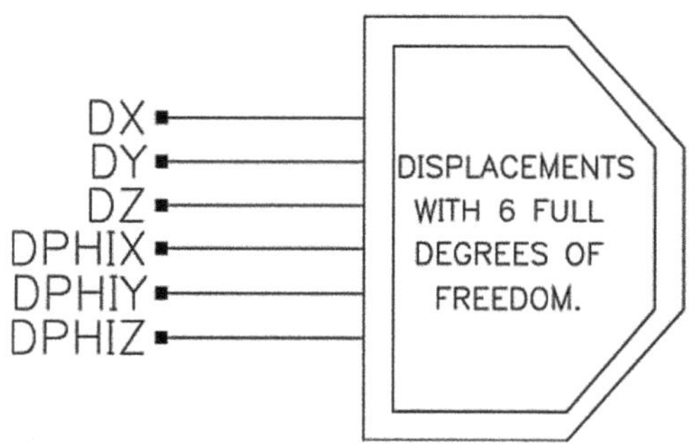

Figure 5.27. Symbol of the 6 DoFs source of linear displacements and rotations.

5.3 Examples of library validation and exploitation in 5G-relevant RF-MEMS components and networks case studies

After the comprehensive discussion around the analytical models developed and implemented in the VerilogA software library of MEMS behavioural elements, it is now time to explore the actual application of such a tool. To this end, the following pages will discuss a few case studies, related to physically fabricated RF-MEMS devices and networks, in which the implemented models are going to be applied to predict the multi-physical characteristics of the observed design concepts, pursuing a twofold target. First, the VerilogA software tool will be validated against experimental data, in order to assess the level of accuracy that can be reached. Second, the ease of parameterisation and simulation of the tool will be capitalised upon, in order to predict the behaviour of RF-MEMS multi-state networks. The targeted characteristics will refer both to the electromechanical coupled behaviour of the devices, i.e. pull-in/pull-out characteristics in quasi-static conditions, as well as the transient dynamic behaviour, along with the S-parameters' performances, in the RF domain, also leveraging the lumped-element network modelling techniques, previously reported in chapter 4. All the physical RF-MEMS devices discussed in the following subsections are fabricated in the technology platform previously mentioned in chapter 4, and discussed in detail in [53, 54].

5.3.1 Electromechanical and RF validation of a clamped–clamped RF-MEMS shunt switched capacitor

The device here at stake is an RF-MEMS switched capacitor, based on the classical clamped–clamped configuration, featuring a square central plate electrostatic transducer suspended with four straight slender beams. A 3D rendering of the physical sample, obtained with an optical profiling system based on interferometry [55, 56], is reported in figure 5.28.

The central rigid transducer is a square plate with sides of 220 μm, populated by openings of 20 μm with a clearance of 20 μm from adjacent ones and from the edges, while the flexible suspensions are 200 μm long and 10 μm wide. Finally, the air gap, defined by the thickness of the photoresist sacrificial layer, for the technology batch from which the measured sample derives, is around 2.4 μm. The following figure 5.29 compares the top view of the profile previously shown in figure 5.28 with the corresponding schematic built with the models discussed in this chapter.

Before stepping into the details of simulations and validation, it has to be stressed that the RF-MEMS physical sample here at stake is purposely chosen with

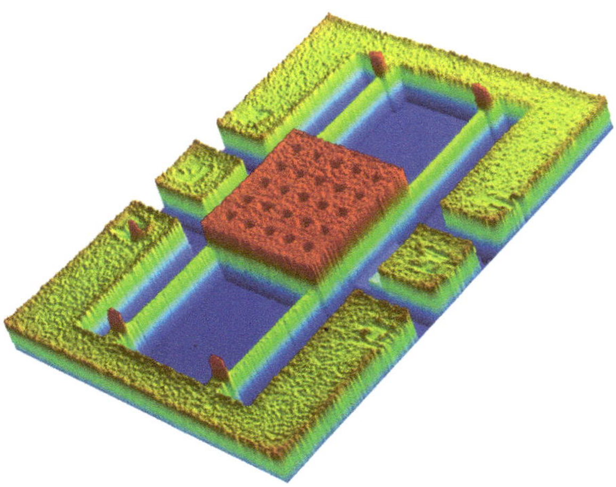

Figure 5.28. Measured 3D profile of an RF-MEMS switched capacitor. The colour scale indicates the elevation along the vertical direction of each device part (red: higher; blue: lower).

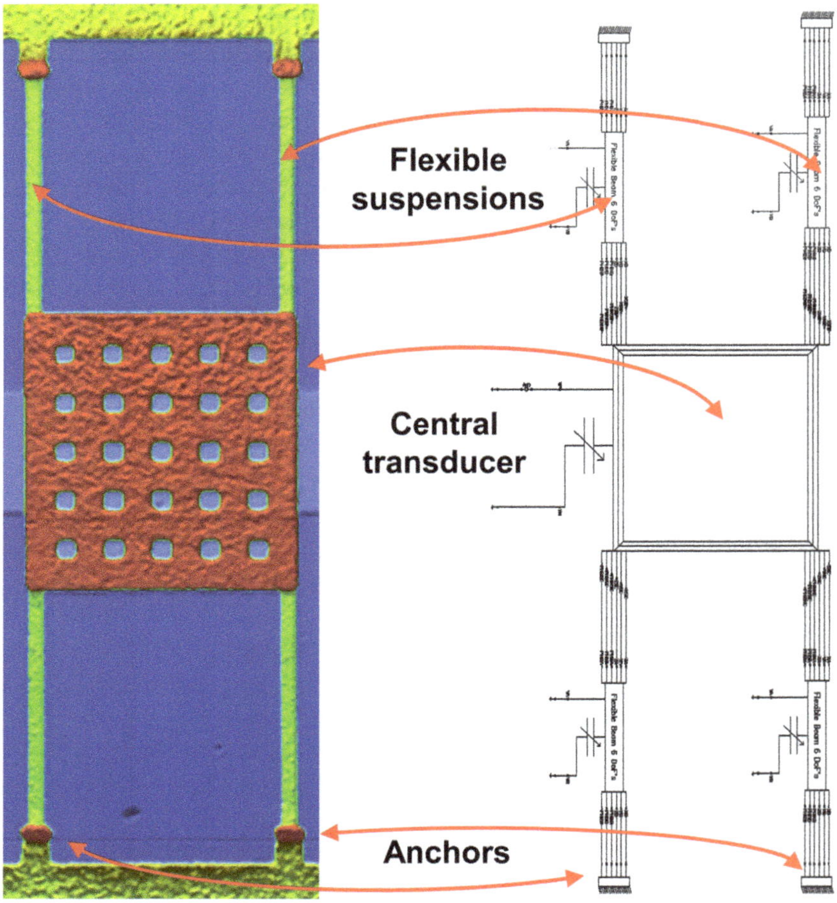

Figure 5.29. Correspondence between the RF-MEMS switched capacitor top-view topology with the schematic for simulations built with the elements included in the VerilogA models library.

non-idealities due to technology spreads that deviate quite importantly in its characteristics and performances from what is expected by design. This is done with the purpose of extending the overview on how the modelling and simulation approach discussed in this chapter can be exploited, as well as for reverse engineering and post-layout analysis purposes. In particular, the device reported in figure 5.28 exhibits two technology non-idealities, the first affecting both the electromechanical and RF behaviour, while the second influencing only the electromagnetic characteristics of the device.

The first technology spread is the non-planarity of the central rigid plate due to the presence of non-uniform residual stress (i.e. a stress gradient distribution) within the material of the MEMS membrane, which leads to the bowing of membranes that are supposed to be flat [57–61]. This is a known issue,that has been already solved in the RF-MEMS fabrication process discussed here, by acting on the temperature of the MEMS suspended membranes release phase [62]. As it is easy to figure out, having a non-planar movable transducer armature affects the electromechanical behaviour, both for what concerns the pull-in/pull-out voltages, along with the capacitance between the two armatures. This technology spread is not visible in the 3D rendering in figure 5.28, as it was preferred to report, for purposes of explanation, the profiling result of a physical device closer to ideality, i.e. with a flat surface of the MEMS central electrostatic transducer. Assuming an homogeneous parabolic distribution of the plate bowing, the way non-planarity is typically modelled in simulations is by considering a constant residual gap of air, limiting the vertical swing of the MEMS membrane at pull-in [63]. The height of such a spurious gap is trimmed, in order to average the mean vertical quote of the bent plate. In the specific case discussed here, the starting experimental data set available is the pull-in/pull-out characteristic, acquired with a dynamic optical profiling system. In particular, the vertical quote of the MEMS rigid transducer is averaged on the entire plate surface, in response to a zero mean value triangular voltage, in the range ±20 V, with a period of 20 Hz, i.e. slow enough to operate the MEMS switch in quasi-static conditions. Given such experimental data, the thickness of the residual air gap is tuned in the schematic in figure 5.29, looking for a good fitting of the measured pull-in, pull-out and extent of vertical displacement. The optimal value is found to be around 600 nm, and the superposition of the measured and simulated pull-in/pull-out characteristics is shown in figure 5.30.

As visible, the pull-in at ±15 V, the pull-out at ±9 V, along with the reduced vertical displacement of around 1.8 μm, are correctly reproduced by the simulation based on the VerilogA models. Concerning the non-planarity of the physical device, it has to be stressed that the averaged quote of the plate measured after pull-in occurs tends to get lower as the applied bias increases. This is because the electrostatic

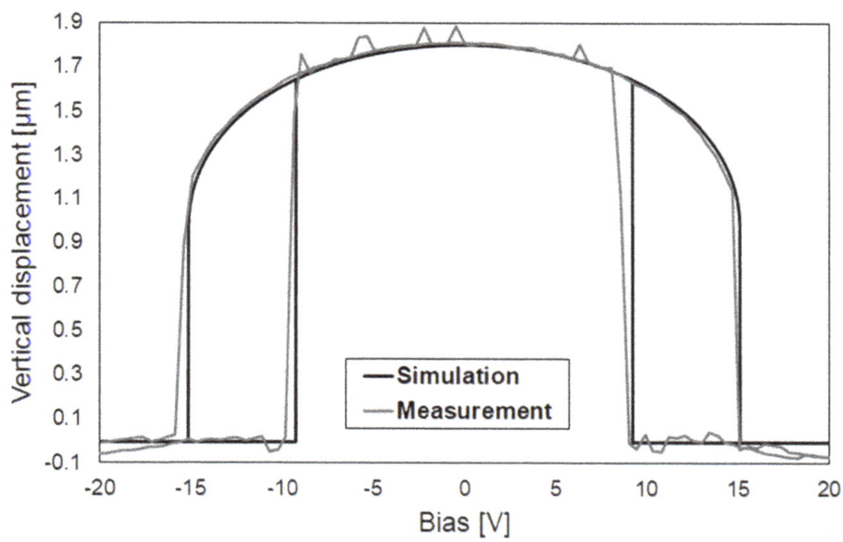

Figure 5.30. Comparison of the quasi-static measured and simulated pull-in/pull-out characteristics of the RF-MEMS switched capacitor in figure 5.28.

attraction force increases, forcing the parts of the MEMS membrane, elevated because of upward bending, to become flatter and closer to the substrate.

The matched results of the just-discussed electromechanical simulation come from a fitting performed on experimental data sets, and, despite complying with all the critical information that can be inferred from experimental data (i.e. pull-in, pull-out, vertical swing), they must be cross-checked against the orthogonal characteristics of the same physical device, in order to be reasonably reinforced and validated. To this purpose, Capacitance versus Voltage (C-V) measurements of the same RF-MEMS physical sample are performed with an LCR meter (Inductance-Capacitance-Resistance). A comparison of the measured and simulated data are reported in figure 5.31.

In particular, the blue solid line is the measured capacitance versus applied bias, and has to be read from the right to the left side of the plot, as the biasing voltage applied started from 25 V and was decreased to 0 V, meaning that the abrupt transition corresponds to the pull-out of the MEMS membrane. The capacitance simulated with the VerilogA models is indicated by the red solid line, and only the pull-out must be considered for comparison. A difference of about 2 V between the measured and simulated pull-out voltage is visible, despite the coarse voltage points measured with the LCR meter. It is more interesting to highlight the good match between the simulated and measured capacitance after the pull-out (C_{OFF}, OFF state capacitance), i.e. below 12 V. Concerning the capacitance of the pulled-in RF-MEMS switched capacitor, the simulated trace marks a constant C_{on} (ON state capacitance) of about 700 fF, while the measured one exhibits a quasi-linear behaviour and a decreasing slope from around 850 fF at 25 V, down to about 550 fF at pull-out. The simulated C_{on} is comparable with the measured value at pull-in, confirming that the strategy of adopting a constant residual air gap is sound and

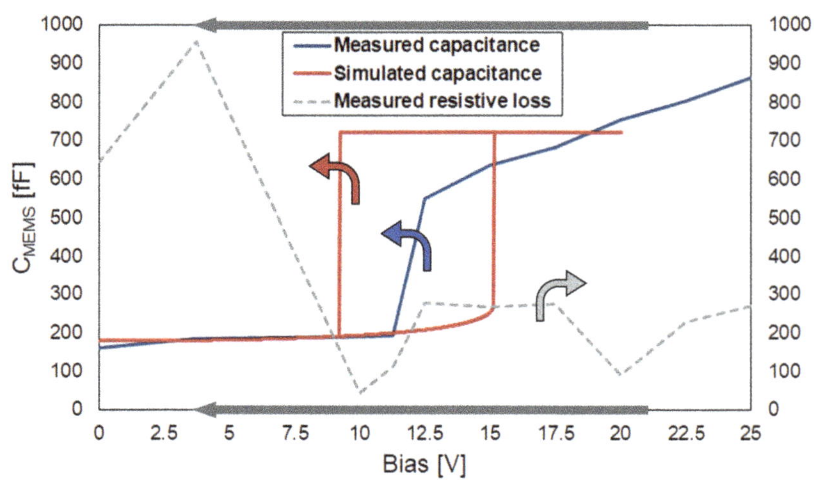

Figure 5.31. Measured versus simulated capacitance (C_{MEMS}) and measured resistive loss (R_{MEMS}) of the RF-MEMS switched capacitor against the applied bias. The measured points are acquired for decreasing bias levels. Therefore, the plot should be observed from the right to the left of the horizontal axis.

that the chosen value correctly predicts the behaviour of the non-ideal physical sample. Moreover, it has to be stressed that the decreasing capacitance in the pulled-in state for decreasing applied bias is fully compatible with the non-planarity of the MEMS membrane. Accordingly, the plot in figure 5.31 includes the resistance R_{MEMS} in parallel to C_{MEMS}, measured by the LCR meter and necessary to assess the losses of the capacitor. As expected, the R_{MEMS} values are rather large, in the order of several hundred GΩ, and fluctuate quite widely, so that it is not necessary to try to predict them with simulations, as it is sufficient choosing a fixed intermediate value. However, what has to be stressed is that R_{MEMS} is significantly smaller (below 300 GΩ) when the MEMS membrane is pulled-in, suggesting larger losses in this state, as it is reasonable to expect [64].

Now that the discussion on the cross-checked electromechanical and electric modelling of the intrinsic RF-MEMS switched capacitor is complete, the equivalent lumped-element network architecture of the whole device can be introduced and detailed. The topology of the whole extracted network is shown in figure 5.32, while the values of each element that compose it are listed in table 5.1.

The network is symmetric, as the RF-MEMS device it describes, and the intrinsic MEMS switched capacitor, i.e. neglecting the surrounding Coplanar Waveguide (CPW) frame, is visible in the centre. It is composed of C_{MEMS}, R_{MEMS} and L_{MEMS}. The first two were discussed above, and the values corresponding to the switched capacitor OFF and ON state are those previously shown in figure 5.31 for an imposed bias of 0 V and 25 V, respectively. On the other hand, L_{MEMS} represents the series inductance due to the MEMS suspensions and anchoring points. It is assessed by means of simple analytical considerations [29, 32], and its value is constant regardless of the switched capacitor configuration.

Moving from the centre towards the right-hand and left-hand sides of the network, the input/output CPW sections, constituting the frame around the intrinsic MEMS actuator (see figure 5.28), are modelled. The topology is the typical one adopted for CPW structures, featuring a series and shunt section. The series section models the resistive loss and the inductance of the line (R_{wse} and L_{wse}, respectively), whilst the shunt section accounts for the capacitive coupling between the central RF

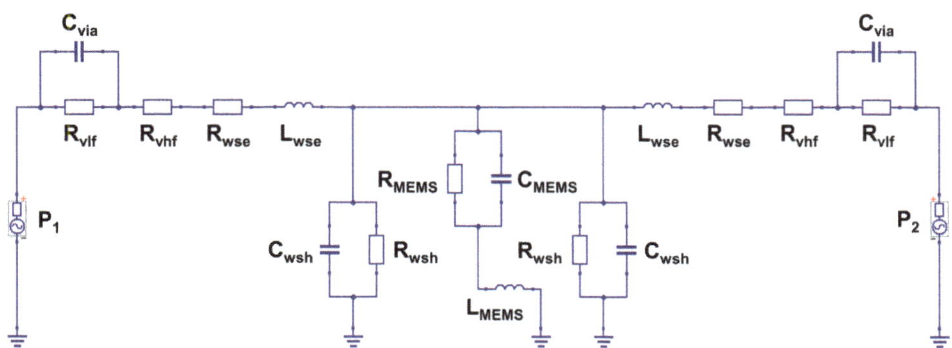

Figure 5.32. Complete equivalent lumped-element network describing the RF behaviour of the device in figure 5.28. The elements values are reported in table 5.1.

Table 5.1. Extracted values of the resistive and reactive elements composing the network in figure 5.32. The components are split into the three network sections, i.e., intrinsic MEMS, input/output CPW and vertical via parasitic effects.

Name	Value	Unit	Network section	Description
C_{MEMS}	160	fF	Intrinsic MEMS	OFF state (CLOSE switch) capacitance
R_{MEMS}	645	GΩ	Intrinsic MEMS	OFF state (CLOSE switch) resistive loss
C_{MEMS}	862	fF	Intrinsic MEMS	ON state (OPEN switch) capacitance
R_{MEMS}	269	GΩ	Intrinsic MEMS	ON state (OPEN switch) resistive loss
L_{MEMS}	15	pH	Intrinsic MEMS	Series inductance of the suspensions and anchors
R_{wse}	100	mΩ	Input/output CPW	Series resistance of the RF line
L_{wse}	122	pH	Input/output CPW	Series inductance of the RF line
C_{wsh}	30	fF	Input/output CPW	RF line capacitive coupling to ground planes
R_{wsh}	840	GΩ	Input/output CPW	RF line resistive losses to the substrate
R_{vlf}	2.82	Ω	Vertical via	Via resistive loss (acting at low frequency)
C_{via}	134	pF	Vertical via	Via parasitic capacitance
R_{vhf}	700	mΩ	Vertical via	Via resistive loss (acting at high frequency)

line and the ground planes, as well as for the resistive losses to the substrate (C_{wsh} and R_{wsh}, respectively).

Moving on towards the input/output ports of the network in figure 5.32, the last sections describing the non-idealities of vertical vias are visible, that is, the second technology spread affecting the RF-MEMS switched capacitor physical sample purposely chosen for this discussion. The mentioned vias are the vertical transitions between the aluminium RF underpass and the above gold metal layers, the latter realising the CPWs and the MEMS itself [53, 54]. Typically, those openings in the oxide, filled with metal, provide a modest contribution in terms of series parasitic resistance that can be neglected as being significantly smaller than the series resistance of the CPW sections and of the MEMS ON state contact resistance. However, in the specific case studied here, some non-ideal conditions occurred during fabrication, yielding an unwanted oxidation of the opened vias (to aluminium), before the gold layer covering and contacting them was evaporated. This led to the formation of a very thin titanium oxide (few nanometres) wherever vias were opened onto the oxide, therefore impairing the desired metal-to-metal (i.e. gold to aluminium) physical contact [31]. The unwanted oxide layer can be effectively modelled in the RF domain as a lossy capacitor, as shown in figure 5.32. In particular, the parasitic capacitor C_{via} has one resistor in parallel (R_{vlf}, i.e. via resistor at low frequency) and one in series (R_{vhf}, i.e. via resistor at high frequency). The via capacitance C_{via}, given the very small thickness of the oxide, has a rather large value, and therefore affects mainly the low-frequency portion of the range, in combination with its resistive losses R_{vlf}. Therefore, as the frequency rises, C_{via} behaves as a short circuit, and just the resistive contribution of R_{vlf} affects the RF characteristics of the device, as discussed in [31].

Before commencing discussion of the S-parameters' characteristics of the network topology in figure 5.32, a couple of considerations have to be stressed first. First, the extracted values listed in table 5.1 are all based on physical and analytical considerations, and are not chosen furthermore simply to fit the measured data sets. Moreover, it has to be noted that the only meaningful parameter that changes in the whole network, depending on the RF-MEMS switched capacitor ON/OFF configuration, is the intrinsic capacitance C_{MEMS}. As mentioned before, R_{MEMS} scores very large values that could also be kept fixed without affecting the accuracy of the model. In addition, all the other parameters, describing the MEMS series inductance, the input/output CPWs and the non-idealities of the vias, remain unchanged. This is in line with what happens in the physical device, as the only varying element, depending on the applied bias, is the MEMS plate transducer.

Now, given the lumped-element network in figure 5.32 and the components values listed in table 5.1, the S-parameters' simulated response of the switched capacitor in the ON and OFF state is compared to the measured data sets, as reported in plots in figure 5.33.

Looking at the S11 and S21 parameters in the CLOSED state, reported in figure 5.33(a) and in figure 5.33(b), respectively, the influence of the vertical vias non-idealities is more evident, as both reflection and losses are small. In particular, the unwanted C_{via} behaves as an open circuit at low frequency, causing an unwanted

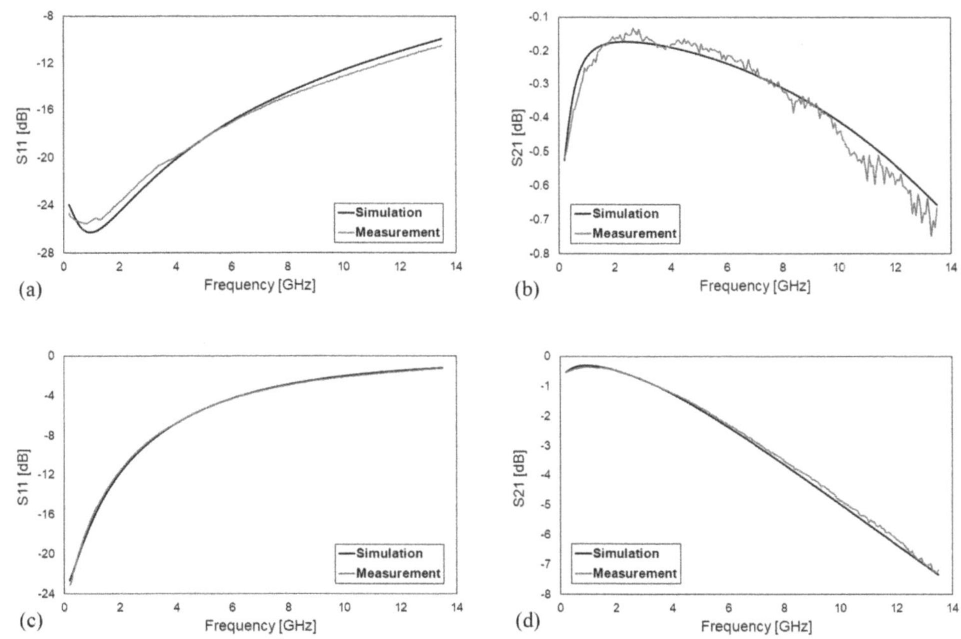

Figure 5.33. Simulated S-parameters' characteristics of the RF-MEMS switched capacitor lumped-element network in figure 5.32 compared against measured data sets up to 13.5 GHz, for what concerns (a) the reflection (S11) in the CLOSED state (MEMS switch OFF), (b) the transmission (S21) in the CLOSED state (MEMS switch OFF), (c) the reflection (S11) in the OPEN state (MEMS switch ON), and (d) the isolation (S21) in the OPEN state (MEMS switch ON).

attenuation of the RF signal through its resistive loss R_{vlf}. On the other hand, given the large value of C_{via}, its behaviour turns into that of a short circuit rather low in the frequency range, that is around 2 GHz. In fact, this is the point where an inversion, both of reflection and transmission, is visible in figure 5.33(a) and in figure 5.33(b), leading to the S11 and S21 characteristics being more similar to what one would expect from such a type of device above 2 GHz.

Stepping now to the S11 and S21 parameters in the OPEN state, reported in figure 5.33(c) and in figure 5.33(d), respectively, the presence of vias non-idealities exerts a smaller impact, especially for what concerns reflection. Nonetheless, in the MEMS transducer ON state, the first technology spread discussed above, that is the non-planarity of the movable membrane, causes significant consequences. In fact, due to the smaller ON state capacitance of the non-planar plate, the reflection (S11) is excessively low for a switch that is supposed to isolate, especially in the low-frequency range, up to around 8 GHz. This is confirmed by the isolation (S21) itself, which is rather poor over the whole span, marking the largest value of about −7.5 dB at 13.5 GHz. In order to have an idea of how an RF-MEMS switched capacitor closer to ideality should behave, figure 1.14 in chapter 1 can be observed.

5.3.2 Electromechanical and RF validation of a clamped–clamped RF-MEMS series ohmic switch

The RF-MEMS device chosen in this section to discuss the validation of the compact modelling approach has a dual behaviour compared to the one populating the previous pages, as it is a series ohmic switch. Its 3D profile, obtained as before with a white-light profiling system based on interferometry, is reported in figure 5.34.

The working principle of the micro-switch is still based on a transversal clamped–clamped membrane, composed by a central rigid plate, and kept suspended by means of four slender beams. This time, however, the input/output RF underpasses are truncated underneath the MEMS plate, and the vias openings are set on their tips, in such a way as to establish a metal-to-metal contact when the rigid plate is

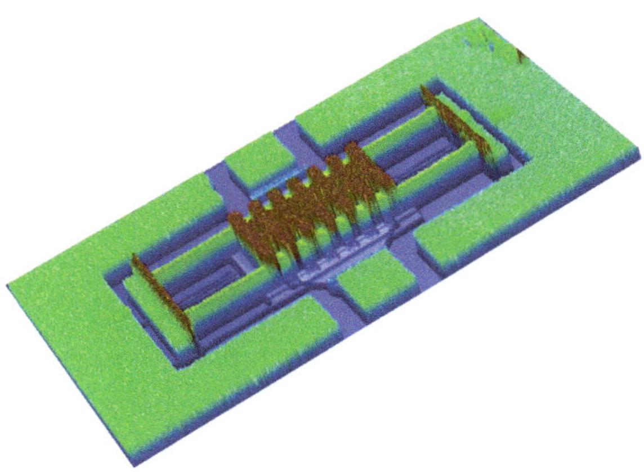

Figure 5.34. Measured 3D profile of an RF-MEMS series ohmic switch. The colour scale indicates the elevation along the vertical direction of each device part (red: higher; blue: lower).

pulled in [32, 65]. Therefore, apart from having an ohmic contact rather than a variable capacitance, the device here at stake is of series type, which means that its behaviour is that of an OPEN switch when the MEMS actuator is OFF, and of a CLOSE switch when the rigid plate is actuated. The central rigid transducer is a rectangular plate with edges of 260 μm and 140 μm, and with square openings of 20 μm, 20 μm distant from adjacent ones and from the plate boundaries. The flexible suspensions are 165 μm long and 10 μm wide. In the wafer the physical sample comes from, the air gap thickness was reduced to about 1.4 μm, while there were negligible issues in terms of residual stress, therefore the membrane is reasonably flat. A schematic similar to that previously reported in figure 5.29 is built with the models available in the software library, and a comparison of the experimental and simulated pull-in/pull-out quasi-static characteristics is shown in figure 5.35.

As visible, the pull-in at ±31 V, the pull-out at ±27 V, along with the reduced vertical displacement of around 1.4 μm, are correctly reproduced by the simulation based on the VerilogA models. It has to be noted that the hysteresis of the device is quite narrow, as the pull-in and pull-out take place at just 4 V one from each other. This is due to the particular geometry of interdigitated input/output ohmic contacts, which are significantly elevated with respect to the biasing fixed electrode, as discussed in [32, 65]. This leads to a rather large residual air gap at pull-in between the collapsed MEMS membrane and the underneath polycrystalline silicon biasing pads, with the input/output RF terminations acting as stoppers. The mentioned gap increases (in magnitude) the pull-out voltage level, moving it closer to the pull-in.

As a further and more extended validation step, the dynamic behaviour of the model is now going to be compared against experiments, looking at the RF-MEMS switch characteristics in the time transient domain. In this case, apart from the

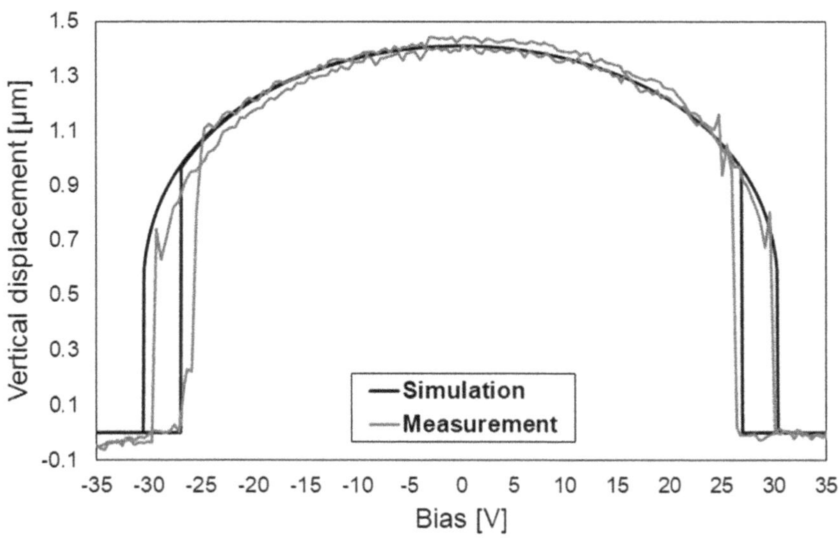

Figure 5.35. Comparison of the quasi-static measured and simulated pull-in/pull-out characteristic of the RF-MEMS series ohmic switch in figure 5.34.

mechanical stiffness characteristics, inertia and viscous damping are also involved. The physical sample is driven by a voltage square pulse (50% duty cycle) with period of 2 ms, 0 V low value and 33 V high value. The vertical displacement is measured by means of a 3D optical profiling system with stroboscopic illumination [55, 56], enabling the reconstruction of the equivalent time periodic behaviour of the device under test. On the other hand, the schematic based on compact models is simulated in the transient regime, applying the same stimulus mentioned above. A comparison of the measured and simulated vertical displacement versus time is reported by the plots in figure 5.36.

Looking at the whole observed period reported in figure 5.36(a), a rather accurate prediction of the dynamic characteristics of the RF-MEMS physical sample is visible. In particular, when magnifying the actuation discontinuity, reported in figure 5.36(b), the first mechanical touchdown at around 35 μs is rather accurately predicted by the model, as well as the qualitative reproduction of the first afterwards bounce, consequent to the impact. Also looking at the close-up around the release, shown in figure 5.36(c), the model predicts accurately the detachment of the MEMS transducer at around 1 ms, i.e. when the applied voltage pulse drops to zero. Also, the subsequent periodicity of the free ringing and amplitude attenuation due to the damping are satisfactorily modelled by the simulated structure. Especially in the latter case, the behavioural features linked to stiffness, inertia and viscous damping all have to be described accurately in order to score a good fitting with experiments.

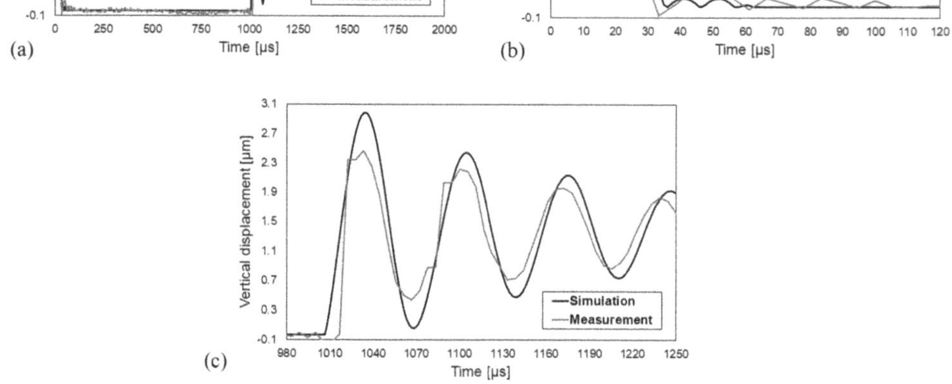

Figure 5.36. Measured versus simulated dynamic response of the RF-MEMS series ohmic switch to a square pulse over (a) the whole observed period, and over narrower intervals corresponding to (b) actuation and (c) release.

Now that the electromechanical behaviour of the device at stake has been validated, its RF characteristics are discussed. To this end, following the same approach previously discussed for the RF-MEMS switched capacitor, an equivalent lumped-element network is extracted for the series ohmic switch in figure 5.34. The complete network architecture is reported in figure 5.37, while the values of the resistive and reactive elements are listed in table 5.2.

Starting as before from the central part of the network, the intrinsic MEMS switch is described with a configuration of elements similar to that of the switched capacitor. The main difference is that, in this case, a series ohmic contact has to be accounted for. To this end, when the micro-switch is actuated (CLOSED), the R_{ON}

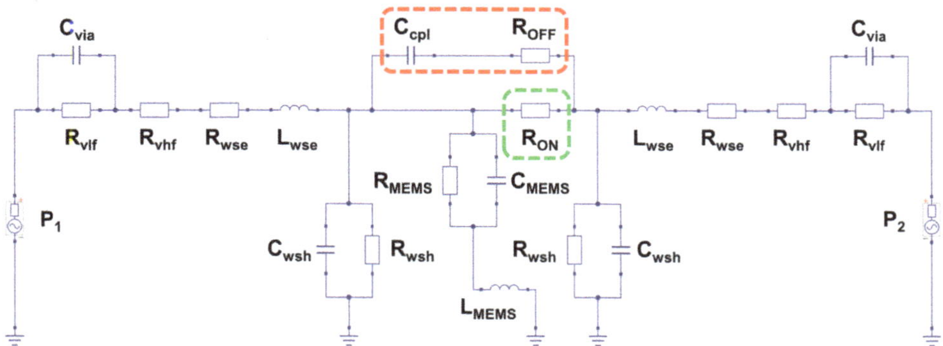

Figure 5.37. Complete equivalent lumped-element network describing the RF behaviour of the device in figure 5.34. The elements values are reported in table 5.2.

Table 5.2. Extracted values of the resistive and reactive elements composing the network in figure 5.37. The components are split into the three network sections, i.e., intrinsic MEMS, input/output CPW and vertical via parasitic effects.

Name	Value	Unit	Network section	Description
C_{cpl}	18	fF	Intrinsic MEMS	OFF state (OPEN switch) input/output capacitance
R_{OFF}	818	Ω	Intrinsic MEMS	OFF state (OPEN switch) input/output resistance
C_{MEMS}	14	fF	Intrinsic MEMS	OFF state (OPEN switch) capacitance to ground
R_{ON}	327	mΩ	Intrinsic MEMS	ON state (CLOSE switch) input/output resistance
C_{MEMS}	116	fF	Intrinsic MEMS	ON state (CLOSE switch) capacitance to ground
R_{MEMS}	7	GΩ	Intrinsic MEMS	Resistance to ground
L_{MEMS}	97	pH	Intrinsic MEMS	Series inductance of the suspensions and anchors
R_{wse}	1200	mΩ	Input/output CPW	Series resistance of the RF line
L_{wse}	137	pH	Input/output CPW	Series inductance of the RF line
C_{wsh}	31	fF	Input/output CPW	RF line capacitive coupling to ground planes
R_{wsh}	400	GΩ	Input/output CPW	RF line resistive losses to the substrate
R_{vlf}	27.5	Ω	Vertical via	Via resistive loss (acting at low frequency)
C_{via}	9.2	pF	Vertical via	Via parasitic capacitance
R_{vhf}	627	mΩ	Vertical via	Via resistive loss (acting at high frequency)

resistance is considered, in order to account for the contact series resistance between the collapsed MEMS bridge and the contacts underneath. Alternatively, when the MEMS is OFF (OPEN switch), the R_{ON} is removed, while the series branch including the C_{cpl} and R_{OFF} elements is inserted. The latter models the series parasitic capacitive coupling through air, between the contacts underneath and the suspended MEMS transducer in the rest position, along with its resistive losses.

Moving towards the edges of the lumped-element network, the input/output CPW branches are visible, along with subnetworks describing the vertical vias non-idealities, still present in the wafer from which the measured sample comes from. The CPW and vertical vias elements exhibit different values if compared to those in table 5.1, as the device dimensions, fabrication batch and technology parameters are not the same as before.

Now, given the lumped-element network in figure 5.37 and the components values listed in table 5.2, the S-parameters' simulated response of the series ohmic switch in the OFF and ON state is compared to the measured data sets, as reported in plots in figure 5.38.

Looking at the reflection (S11) and isolation (S21) parameters in the OPEN state, in figure 5.38(a) and in figure 5.38(b), respectively, as well as at the reflection (S11) and transmission (S21) parameters in the CLOSED state, in figure 5.38(c) and in figure 5.38(d), respectively, a good match between the simulated and measured traces is observable, completing the validation of the RF-MEMS series ohmic switch discussed in this subsection.

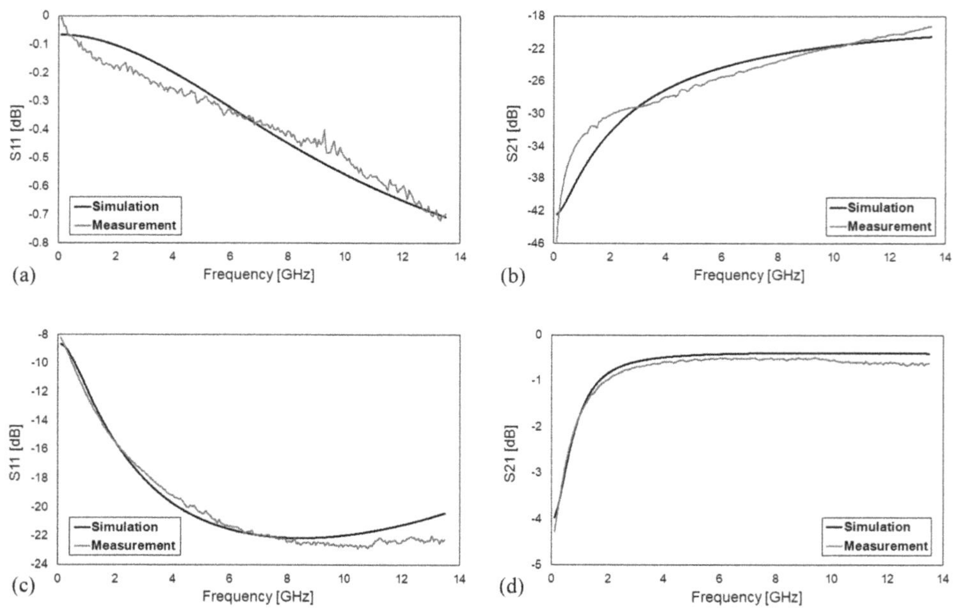

Figure 5.38. Simulated S-parameters' characteristics of the RF-MEMS series ohmic switch lumped-element network in figure 5.34 compared against measured data sets up to 13.5 GHz, for what concerns (a) the reflection (S11) in the OPEN state (MEMS switch OFF), (b) the isolation (S21) in the OPEN state (MEMS switch OFF), (c) the reflection (S11) in the CLOSED state (MEMS switch ON), and (d) the transmission (S21) in the CLOSED state (MEMS switch ON).

5.3.3 Electromechanical validation of a cantilever-based RF-MEMS series ohmic switch

This subsection develops a brief discussion around the electromechanical behaviour of a cantilever-type RF-MEMS series ohmic switch that will be employed in a multi-state RF power attenuator network, whose RF behaviour is going to be covered in the following subsection. The experimentally measured 3D topology of the device here at stake is reported in figure 5.39.

The cantilever is placed in a transversal fashion with respect to the input/output RF branches. Two contact fingers are hinged to the central transducer in correspondence to its free end, and are placed above the contact areas underneath. When the MEMS is pulled-in, such fingers realise an ohmic contact with the RF terminations, thus closing the contact between the input and output. The MEMS transducer is connected to short deformable suspensions on its anchoring side, ensuring downward mechanical deformation when the bias voltage is imposed [66]. The bottom image in figure 5.39 shows a schematic built with the compact models implemented in VerilogA. The comparison of the measured and simulated pull-in/pull-out quasi-static characteristics is reported in figure 5.40.

Once again, a good match between simulations and measurements is attained. The pull-in at ±26 V and the pull-out at ±22 V mark a rather narrow hysteresis, as can be expected by a cantilevered micro-switch.

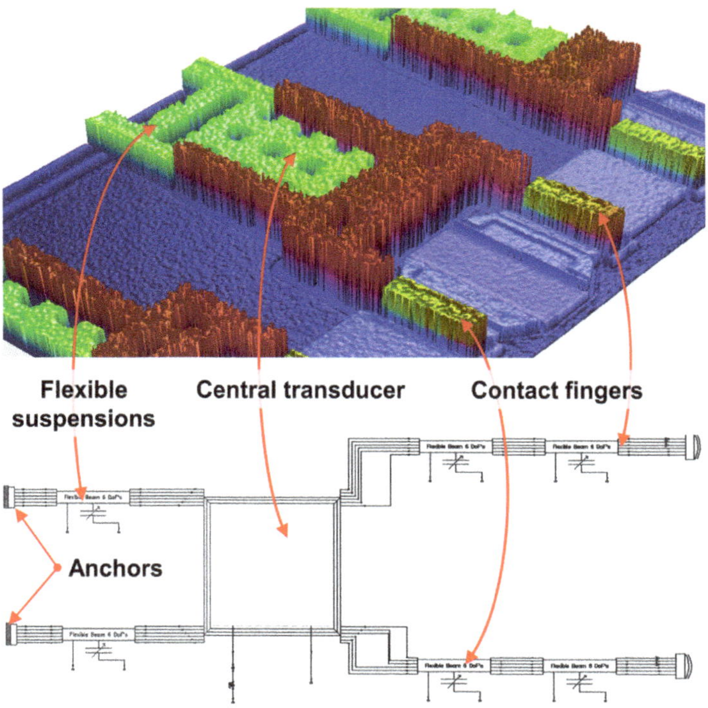

Figure 5.39. Measured 3D profile of a cantilever-based RF-MEMS series ohmic switch (top image). The colour scale indicates the vertical quote of each device part (red: higher; blue: lower). Schematic for simulations built with VerilogA compact models (bottom image).

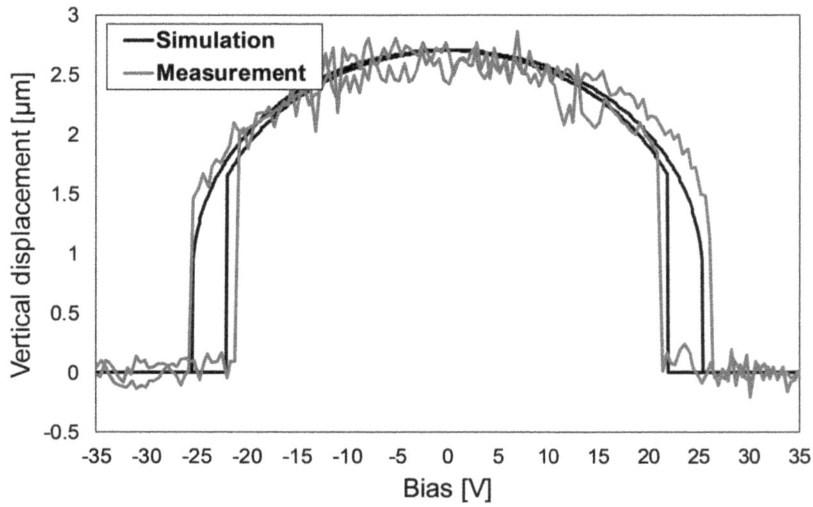

Figure 5.40. Comparison of the quasi-static measured and simulated pull-in/pull-out characteristics of the RF-MEMS cantilever-type series ohmic switch in figure 5.39.

5.3.4 RF simulation of a multi-state RF-MEMS power attenuator

Starting from the cantilevered series ohmic switch discussed in the previous subsection, its employment in a multi-state RF-MEMS power attenuator is going to be covered below. The network features four stages (bits) that can be reconfigured by cantilevered ohmic switches. The level of attenuation is realised by buried polycrystalline silicon resistive loads, shorted or inserted on the RF line depending on the ON/OFF configuration of the switches [67]. A microphotograph of a physical network sample is reported in figure 5.41.

Starting from the bottom side of the network, the RF signal line splits into two branches. The first stage of left/right RF-MEMS switches simply selects one or both branches, depending on their ON/OFF state, as well as isolates the input/output terminations when are both OFF. For this reason, the mentioned switches can be independently driven. Moving towards the top, three buried resistors of 25 Ω, 100 Ω and 500 Ω, respectively, are placed on both the RF branches. The left/right resistors of each size load the RF line or are shorted when the corresponding switches are OFF or ON, respectively. Having said that, the left/right switches of each of the three resistive stages are not independent, but controlled by the same biasing signal. This way, the three resistive loads realise eight different values of resistance, while the lower stage, depending on if one or both the RF branches are selected, doubles the total resistance values, therefore enabling 16 different configurations of the whole network (excluding the OFF state and the left or right branch redundant configuration). A complete schematic of the network in figure 5.41, realised with the compact models developed in VerilogA, is shown in figure 5.42.

The schematic comprises eight RF-MEMS switches, like the one reported in figure 5.39 (bottom image), arranged according to two RF line branches in parallel. Each micro-relay is connected to a two-state resistance element that switches

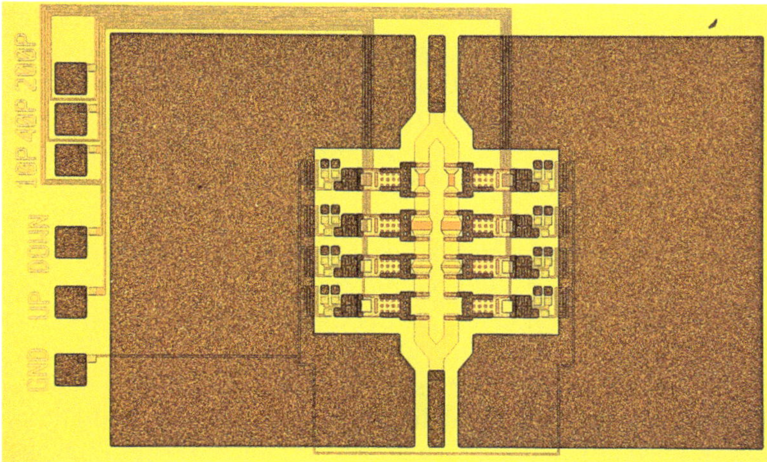

Figure 5.41. Microphotograph of a RF-MEMS multi-state power attenuator featuring four bits and implementing 16 different configurations.

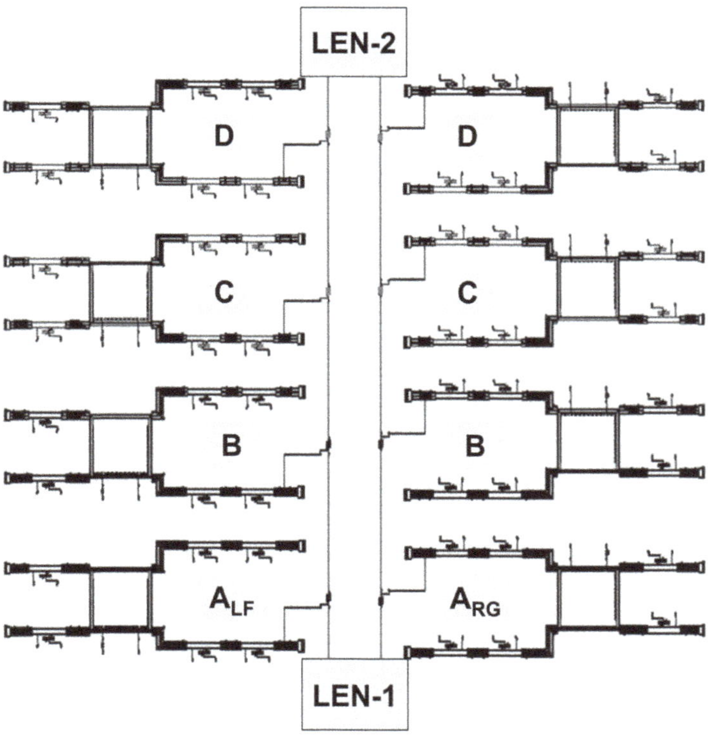

Figure 5.42. Complete schematic of the multi-state RF-MEMS power attenuator.

between two values, depending on if the MEMS device is OFF or ON. In the case of the branch-selecting switches, indicated as A_{LF} (stage A left) and A_{RG} (stage A right), the resistance is set to 10 kΩ when the switch is OFF (OPEN state), and to 1 Ω when the switch is ON (CLOSED state). The other stages, alternatively, realise the actual attenuation levels. The switches belonging to the stage B load each branch of the RF line with 25 Ω when ON, and with 1 Ω when OFF (i.e. shorted polycrystalline resistors), while those of stages C and D do the same, yet with 100 Ω and 500 Ω when ON, respectively, and with 1 Ω when OFF. The network in figure 5.42 is completed by two blocks, labelled as LEN-1 and LEN-2 (Lumped-Element Network), placed near port 1 and port 2, respectively. The topology of the input/output networks is sketched in figure 5.43.

The network models both the non-ideality of the vertical vias and the CPW sections of the physical device, along with providing the splitting of the RF line into two branches (LEN-1) and vice versa (LEN-2). The extracted values of each single element of the network are listed in table 5.3.

A comparison of the measured and simulated attenuation (S21) of the network in figure 5.41 obtained with the schematic in figure 5.42 up to 30 GHz, is shown in figure 5.44, for a few different configurations, including the maximum (A_{LF} = ON) and no attenuation (A_{LF}, B, C, D = ON) state.

As before, a good match between the measured and simulated traces is visible, as well as the undesired effects of vertical vias parasitic effects, affecting mainly the low-frequency-range portion (up to around 4 GHz).

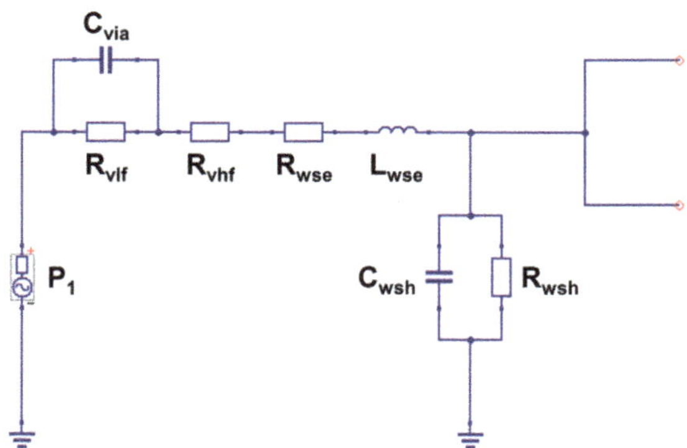

Figure 5.43. Arrangement of the LEN-1 and LEN-2 (Lumped-Element Network) included in the complete schematic in figure 5.42.

Table 5.3. Extracted values of the resistive and reactive elements composing the network in figure 5.43. The components are split into the two network sections, i.e. input/output CPW and vertical via parasitic effects.

Name	Value	Unit	Network section	Description
R_{wse}	1	Ω	Input/output CPW	Series resistance of the RF line
L_{wse}	300	pH	Input/output CPW	Series inductance of the RF line
C_{wsh}	20	fF	Input/output CPW	RF line capacitive coupling to ground planes
R_{wsh}	1	GΩ	Input/output CPW	RF line resistive losses to the substrate
R_{vlf}	15	Ω	Vertical via	Via resistive loss (acting at low frequency)
C_{via}	10	pF	Vertical via	Via parasitic capacitance
R_{vhf}	15	Ω	Vertical via	Via resistive loss (acting at high frequency)

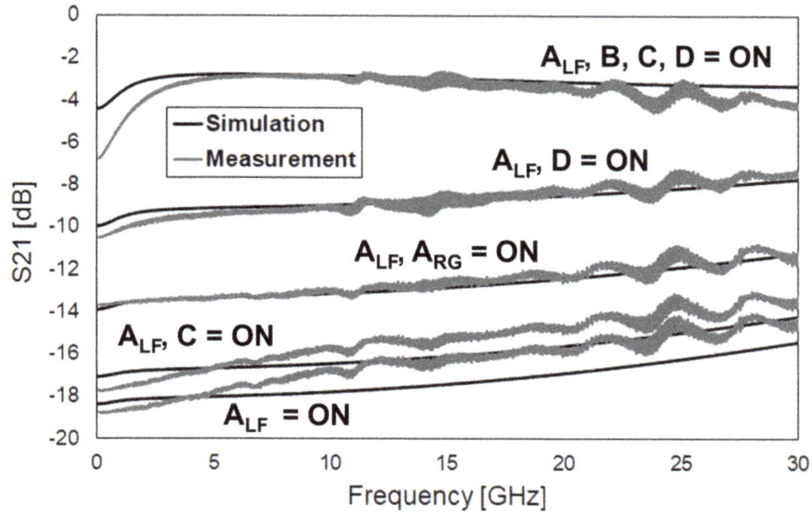

Figure 5.44. Comparison of the measured and simulated attenuation (S21) for a few different configurations of the network in figure 5.41 up to 30 GHz.

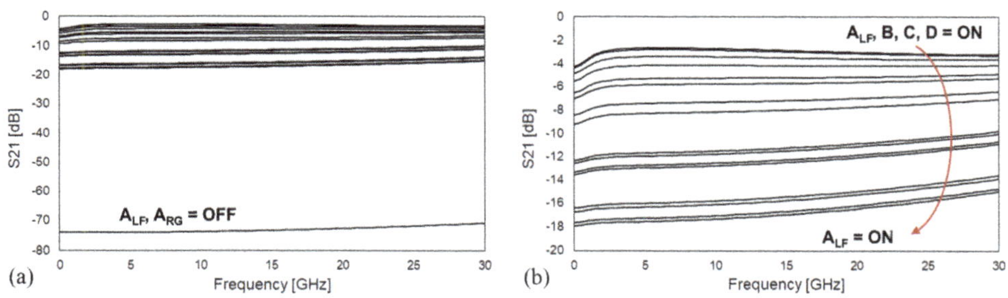

Figure 5.45. Simulated attenuation (S21) up to 30 GHz of the multi-state power attenuator, including the non-ideal behaviour of vias, in (a) all 17 possible configurations, and in (b) all 16 attenuation configurations excluding the OPEN state.

As a final analysis, all the possible 17 non-redundant configurations of the RF-MEMS multi-state power attenuator are simulated, both in the real case of vertical vias non-idealities, as well as assuming the behaviour of vertical transitions closer to ideality. In the latter case, the subnetwork modelling the parasitic effects of vertical vias, included in the architecture previously shown in in figure 5.43, is simply ignored. Having said that, the attenuator S21 traces, corresponding to all the configurations accounting for the parasitic effects of vias, are shown in figure 5.45.

The plot in figure 5.45(a) shows the large gap in terms of S21 existing between the isolation configuration (A_{LF}, A_{RG} = OFF) and all the attenuation states, in the latter case with the S21 ranging between around −18 dB (maximum attenuation) and −3 dB (no attenuation). In order to ease the interpretation of the 16 attenuation levels, the trace corresponding to the OPEN configuration is removed in figure 5.45(b). In this case, the flatness of all the S21 curves, as well as the impact

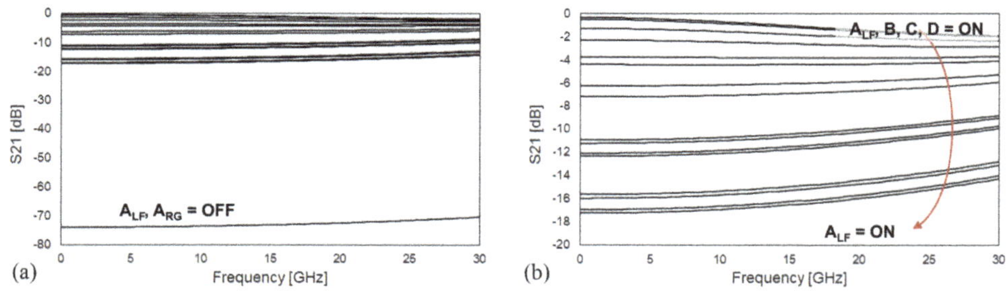

Figure 5.46. Simulated attenuation (S21) up to 30 GHz of the multi-state power attenuator, neglecting the non-ideal behaviour of vias, in (a) all 17 possible configurations, and in (b) all 16 attenuation configurations excluding the OPEN state.

of vias non-idealities up to around 4 GHz, are clearly visible features. On the other hand, the plot in figure 5.46(a) reports all the simulated configurations when vertical vias behave closer to ideality.

In particular, when looking at the plot in figure 5.46(b), where the OPEN state is not included, the more pronounced flatness of the S21 in all the possible configurations is evident. To this end, it must be highlighted that in the worst case, the S21 shift over the whole frequency span is within 2 dB. Also relevantly, it has to be noted that the loss, in the case of no attenuation (A_{LF}, B, C, D = ON), is better than −1.4 dB up to 30 GHz. The latter performances are realistically scored by physical devices, when the vertical vias have less intrinsic parasitic effects, as already published and discussed in [68–70].

5.4 Conclusion

This chapter developed a focus around the multi-physical modelling and simulation of RF-MEMS lumped components and networks, leveraging on-purpose compact analytical models in order to enable a fast, accurate and easily parameterisable tool, helping the designer especially in the early optimisation phase of novel MEMS-based RF passive design concepts, when a plethora of DoFs have to be handled.

At first, a brief review of the most common modelling and simulation approaches and techniques has been established. Starting from a full analytical description of the problem, which in the case of RF-MEMS involves structural mechanics, electrostatics and electromagnetism, two fundamental approaches can be utilised. On one side, numerical approaches are available, among which the most common are the Finite Element Method (FEM), Boundary Element Method (BEM) and Finite Difference Method (FDM), all being based on some form of discretisation of the 2D or full-3D structure to be analysed. On the other hand, compact and simplified approaches are also viable. In this case, the mathematical problem description itself is simplified, in order to address the most suitable trade-off between model complexity and accuracy of results. Among compact and simplified approaches, basic analytical models, compact analytical models and Model Order Reduction

(MOR), among others, are mentioned. The introductory section is completed by a brief comparison of the pros and cons of each listed technique and approach.

Subsequently, the discussion progressed to the specific technique covered by the chapter, based on compact analytical models. The developed tool features a set of basic elements that, wired together at schematic level, enable one to build typical RF-MEMS passives, like switches and reconfigurable networks. A software library is implemented in the VerilogA programming language, i.e. one of the various types of Hardware Description Languages (HDLs), within the Cadence Integrated Circuits (ICs) development environment, although it can be ported to other simulation tools supporting VerilogA, like, for instance, the Keysight ADS.

Mathematical models of the main elements included in the library, i.e. a rigid plate suspended electrostatic transducer (air gap) and a flexible straight suspension, were reported in detail, focusing in particular on the electrostatic and electromechanical transduction, on the description of mechanical deformations (flexible beam), on inertia effects, as well as on viscous damping. The VerilogA software library also features mechanical Boundary Conditions (BCs), like mechanical anchoring points and linear/angular force/torques and displacements/rotations.

After completing the discussion of the VerilogA compact models library, the tool was exploited in practical examples for validation and prediction purposes. In particular, two fundamental RF-MEMS basic components were chosen to carry out multi-physical validation of the results produced with the compact models, both for what concerns the electromechanical behaviour as well as with respect to the Radio Frequency (RF) Scattering parameters' (S-parameters) characteristics. In the latter case, equivalent lumped-element networks were extracted from experimental data and wrapped around the intrinsic MEMS transducers. The two mentioned devices are an electrostatically controlled switched capacitor in shunt configuration and a series ohmic switch. Physical samples with significant non-idealities due to technology spreads were purposely selected, in order to make the modelling more articulated and to demonstrate the flexibility of the on-purpose simulation tool.

Finally, a parametric simulation of a RF-MEMS complex network, that is, of an RF power attenuator with 16 different configurations, was reported, and the differences in terms of performance between a sample with technology non-idealities and one closer to nominal characteristics were remarked upon.

References

[1] Timoshenko S P and Gere J 1961 *Theory of Elastic Stability* 2nd edn (New York: McGraw-Hill)

[2] Wang Y G and Lin W H 2010 Pull-in voltage analysis for an electrostatically actuated extensional microbeam with large deflection *Z Angew. Math. Mech.* **90** 211–8

[3] Pozar D M 2012 *Microwave Engineering* (New York: Wiley) pp 6–158

[4] Zhu B 2017 *The Finite Element Method: Fundamentals and Applications in Civil, Hydraulic, Mechanical and Aeronautical Engineering* (Singapore: Wiley)

[5] Dick E 2009 Introduction to finite element methods in computational fluid dynamics *Computational Fluid Dynamics* ed J F Wendt (Berlin: Springer) pp 235–301

[6] Sauter S and Schwab C 2011 *Boundary Element Methods* (Berlin: Springer)

[7] Dimov I, Farago I and Vulkov L 2019 Finite difference methods: theory and applications *7th Int. Conf. FDM 2018*

[8] Rocha L A, Cretu E and Wolffenbuttel R F April 2004 Analysis and analytical modeling of static pull-In with application to MEMS-based voltage reference and process monitoring *J. Microelectromech. Syst.* **13** 342–54

[9] Németh M, Jani L and Poppe A 2016 Compact modeling approach for microchannel cooling aimed at high-level thermal analysis of 3D packaged ICs *2016 Symp. on Design, Test, Integration and Packaging of MEMS/MOEMS (DTIP)* pp 1–6

[10] Dawale H E, Jourdan G, Sibeud L, Lamberti F, Hentz S and Badets F 2020 Analytical compact model for opto-mechanical sensor *2020 IEEE 33rd Int. Conf. on Micro Electro Mechanical Systems (MEMS)* pp 869–72

[11] Galisultanov A, Perrin Y, Fanet H, Hutin L and Pillonnet G 2018 Compact MEMS modeling to design full adder in capacitive adiabatic logic *2018 48th European Solid-State Device Research Conf. (ESSDERC)* pp 174–7

[12] Bazigos A *et al* June 2014 Analytical compact model in verilog-a for electrostatically actuated ohmic switches *IEEE Transactions on Electron Devices* **61** 2186–194

[13] Senturia S D 2010 *Microsystem Design* (Berlin: Springer)

[14] Min Q, Li E, Zhuo C, Li Y, Zhou S and Jin J 2017 Electrical-thermal co-analysis of through silicon via with equivalent circuit model *2017 IEEE Electrical Design of Advanced Packaging and Systems Symp. (EDAPS)* pp 1–4

[15] Yaren T, Kizir S and Yildiz A B 2019 Electrical equivalent circuit based analysis of double pendulum system *2019 6th Int. Conf. on Electrical and Electronics Engineering (ICEEE)* pp 258–62

[16] Nazdrowicz J and Napieralski A 2018 Electrical equivalent model of MEMS accelerometer in Matlab/SIMULINK environment *2018 XIV-th Int. Conf. on Perspective Technologies and Methods in MEMS Design (MEMSTECH)* pp 69–72

[17] Homayouni S M, Schreurs D and Nauwelaers B 2009 Wide-band multi-bias equivalent circuit extraction for FinFET transistors accommodating high-frequency kink-behaviours *Proc. Europ. Microw. Integrated Circuits Conf. EuMIC (Rome, Sept. 2009)* pp 85–8

[18] Wada T, Nakajima R, Obiya H, Ogami T, Koshino M, Kawashima M and Nakajima N 2014 A miniaturized broadband lumped element circulator for reconfigurable front-end system *Proc. IEEE MTT-S Int. Microw. Symp. IMS2014 (Tampa, FL, June 2014)* pp 1–3

[19] Essaadali R, Jarndal A, Kouki A B and Ghannouchi F M 2016 A new GaN HEMT equivalent circuit modeling technique based on X-parameters *IEEE Trans. Microw. Theory Tech.* **64** 2758–77

[20] Arcioni P, Castello R, Perregrini L, Sacchi E and Svelto F 1998 An improved lumped-element equivalent circuit for on silicon integrated inductors *Proc. IEEE Radio and Wireless Conf. RAWCON (Colorado Springs, CO, Aug. 1998)* pp 301–4

[21] Kang J, Sun L, Wen J and Zhao M 2009 An equivalent lumped-circuit model for on-chip symmetric intertwined transformer *Proc. IEEE Int. Conf. on ASIC (Changsha, Hunan, Oct. 2009)* pp 674–7

[22] Horng T S, Jau J K, Tsai Y S and Huang C S A decomposition and reconstruction scheme for broadband modeling of on-chip passive components using the modified T-equivalent circuit topology *Proc. IEEE Radio Frequency integrated Circuits (RFIC) Symp (Long Beach, CA, June 2005)* pp 299–302

[23] Hwang J, Jung W and Kim S 2015 Coupling analysis and equivalent circuit model of the IC stripline method *Proc. Asia-Pacific Symp. on Electromag. Compatibility APEMC (Taipei, May 2015)* pp 650–3

[24] Zhou Y and Chen Y 2008 Lumped-element equivalent circuit models for distributed microwave directional couplers *Proc. Int. Conf. on Microw. and Millimeter Wave Technol. (Nanjing, April 2008)* pp 131–4

[25] Yang K, Yin W Y and Mao J F 2007 Wideband lumped element model for on-chip (a) symmetrical coupled interconnects on lossy silicon substrate *Proc. Asia-Pacific Microw. Conf. (Bangkok, Dec. 2007)* pp 1–4

[26] Chen K L, Salvia J, Potter R, Howe R T and Kenny T W 2009 Performance evaluation and equivalent model of silicon interconnects for fully-encapsulated RF MEMS devices *IEEE Trans. Adv. Packag.* **32** 402–9

[27] Grinde C, Nygaard K H, Due-Hansen J and Fjeldly T A 2009 Lumped modeling of a novel RF MEMS double-disk resonator system *Proc. Symp. on Design, Test, Integration & Packag. of MEMS/MOEMS DTIP (Rome, April 2009)* pp 14–8

[28] Unlu M, Topalli K, Demir S, Civi O A, Koc S and Akin T 2004 A parametric modeling study on distributed MEMS transmission lines *Proc. Europ. Microw. Conf. EuMW (Amsterdam, Oct. 2004)* pp 1157–60

[29] Iannacci J 2010 Mixed-domain fast simulation of RF and microwave MEMS-based complex networks within standard IC development frameworks *Advanced Microwave Circuits and Systems* ed V Zhurbenko (London: InTechOpen)

[30] Iannacci J, Gaddi R and Gnudi A 2007 Non-linear electromechanical RF model of a MEMS varactor based on VerilogA© and lumped-element parasitic network *2007 European Microwave Conf.* pp 1342–5

[31] Iannacci J, Giacomozzi F, Colpo S, Margesin B and Bartek M 2009 A general purpose reconfigurable MEMS-based attenuator for radio frequency and microwave applications *IEEE EUROCON* pp 1197–205

[32] Iannacci J, Gaddi R and Gnudi A June 2010 Experimental validation of mixed electro-mechanical and electromagnetic modeling of RF-MEMS devices within a standard IC simulation environment *J. Microelectromech. Syst.* **19** 526–37

[33] Schilders W H, van der Vorst H A and Rommes J (ed) 2008 *Model Order Reduction: Theory, Research Aspects and Applications* (Berlin: Springer)

[34] Niessner M, Schrag G, Iannacci J and Wachutka G 2009 Experimentally validated and automatically generated multi-energy domain coupled model of a RF-MEMS switch *EuroSimE 2009–10th Int. Conf. on Thermal, Mechanical and Multi-Physics Simulation and Experiments in Microelectronics and Microsystems* pp 1–6

[35] Del Tin L *et al* 2007 Non linear compact modeling of RE-MEMS switches by means of model order reduction *TRANSDUCERS 2007–2007 Int. Solid-State Sensors, Actuators and Microsystems Conf.* pp 635–8

[36] Sanginario A, Schropfer G, Zerbini S, Ekwinska M, Houlihan R and Demarchi D 2014 A MEMS design methodology for model-order-reduction, based on high-order parametric elements *The Tenth Int. Conf. on Advanced Semiconductor Devices and Microsystems* pp 1–4

[37] Chen J and Kang S M 2000 An algorithm for automatic model-order reduction of nonlinear MEMS devices *2000 IEEE Int. Symp. on Circuits and Systems (ISCAS)* **2** 445–8

[38] Parent A, Krust A, Lorenz G and Piirainen T 2015 A novel model order reduction approach for generating efficient nonlinear verilog-a models of mems gyroscopes *2015 IEEE Int. Symp. on Inertial Sensors and Systems (ISISS) Proc.* pp 1–4

[39] Parent A, Krust A, Lorenz G, Favorskiy I and Piirainen T 2015 Efficient nonlinear simulink models of MEMS gyroscopes generated with a novel model order reduction method *2015 Transducers – 2015 18th Int. Conf. on Solid-State Sensors, Actuators and Microsystems (TRANSDUCERS)* pp 2184–7

[40] Agilent Technologies *Verilog-A Reference Manual* https://literature.cdn.keysight.com/litweb/pdf/ads2004a/pdf/verilogaref.pdf

[41] Agilent Technologies *Verilog-A and Verilog-AMS Reference Manual* https://literature.cdn.keysight.com/litweb/pdf/ads2006update3/pdf/verilogaref.pdf

[42] Open Verilog Int. *Verilog-A Language Reference Manual* https://siue.edu/~gengel/ece585WebStuff/OVI_VerilogA.pdf)

[43] Massachusetts Institute of Technology *Verilog-A Reference Manual* https://lost-contact.mit.edu/afs/cnf.cornell.edu/windows/Program%20Files/LEdit/program%20files/Tanner%20EDA/Tanner%20Tools%20v13.0/Docs/VerilogA_Reference.pdf

[44] Sutherland S 2002 *Verilog—2001. A Guide to the New Features of the Verilog® Hardware Description Language* (The Springer Interntional Series in Engineering and Computer Science vol 652) (New York: Springer US)

[45] Thomas D E and Moorby P R 1991 *The Verilog Hardware Description Language* (Berlin: Springer)

[46] Boute (Auth.) R and Mermet J P (ed) 1993 *Fundamentals and Standards in Hardware Description Languages* (NATO ASI Series vol 249) (Berlin: Springer)

[47] Hartenstein R W 1988 *Hardware Description Languages (Advances in CAD for VLSI)* (Amsterdam: Elsevier)

[48] Przemieniecki J S 1968 *Theory of Matrix Structural Analysis* (New York: McGraw-Hill)

[49] Halliday D, Resnick R and Walker J 2004 *Fundamentals of Physics* (New York: John Wiley and Sons)

[50] Gary M, Pornin J, Cibrario G and Jaffard C 2013 Release holes: physical implementation checks for thin films packaging *2013 Eurpoean Microelectronics Packaging Conf. (EMPC)* pp 1–5

[51] Iannacci J 2013 *Practical Guide to RF-MEMS.* (Weinheim: Wiley-VCH Verlag GmbH & Co. KGaA)

[52] Öchsner A 2021 Euler–Bernoulli beam theory *Classical Beam Theories of Structural Mechanics* (Dordrecht: Springer)

[53] Giacomozzi F, Mulloni V, Colpo S, Iannacci J, Margesin B and Faes A 2011 A flexible fabrication process for RF MEMS devices *Rom. J. Inf. Sci. Technol.* **14**

[54] Giacomozzi F, Mulloni V, Colpo S, Iannacci J, Margesin B and Faes A 2011 A flexible technology platform for the fabrication of RF-MEMS devices *Proc. of the Int. Semiconductor Conf., CAS* **1** 155–8

[55] Conway J A, Osborn J V and Fowler J D June 2007 Stroboscopic imaging interferometer for MEMS performance measurement *J. Microelectromech. Syst.* **16** 668–74

[56] Xie Y, Liu S, Shi T, Wang H and Zhang W 2006 Measurement system for MEMS dynamics characterization with environmental control facility *2006 1st IEEE Int. Conf. on Nano/Micro Engineered and Molecular Systems* pp 1055–9

[57] Dutta S, Imran M, Pal R, Jain K K and Chatterjee R 2011 Effect of residual stress on RF MEMS switch *Microsyst. Technol.* **17** 1739–45
[58] Dutta S and Pandey A 2021 Overview of residual stress in MEMS structures: its origin, measurement, and control *J. Mater. Sci.: Mater. Electron.* **32** 6705–41
[59] Chen K-S 2007 Techniques in residual stress measurement for MEMS and their applications *MEMS/NEMS* (Boston, MA: Springer)
[60] Tilli M, Motooka T, Airaksinen V M, Franssila S, Paulasto-Kröckel M and Lindroos V 2015 *Handbook of Silicon Based MEMS Materials and Technologies* 2nd edn (New York: Elsevier)
[61] Kawdi B and Vrushabha C M 2019 Residual stress on a MEMS structure: a survey *Int. J. Mod. Stud. Mech. Eng.* **5** 37–40
[62] Iannacci J, Masotti D, Kuenzig T and Niessner M 2011 A reconfigurable impedance matching network entirely manufactured in RF-MEMS technology *Proc. of SPIE* **8066** 80660X
[63] Iannacci J, Repchankova A, Macii D and Niessner M 2009 A measurement procedure of technology-related model parameters for enhanced RF-MEMS design *2009 IEEE Int. Workshop on Advanced Methods for Uncertainty Estimation in Measurement* pp 44–9
[64] Llamas M A *et al* 2009 Capacitive and resistive RF-MEMS switches 2.5D & 3D electromagnetic and circuit modelling *2009 Spanish Conf. on Electron Devices* pp 451–4
[65] Gaddi R *et al* 2005 Reconfigurable MEMS-enabled LC-tank for multi-band CMOS oscillator *IEEE MTT-S Int. Microwave Symp. Digest* **2005** 1353–6
[66] Iannacci J, Faes A, Mastri F, Masotti D and Rizzoli V 2010 A MEMS-based wide-band multi-state power attenuator for radio frequency and microwave applications *Nanotechnology 2010: Electronics, Devices, Fabrication, MEMS, Fluidics and Computational—Technical Proc. of the 2010 NSTI Nanotechnology Conf. and Expo, NSTI-Nanotech 2010* 2010 **2** 328–31
[67] Iannacci J, Faes A, Kuenzig T, Niessner M and Wachutka G 2011 Electromechanical and electromagnetic simulation of RF-MEMS complex networks based on compact modeling approach *Technical Proc. of the 2011 NSTI Nanotechnology Conf. and Expo, NSTI-Nanotech 2011* **2**
[68] Iannacci J, Huhn M, Tschoban C and Pötter H 2016 RF-MEMS technology for 5G: series and shunt attenuator modules demonstrated up to 110 GHz *IEEE Electron Device Lett.* **37** 1336–9
[69] Iannacci J, Reyes J, Maaß U, Huhn M, Ndip I and Pötter H 2016 RF-MEMS for 5G mobile communications: a basic attenuator module demonstrated up to 50 GHz *Proc. IEEE SENSORS (Orlando, FL, Oct.–Nov. 2016)* pp 1–3
[70] Iannacci J, Huhn M, Tschoban C and Pötter H Dec. 2016 RF-MEMS technology for future mobile and high-frequency applications: reconfigurable 8-bit power attenuator tested up to 110 GHz *IEEE Electron Device Letters* **37** 1646–9

Appendix A

Moving MEMS: Dynamics down in the micro-world

A.1 Introduction

The aim of this appendix is to provide a brief insight into Micro-ElectroMechanical Systems (MEMS) and Radio Frequency Micro-ElectroMechanical Systems (RF-MEMS) dynamics, by reporting and explaining a few video animations of micro-system devices recorded during their operation. The material will be based both on experimental measurements as well as Finite Element Method (FEM) simulations. The way sections are split is based upon the type of measurement setup or simulation exploited to obtain visual dynamic data. According to this arrangement, the necessary background on the observation procedure is provided first, in such a way as to ease understanding.

A.2 Measurements based on interferometric microscopy

The experimental material that is going to be reported in this section is acquired by means of a setup based on optical interferometry. This particular type of microscopy analysis exploits the constructive/destructive interference of the light coming back from the Device Under Test (DUT) with that reflected by a mirror, in order to reconstruct information about the vertical dimension of the DUT itself [1, 2]. In other words, optical interferometry allows one to reconstruct a 3D profile of the DUT, starting from single microscopy observation. Proper software-based post-processing of the measured data enables 3D images rendering of the DUT, in which a colour scale is typically used to provide information about the elevation along the vertical direction of each measured spot. The image reported in figure A.1 shows the 3D topology of an RF-MEMS two-state capacitor obtained with an optical interferometer based on white light. The DUT is a typical MEMS suspended electrostatically controlled plate, framed within a Coplanar Waveguide (CPW) structure.

doi:10.1088/978-0-7503-4199-8ch6

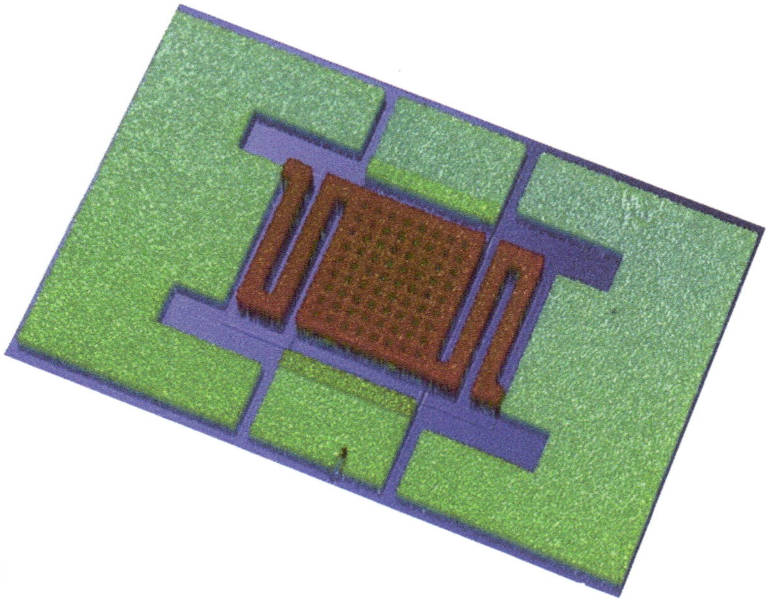

Figure A.1. 3D image of an RF-MEMS two-state capacitor obtained by means of optical interferometry. The colour scale indicates the vertical quote of the DUT.

Looking at the colour scale of the image in figure A.1, three main reference quotes can easily be identified. The lower plane, depicted in blue, is the underlying substrate. The intermediate quote, reported in green, is that of the electrodeposited gold defining the CPW line and ground planes. Finally, the higher level, shown in red, is the elevated MEMS membrane. The central suspended plate with holes and the meandered suspensions are clearly visible. Any additional information about the technology is available in figure 4.2 and table 4.1.

Besides white-light static 3D profiling, interferometric microscopy also enables measurements in dynamic conditions. In this case, the DUT is illuminated with stroboscopic (pulsed) light. This type of measurement can be effectively exploited just to observe periodic behavioural characteristics. The DUT is stimulated by a periodic waveform (e.g. sine, pulse, etc) at a certain frequency. On the other hand, the stroboscopic light is pulsed at the same frequency. The phase delay of the pulse feeding the light source with respect to the bias driving the DUT is increased, period after period. This way, at each iteration of the periodic behaviour, an image of the DUT at a different point of its characteristic is acquired. Finally, by putting together all the collected snapshots, the behaviour of the observed sample over an entire period is reconstructed [3–6].

The RF-MEMS sample in figure A.1 is biased with a zero mean value triangular waveform ranging between −20 V and +20 V. The absolute peak value is selected in order to be larger than the pull-in voltage. Moreover, its frequency is kept low (0.05 Hz) in order to perform the measurement in quasi-static conditions. The bias waveform versus time, with markers in the corresponding pull-in and pull-out transitions, is reported in figure A.2.

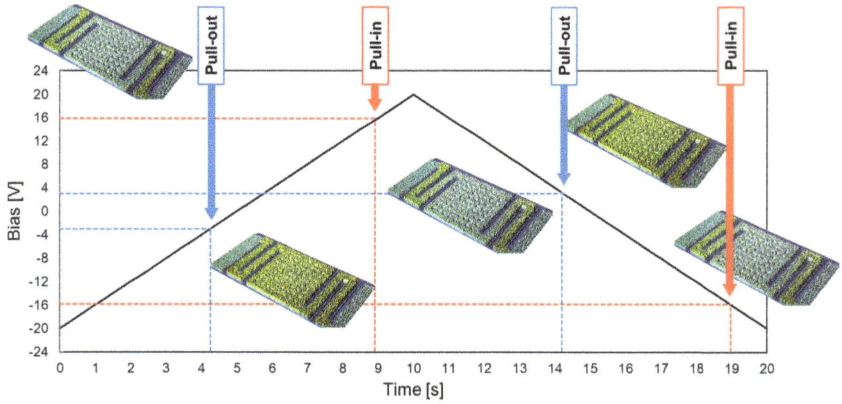

Figure A.2. Triangular waveform versus time biasing the DUT. Pull-in and pull-out levels are indicated along with measured images of the RF-MEMS when actuated (membrane colour closer to green) and not actuated (membrane colour closer to red).

Video A.1. Video animation of the RF-MEMS in figure A.1 when biased by the waveform in figure A.2, as reconstructed by 3D stroboscopic light optical dynamic interferometry. Video available at https://iopscience.iop.org/book/978-0-7503-4199-8.

Pull-in takes place at ±16 V, while pull-out happens at ±3 V. With the waveform centred around zero, two actuations and releases are observed over one period. The small measured snapshots of the RF-MEMS movable membrane help one to understand what it looks like (in terms of colours) when actuated and not actuated. Finally, an animation of the RF-MEMS, as reconstructed by the 3D stroboscopic light optical interferometric systems, is reported in video A.1.

A dynamic characterisation performed on a different DUT by means of the same measurement procedure discussed above is now going to be reported. The observed RF-MEMS device is a multi-state power attenuator comprising six cantilevered ohmic switches, exploited to select or short resistive loads on a split RF line [7, 8]. The 3D topology of the physical sample, obtained by means of white-light interferometry, is shown in figure A.3.

The DUT is stimulated with a waveform similar to that previously discussed in figure A.2. Transitions of cantilevered switches, corresponding to pull-in and pull-out, are visible in the animation, reconstructed by the 3D stroboscopic light optical interferometric systems, in video A.2.

Figure A.3. 3D image of the RF-MEMS power attenuator discussed in [7, 8], obtained by means of optical interferometry. The colour scale indicates the vertical quote of the DUT.

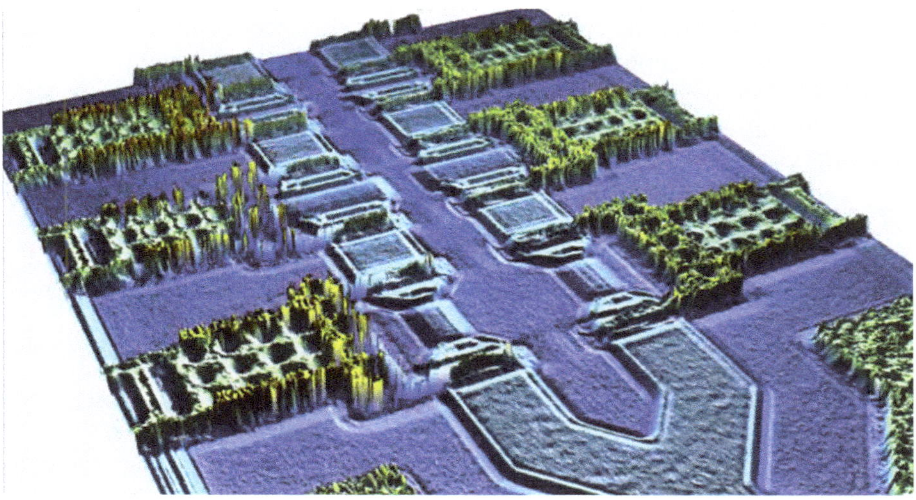

Video A.2. Video animation of the RF-MEMS attenuator in figure A.3 when biased by a waveform similar to that in figure A.2, as reconstructed by 3D stroboscopic light optical dynamic interferometry. Video available at https://iopscience.iop.org/book/978-0-7503-4199-8.

A.3 Measurements based on laser Doppler vibrometer (LDV) microscopy

In this section a different measurement technique, able to capture the dynamic characteristics of RF-MEMS, is exploited. It is called Laser Doppler Vibrometer (LDV) microscopy, and exploits the Doppler effect in order to reconstruct the movement of a certain spot pointed by a laser light source, i.e. if getting closer or farther with respect to the source [9–11]. Differing from dynamic interferometry, LDV microscopy enables one to capture the real-time evolution of the DUT. In other words, there is no need to cycle the sample many times to build its dynamics over one period. Nonetheless, the laser spot monitors just one point over the DUT surface. Therefore, if one wants to observe the dynamic characteristic of a certain sample portion, the biasing waveform has to be cycled several times, in order for the laser to acquire a different spot, period after period, and then reconstruct the evolution of a surface rather than of a single point.

The DUT example selected for LDV microscopy investigation is an RF-MEMS toggle-type switch. This particular geometry employs two suspended electrostatic actuators, anchored in their central point. Each actuator is provided with two distinct counter-electrodes for DC/AC biasing. Depending on whether the inner or outer electrodes are driven (pull-down and pull-up, respectively), the central plate can be displaced downwards or upwards, thus extending the vertical range it can span [12–14]. The layout of the RF-MEMS toggle switch exploited for LDV microscopy dynamic measurement is shown in figure A.4.

In the experimental setup, the pull-down and pull-up electrodes are alternately biased, so that the actuators can tilt both clockwise and counter clockwise around the axis passing through flexible suspensions and anchoring points. The dynamic response captured by means of LDV microscopy is reported in video A.3. In the animation, the surface measured with the laser source (i.e. the moving one) is reported in colours, and the colour scale indicates the vertical displacement extent. Such a surface is superimposed on the grayscale measurement of the DUT. Since the entire RF-MEMS toggle switch is larger than the maximum surface observable with the experimental setup, it was not possible to display the whole structure in the animation.

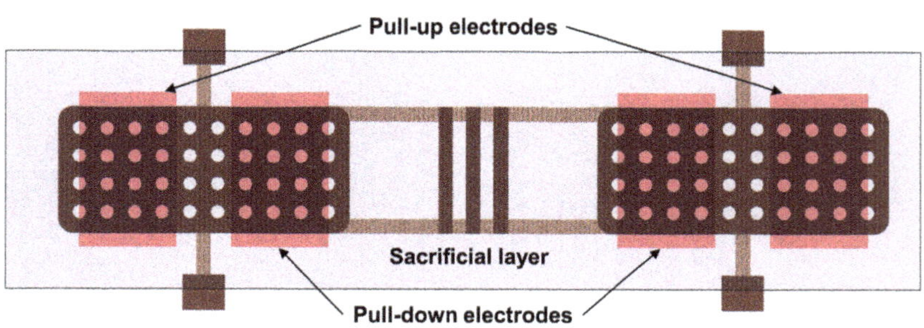

Figure A.4. Layout of the RF-MEMS toggle switch exploited for dynamic characterisation based on LDV microscopy.

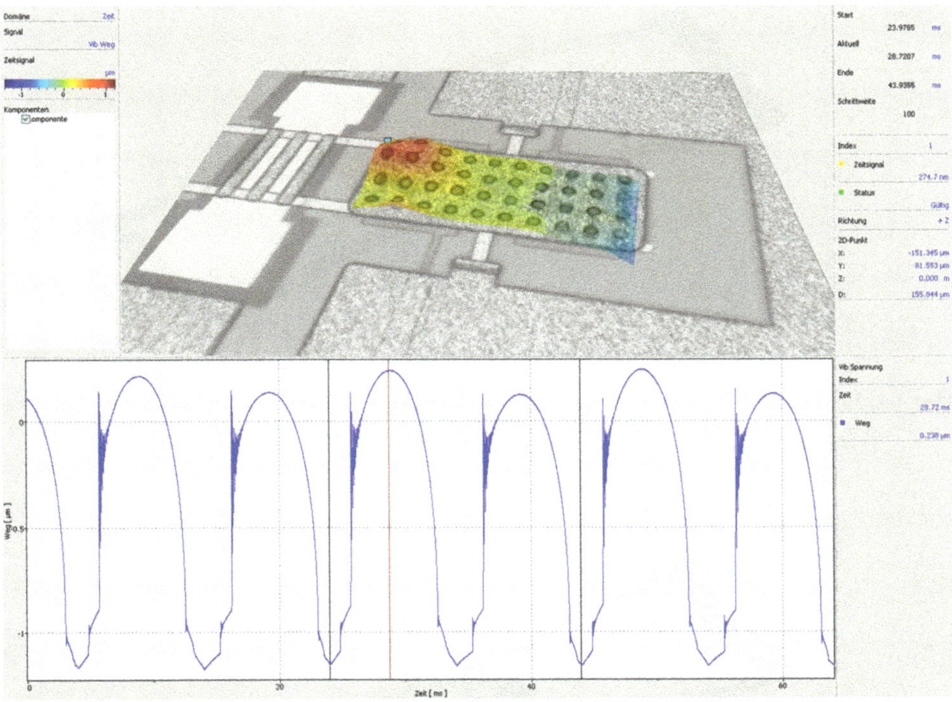

Video A.3. Video animation of the RF-MEMS toggle-switch in figure A.4 when pull-down and pull-up electrodes are alternately biased, as reconstructed by LDV microscopy dynamic measurement. Video available at https://iopscience.iop.org/book/978-0-7503-4199-8.

A.4 Simulated animations

In this section, simulated animations of an RF-MEMS switch are briefly reported. The DUT is the series ohmic micro-relay already discussed in chapter 4 and reported in figures 4.3 and 4.5. Simulations are performed in Ansys Workbench, and the type of analysis is modal eigenfrequency, previously mentioned in section 4.3.2. The predicted resonant modes, from the fundamental (first mode) to the fourth one, are reported in the animations in video A.4, video A.5, video A.6 and video A.7, respectively.

A.5 Conclusion

In this appendix, a few significant examples of moving MEMS were reported. The reviewed data come both from experimental measurements of physical samples, as well as from Finite Element Method (FEM) simulations. Before showing any Devices Under Test (DUTs) animated dynamics, the specific measurement/simulation procedure was described, in order to ease understanding of the presented material. The content of this appendix complements the overall discussion developed in the previous chapters of this work.

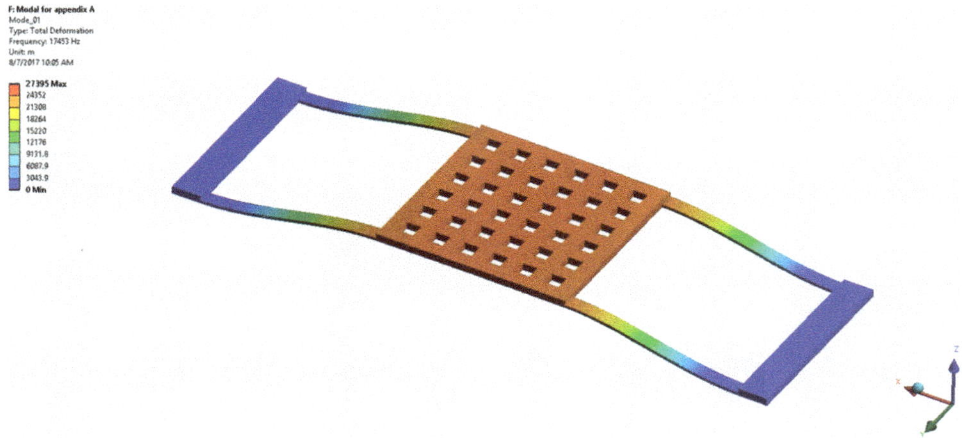

Video A.4. Fundamental resonant mode (1st mode) of the RF-MEMS ohmic switch in figure 4.5. Resonance is predicted by Workbench at 17.4 kHz. Video available at https://iopscience.iop.org/book/978-0-7503-4199-8.

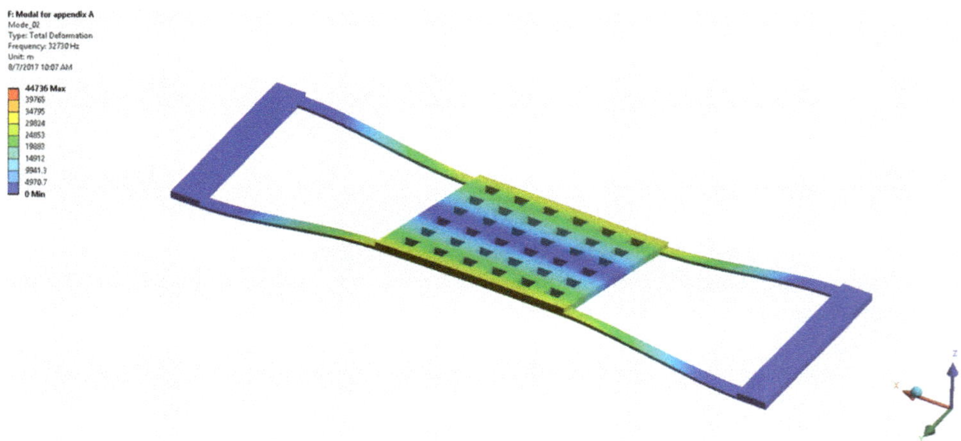

Video A.5. 2nd resonant mode of the RF-MEMS ohmic switch in figure 4.5. Resonance is predicted by Workbench at 32.7 kHz. Video available at https://iopscience.iop.org/book/978-0-7503-4199-8.

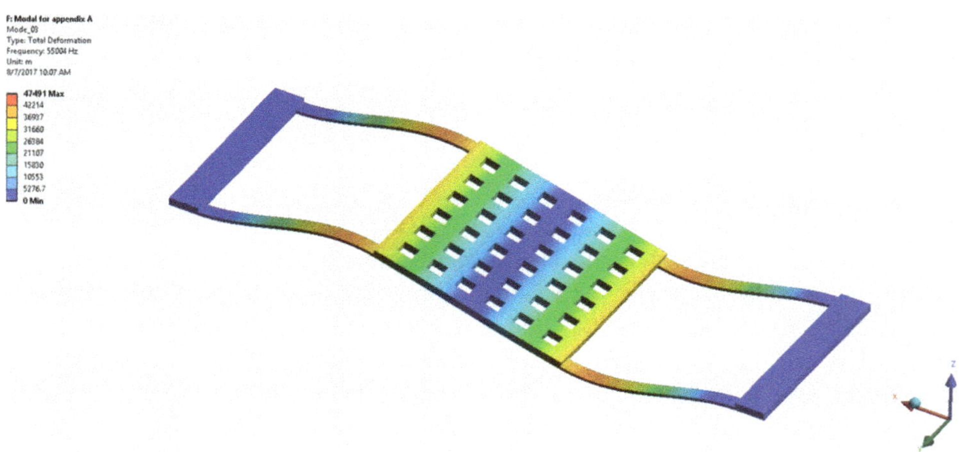

Video A.6. 3rd resonant mode of the RF-MEMS ohmic switch in figure 4.5. Resonance is predicted by Workbench at 55 kHz. Video available at https://iopscience.iop.org/book/978-0-7503-4199-8.

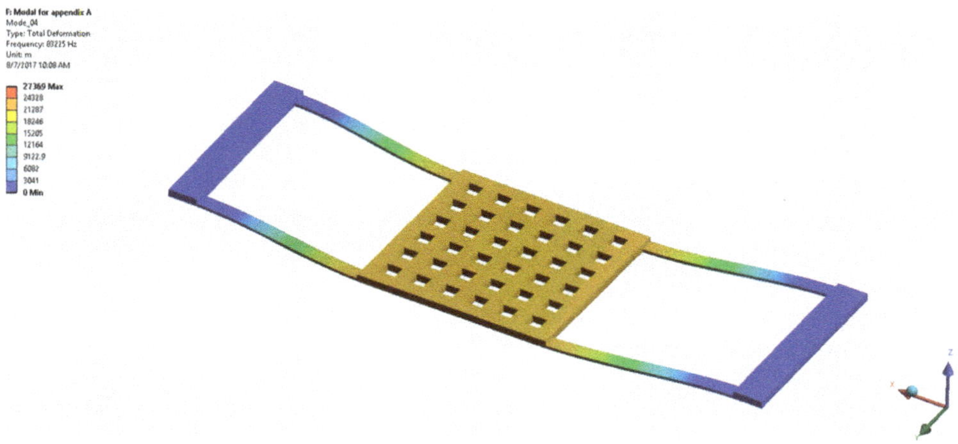

Video A.7. 4th resonant mode of the RF-MEMS ohmic switch in figure 4.5. Resonance is predicted by Workbench at 83.2 kHz. Video available at https://iopscience.iop.org/book/978-0-7503-4199-8.

References

[1] Wang S H, Quan C, Tay C J, Reading I and Fang Z P 2003 Deformation measurement of MEMS components using optical interferometry *Meas. Sci. Technol.* **14** 909–15

[2] Blackshire J L and Sathish S 2002 Characterization of MEMS transducer performance using near-field scanning interferometry *IEEE Trans. Ultrason. Ferroelectr. Freq. Control* **49** 669–74

[3] Kim B, Razavi H A, Degertekin F L and Kurfess T R 2002 Micromachined interferometer for measuring dynamic response of microstructures *Proc. ASME Int. Mech. Eng. Congress and Exposition, MEMS Symp. (New Orleans, LA, Nov. 2002)*

[4] Graebner J E 2000 Optical scanning interferometer for dynamic imaging of high-frequency surface motion *Proc. IEEE Ultrasonics Symp. (San Juan, Oct. 2000)* pp 733–6

[5] Bosseboeuf A and Petitgrand S 2003 Characterization of the static and dynamic behaviour of M(O)EMS by optical techniques: Status and trends *J. Micromech. Microeng.* **13** 23–33

[6] Chen L-C, Huang Y-T and Chang P-B 2006 High-bandwidth dynamic full-field profilometry for nano-scale characterization of MEMS *J. Phys. Conf. Ser.* **48** 1058–62

[7] Iannacci J, Faes A, Mastri F, Masotti D and Rizzoli V 2010 A MEMS-based wide-band multi-state power attenuator for radio frequency and microwave applications *Proc. of TechConnect World, NSTI-Nanotech (Anaheim, CA, June 2010)* pp 328–31

[8] Iannacci J, Faes A, Kuenzig T, Niessner M and Wachutka G 2011 Electromechanical and electromagnetic simulation of RF-MEMS complex networks based on compact modeling approach *Proc. of TechConnect World, NSTI Nanotech 2011 (Boston, MA, June 2011)* pp 591–94

[9] Lawrence E *Optical Measurement Techniques for Dynamic Characterization of MEMS Devices* http://polytec.com/fileadmin/user_uploads/Applications/Micro_Nano_Technology/Documents/OM_TP_MEMS_Whitepaper_2012_07_E.pdf

[10] Kim M G, Jo K, Kwon H S, Jang W, Park Y and Lee J H 2009 Fiber-optic laser Doppler vibrometer to dynamically measure MEMS actuator with in-plane motion *IEEE J. Microelectromech. Syst.* **18** 1365–70

[11] Liu J-W, Huang Q-A, Song J and Tang J-Y 2008 Detection of stiction of suspending structures in MEMS by a laser Doppler vibrometer systems *Proc. Int. Conf. on Solid-State and Integrated-Circuit Technology (Beijing, Oct. 2008)* pp 2472–5

[12] Schauwecker B, Strohm K A, Simon W, Mehner J and Luy J F 2002 Toggle-switch—a new type of RF MEMS switch for power applications *Proc. IEEE MTT-S Int. Microw. Symp. Digest (Seattle, WA, June 2002)* pp 219–22

[13] Solazzi F, Tazzoli A, Farinelli P, Faes A, Mulloni V, Meneghesso G and Margesin B 2010 Active recovering mechanism for high performance RF MEMS redundancy switches *Proc. Europ. Microw. Conf. (Paris, Sept. 2010)* pp 93–6

[14] Schauwecker B, Strohm K M, Mack T, Simon W and Luy J F 2003 Single-pole-double-throw switch based on toggle switch *IET Electron. Lett.* **39** 668–70

IOP Publishing

RF-MEMS Technology for High-Performance Passives (Second Edition)
5G applications and prospects for 6G
Jacopo Iannacci

Appendix B

How to set up, configure and exploit the MEMS behavioural models software library

B.1 Introduction

The target of this appendix is that of gathering all the necessary information to set up and exploit the Micro-ElectroMechanical Systems (MEMS) compact model library discussed in the previous chapter 5, according to a user-guide arrangement. The software library is made available for download together with this book, and is implemented in two different Hardware Description Language (HDL)-based codes, i.e. VerilogA [1, 2] and SpectreHDL [3], within the Cadence Integrated Circuits (ICs) development environment (www.cadence.com), for the Spectre Circuit Simulator [4, 5]. As mentioned before, the tool can be ported in other simulation environments, i.e. alternative to Cadence, as long as HDL codes are supported.

All the most important issues, including the proper configuration of the library itself within the Cadence environment, the properties settings of each model together with the practical use of the tool, e.g. simple examples of MEMS structures and clues about how to face convergence difficulties, will be covered through the following pages.

Moreover, the library being available in two different HDL based codes, different details depending on the particular implementation will be kept and explained separately, both for the VerilogA and the SpectreHDL version.

B.2 How to configure the software library within the Cadence environment

The MEMS compact model library is treated by Cadence as a standard design kit. This means that it has to be copied in a proper location of the file system, which we define here to be **/users/OPUS**. If we now name the two libraries as **MEMS_MODELER_CONTAINER_VerilogA** and **MEMS_MODELER_CONTAINER_SpectreHDL**, the components within each of them will be seen as subdirectories

doi:10.1088/978-0-7503-4199-8ch7

addressed by the following paths: **/users/OPUS/MEMS_MODELER_CONTAINER_VerilogA** and **/users/OPUS/MEMS_MODELER_CONTAINER_SpectreHDL**.

The previous absolute paths must then be added to the **Library Path Editor** [5] within Cadence, in order to make the two libraries visible and usable within the software, together with the other pre-existing libraries and design kits, as reported in figure B.1.

The MEMS compact models deal with mixed-domain magnitudes, i.e. electrical and mechanical. Proper definitions for the electrical and mechanical magnitudes are declared within the libraries. For each of the two parameters, i.e. **abstol** and **blowup**

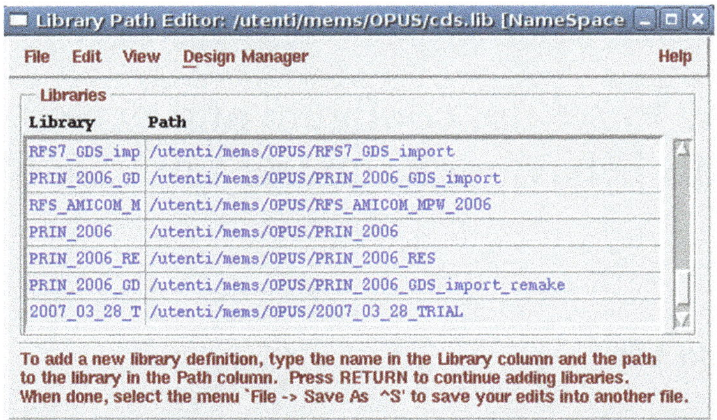

Figure B.1. Library Path Editor window in Cadence. Once the library definitions are added, the new configuration must be saved in the **cds.lib** file.

Table B.1. Summary of all the electrical and mechanical magnitudes' definitions within the VerilogA and SpectreHDL library versions.

Name within Cadence	Physical meaning	Unit	abstol	blowup
disp	Mechanical linear displacement	Meter (m)	1.0e−9	
frc	Mechanical linear force	Newton (N)	1.0e−9	
vel	Mechanical linear velocity	Meter/Second (m s^{-1})	1.0e−3	1.0e+12
acc	Mechanical linear acceleration	Meter/Second^2 (m s^{-2})	1.0e−3	1.0e+12
ang	Mechanical rotation	Radiant (rad)	1.0e−3	
trq	Mechanical torque	Newton*Meter (N*m)	1.0e−3	
avel	Mechanical angular velocity	Radiant/Second (rad s^{-1})	1.0e−3	1.0e+12
aacc	Mechanical angular acceleration	Radiant/Second^2 (rad s^{-2})	1.0e−3	1.0e+12
V	Voltage	Volts (V)	1.0e−6	1.0e+12
I	Current	Ampere (I)	1.0e−6	1.0e+12

[2, 3], threshold values are defined for all the magnitudes that are necessary to the Spectre simulator to discriminate convergence and non-convergence conditions. All the magnitudes with their physical meaning, name within Cadence, unit and parameter(s) value(s) are listed in table B.1.

In the VerilogA library the implementation definitions listed in table B.1 are included in the file **/users/OPUS/MEMS_MODELER_CONTAINER_VerilogA/discipline/discipline.h** and no special settings are required to properly link such a file.

In the SpectreHDL the implementation information listed in table B.1 are defined within the file **/users/OPUS/MEMS_MODELER_CONTAINER_SpectreHDL/quantity.spectre**.

This file must be linked to make it visible during simulations. To do so, a line containing the information on where the file is located must be added to the **cds.lib** file (absolute path: **/users/OPUS/ cds.lib**) before the lines referring to the actual libraries and loaded design kits. In this case, the line has to be as follows: **DEFINE MEMS_MODELER_CONTAINER_SpectreHDL /users/OPUS/MEMS_MODELER_CONTAINER_SpectreHDL**.

B.3 How to instance the available components

Each component within the library is identified by a schematic symbol. This allows one to compose the object to be simulated by adding instances within a Cadence Composer Schematic window, as is usually done with standard libraries of semiconductor devices, and as is shown in figures B.2 and B.3.

Then, the components instanced within the schematic are connected together by means of wires, as reported in figure B.4.

B.4 The flexible straight deformable beam

This section describes how to set all the parameters of the flexible straight beam model, previously discussed in section 5.2.2. Custom configurations of the components are possible, depending on the type of simulation that must be performed. The selection of features is going to be shown below. Any time a change is performed, the component code must be recompiled, in order to allow Spectre to generate the

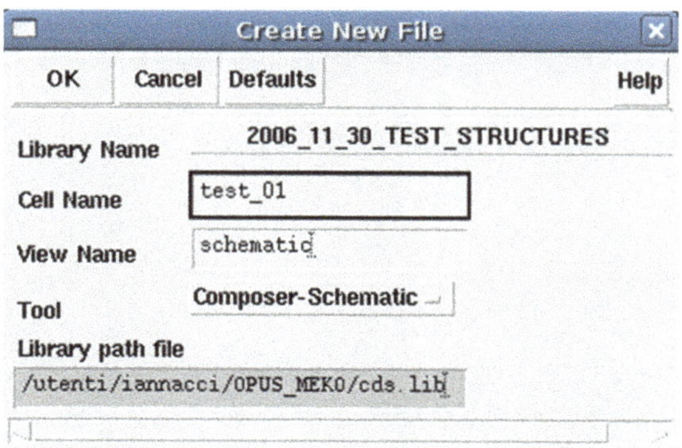

Figure B.2. Definition of a new Composer Schematic file within an existing library.

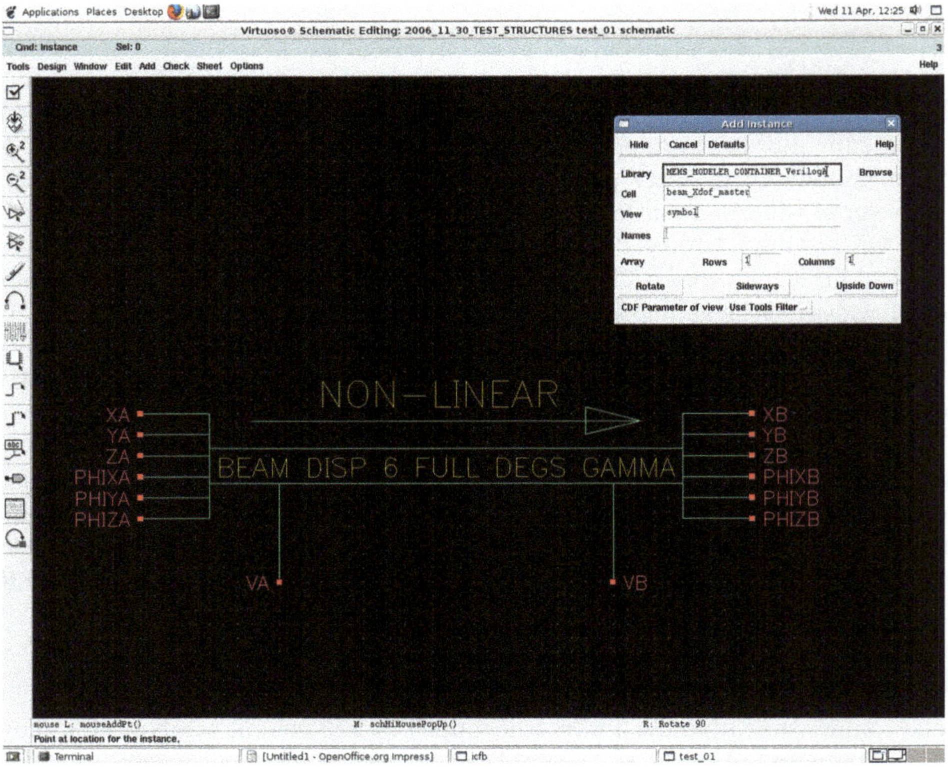

Figure B.3. Instantiation of MEMS elementary components within the schematic window.

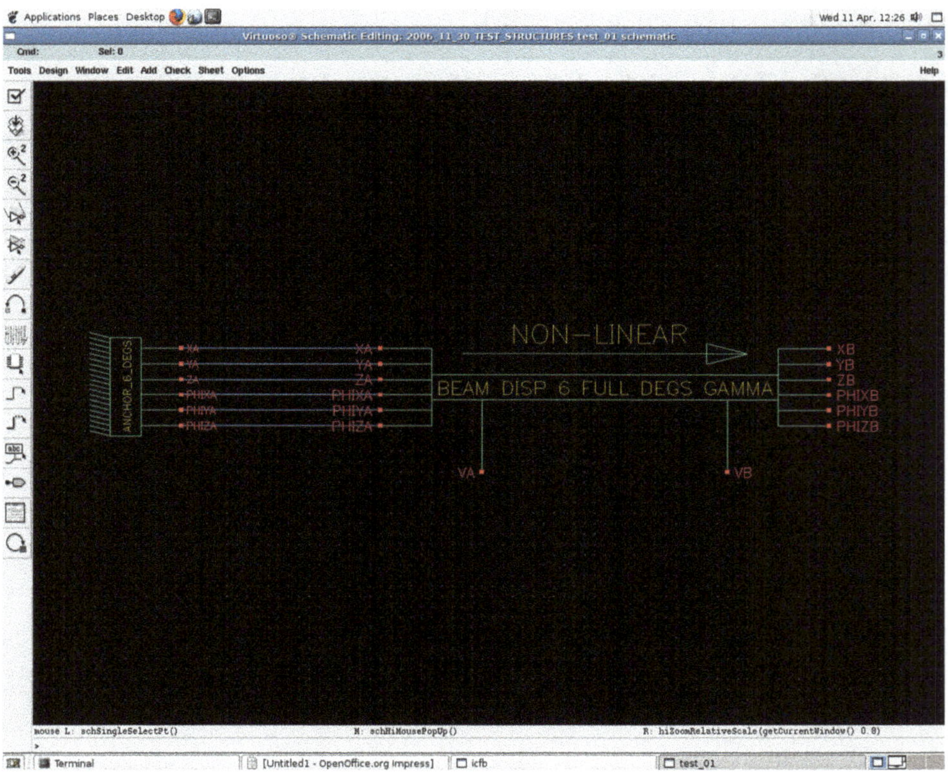

Figure B.4. Connection of single components through wires.

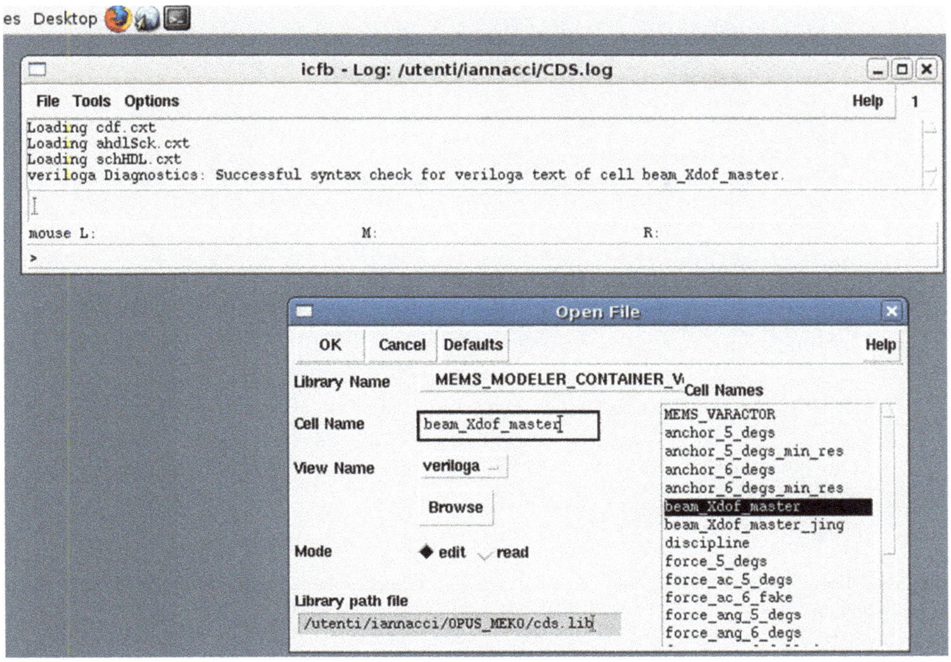

Figure B.5. Opening of the main code file for the 12 DoF flexible beam. In the case of the SpectreHDL implementation the **View Name ahdl** has to be selected.

proper netlist, necessary for the simulation. This means that the main code file must be opened and saved within Cadence. Each component is defined in a subdirectory within the library. The 12 Degrees of Freedom (DoF) straight beam subfolder is named **beam_Xdof_master** for both the library implementations. The main code file for the VerilogA and SpectreHDL is respectively located in the subsequent absolute paths: **/users/OPUS/MEMS_MODELER_CONTAINER_VerilogA/beam_Xdof_master/veriloga/veriloga.va** and **/users/OPUS/MEMS_MODELER_CONTAINER_SpectreHDL/beam_Xdof_master/ahdl/ahdl.def**.

Such files can be opened from the main Cadence window (**File → Open**) selecting the proper **View Name** (**veriloga** or **ahdl**) depending on the implementation. The case of the VerilogA implementation is shown in figure B.5.

In response, a text editor window will open (usually the **vi** Unix Editor). By saving and closing such a file, the compilation will be automatically launched. In the case of **vi** the command to be typed is **:wq + Enter**, i.e. write and quit, as shown in figure B.6.

B.4.1 Features definitions

The list of all the possible features is contained in the file with absolute path: **/users/OPUS/MEMS_MODELER_CONTAINER_VerilogA/beam_Xdof_master/main/main_beam.rgd**, and its content is reported below:

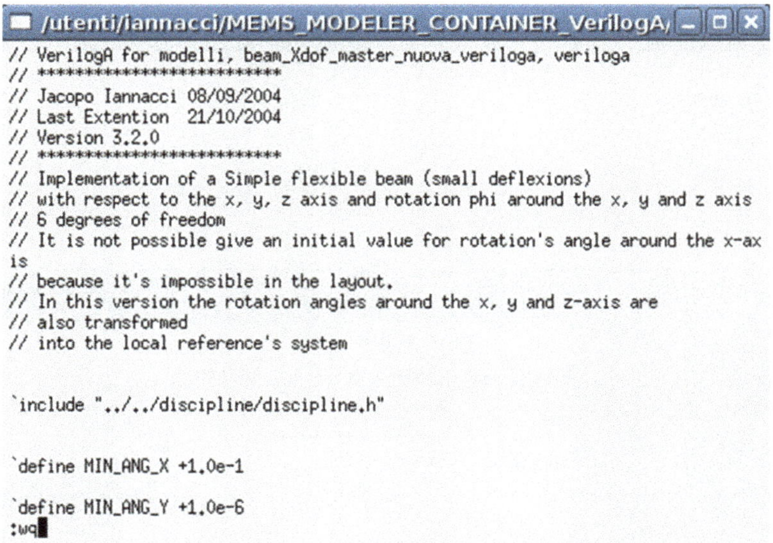

Figure B.6. The command **:wq + Enter**, i.e. write and quit, launches the compilation of the file with the new features selection.

```
/// ********************************************************* ///
/// MODULES INCLUSION. DEFINITION OF A CUSTOM BEAM MODEL FOR
SIMULATIONS. ///
/// ********************************************************* ///
// If parameters values are set to 1, the corresponding
module is included
// in the beam model, otherwise, if the value is equal to 0,
it is not included
//`define Rotation_Matrix_Modified
//`define Non_Linear_Deformations
//`define DC_Simulation_Model_Reduction
`define Shear_Effects
`define Contact_Effects
`define Electrostatic_Force
`define Capacitance
`define Apart_Electrodes
//`define No_Torsion_Around_X
`define Viscous_Damping
/// ********************************************************* ///
///———————— END OF THE MODULES INCLUSION ———————— ///
/// ********************************************************* ///
```

The file contains a series of **macro** variables definition. When a variable is defined, it assumes a **TRUE** value (which means that the corresponding line is not commented, i.e. it does not begin with the comment operator **//**), and the corresponding feature is included in the code compilation. Whereas when a variable is commented it is not defined, and consequently the compiler query returns a **FALSE** value. In this case, the feature defined by that variable will not be included in the model.

For the SpectreHDL implementation, the file is located at the following absolute path: **/users/OPUS/MEMS_MODELER_CONTAINER_SpectreHDL/beam_Xdof_master/main/main_beam.rgd** and its content is reported below:

```
/// ************************************************************ ///
/// MODULES INCLUSION. DEFINITION OF A CUSTOM BEAM MODEL FOR
SIMULATIONS. ///
/// ************************************************************ ///
// If parameters values are set to 1, the corresponding
module is included
// in the beam model, otherwise, if the value is equal to 0,
it is not included
#define Rotation_Matrix_Modified 0
#define Non_Linear_Deformations 0
#define DC_Simulation_Model_Reduction 0
#define Shear_Effects 1
#define Contact_Effects 1
#define Electrostatic_Force 1
#define Capacitance 1
#define Apart_Electrodes 1
#define No_Torsion_Around_X 0
#define Viscous_Damping 1
/// ************************************************************ ///
/// ---------- END OF THE MODULES INCLUSION ----------- ///
/// ************************************************************ ///
```

In this case, a numeric value is associated to each **macro** variable definition, i.e. **0 = FALSE** and **1 = TRUE**. It has to be stressed that, in order to simplify the explanation, the configuration of the main inclusion file is the same for the two VerilogA and SpectreHDL implementations of the 12 DoF beam. The variables are now discussed and explained in detail, one by one, as listed in table B.2.

When the features selection corresponds to the one shown above in the **main_-beam.rgd** file (for both the VerilogA and SpectreHDL implementations), the parameters accessible through the **Edit Object Properties** windows (reachable by selecting the beam schematic and typing **Q**), are reported in figure B.7.

A detailed explanation of each parameter reported in figure B.7 is provided in the table B.3.

NOTE: Misunderstandings can rise among orientation angles values within the mask in figure B.7 and the position of the beam schematic in the Composer windows of Cadence (previously reported in figure B.3). To this end, it has to be stressed that whatever rotation of the schematic does not imply a corresponding rotation of the beam

Table B.2. Summary of all the features that can be modified in the 12 DoF beam model.

Variable name	Corresponding feature
Rotation_Matrix_Modified	Defines a different analytical implementation for the rotation matrix compatible with large deformations [6]. **NOTE**: This feature is not tested and the variable should be left to the **FALSE** value.
Non_Linear_Deformations	Defines an analytical implementation accounting for the material non-linearities when large deformations occur [6]. **NOTE**: This feature is not tested and the variable should be left to the **FALSE** value. In most cases, the beam deformations are small enough to make the linear mechanical model still suitable to reach a good predictive accuracy.
DC_Simulation_Model_ Reduction	When set to **TRUE**, it cuts out the compilation of all the dynamic nodes' definitions (i.e. time derivatives). **NOTE**: This feature makes the model more robust against convergence problems. However, it can be used only for static DC simulations, as all the dynamics due to inertia and viscous damping are ignored.
Shear_Effects	When set to **TRUE**, it includes the shear effects in the linear mechanical model [7]. **NOTE**: It should be always set **TRUE**.
Contact_Effects	When set to **TRUE**, it includes the presence of the substrate underneath the beam, i.e. includes the collapse (pull-in) of the beam onto the substrate. **NOTE**: If set to **FALSE**, none of the features that are following can be included.
Electrostatic_Force	When set to **TRUE**, it includes the electrostatic attractive force between the beam and the substrate, when a bias voltage is applied between them.
Capacitance	When set to **TRUE**, it enables the calculation of the capacitance between the suspended beam and the underlying substrate.
Apart_Electrodes	When set to **TRUE**, enables the definition of separated electrodes along the beam length (refer to [8], their appendix B, for more details).
No_Torsion_Around_X	When set to **TRUE**, it ignores the torsion around the X axis along the beam length, both in the electrostatic magnitudes calculations and in the viscous damping effect model. **NOTE**: It can be reasonably left **FALSE** as torsions around the X axis are usually small, reducing the computational complexity of the model.
Viscous_Damping	When set to **TRUE**, it includes the dynamic effect of the squeeze film damping, which counteracts the beam movements [9].

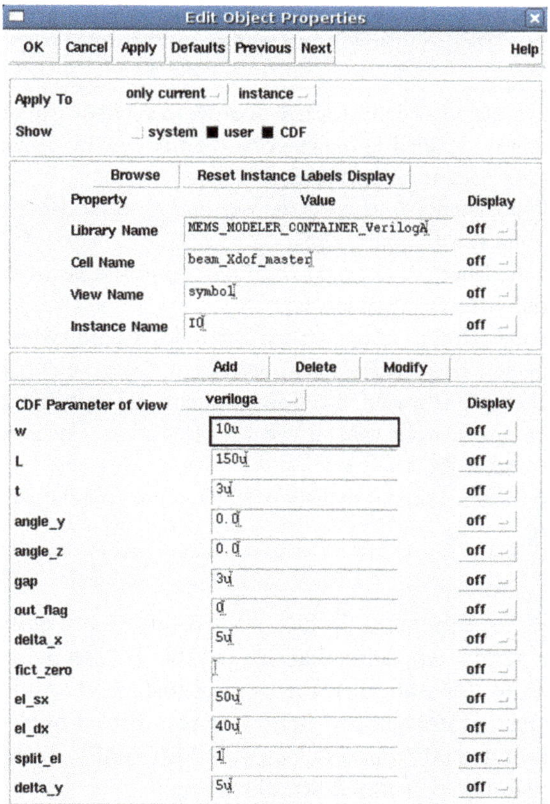

Figure B.7. Object Properties mask of the 12 DoF flexible beam with the features described above in the **main_beam.rgd** file. In the case of SpectreHDL, the **CDF Parameter of view ahdl** must be chosen.

Table B.3. Summary of all the parameters in the 12 DoF beam properties mask.

Parameter	Meaning
w	Beam width along the *Y* axis (see section 5.2.2). In the case, as reported in figure B.7, it is 10 μm.
L	Beam length along the *X* axis.
t	Beam thickness along the *Z* axis.
angle_y	Initial beam orientation angle (in degrees) around the *Y* axis.
angle_z	Initial beam orientation angle (in degrees) around the *Z* axis.
gap	Vertical distance between the suspended beam and the underlying substrate (i.e. air gap).
out_flag	Boolean flag. If set to **1** (**TRUE**), real-time information about the electrostatic force and capacitance values are displayed during the simulation.
delta_x	In the electrostatic magnitudes calculation the beam length is discretised into an integer number of rigid non-parallel plates (see section 5.2.2). The parameter **delta_x** defines the length of each of these elements.

(*Continued*)

Table B.3. (*Continued*)

Parameter	Meaning
fict_zero	Value considered as fictitious zero in order to avoid division by zero. **NOTE**: When nothing is specified in the mask, it is set to 1e-6 by default. This value usually does not need to be modified.
el_sx	Defines the length of the left-hand portion of the electrode (refer to [8], their appendix B, for more details).
el_dx	Defines the length of the right-hand portion of the electrode (refer to [8], their appendix B, for more details).
split_el	Defines the splitting scheme of the electrodes determined by the previous two parameters (refer to [8], their appendix B, for more details). When **split_el = 1**, the electrode is placed in the area not covered both from **el_sx** and **el_dx**. Alternatively, when **split_el = 0**, the electrode is placed on the area covered by **el_sx** and **el_dx**.
delta_y	This parameter is similar to **delta_x**, but it defines the sub-elements length along the *Y* axis.

within the structure to be simulated. In fact, such rotations are defined by means of the **angle_y** and **angle_z** parameters defined above in table B.3. In light of this, maintaining a logic link between the orientation of the symbol instanced within the schematic and the values of the latter parameters can help. For instance, if **angle_z = 90°**, the beam schematic should be rotated 90° counter clockwise. Differently, if **angle_z = −90° = 270°**, the beam schematic should be rotated 90° clockwise.

B.4.2 Materials' characteristics

Materials' properties, likewise several other constants, can be modified to adapt the beam model to the technology specifications. In both the VerilogA and SpectreHDL implementations, a header file exists defining which other files containing the technology and constants definitions must be included. The absolute paths of such header files are: **/users/OPUS/MEMS_MODELER_CONTAINER_VerilogA/qunatities/quantities.tf** and **/users/OPUS/MEMS_MODELER_CONTAINER_SpectreHDL/qunatities/quantities.tf**.

In the following, only the case of the VerilogA implementation is going to be shown, as the approach is the same for the SpectreHDL version. The content of the file **quantities.tf** follows:

```
// **************** //
// —CONSTANTS— //
// **************** //
`include "constants.tf"
// **************** //
// ———END——— //
// **************** //
// ************************** //
```

```
// —PHYSICAL QUANTITIES— //
// *************************** //
`include "air.tf"
// *************************** //
// ————————END———————— //
// *************************** //
// ********************** //
// —TECHNOLOGY FILE— //
// ********************** //
//`include "./tech/irst.tf"
//`include "./tech/poly.tf"
`include "./tech/irst_new.tf"
//`include "./tech/philips_04_06_2004.tf"
// ********************** //
// ———————END——————— //
// ********************** //
```

The first two inclusions are for the **constants.tf** and **air.tf** files, which are located in the same subfolder of **quantities.tf**. They define some constants and the air's characteristic values (e.g. viscosity and permittivity), respectively. For instance, if one is interested in simulating MEMS structures within a nitrogen atmosphere instead of air, the values within **air.tf** are to be modified, or a new file (e.g. **nitrogen.tf**) containing the proper values has to be created within the current subfolder and then included in **quantities.tf** (with the line **'include 'nitrogen.tf'**).

Concerning the technology file with materials information, four of them are declared within **quantities.tf**, despite just one of them being chosen (in this case **irst_new.tf**). The technology files are located in the following path: **/users/OPUS/MEMS_MODELER_CONTAINER_VerilogA/quantities/tech/**. The content of **'irst_new.tf'** is the following:

```
////////////////////////// GOLD VALUES //////////////////////
rho= 0.005;// Structural layer sheet resistance ohm/square.
E=98.5e9;// Young's modulus (Pascal).
dens=19300.0;// Density (kg/m^3).
Poi_nu=0.42;// Poisson's ratio.
thox=100.0e-9;// Oxide thickness above the bottom plate.
erox=5.7;// Oxide dielectric constant.
rough_air_spacer=0;// Extra air gap due to the oxide
roughness.
K_touch=10.0G;// Fictitious elastic parameter when collision
occurs.
min_dist = 10.0e-6;// Minimum distance of two hole on the
plate surface.
Stat_damp = 9.0e-2;// Static damping coefficient.
Dyna_damp = 2.0e-2;// Dynamic damping coefficient.
/////////////////////////////////////////////////////////////
```

Table B.4. Summary of all the technology parameters within the **irst_new.tf** file.

Parameter name	Definition and physical magnitude
rho	Sheet resistance of the gold structural layer (Ω/sq).
E	Young's modulus of gold (Pa).
dens	Density of gold (Kg/m^3).
Poi_nu	Poisson's ratio of gold.
thox	Thickness of the insulating silicon oxide layer deposited above the bottom electrode (m).
erox	Relative dielectric constant of the silicon oxide (F/m).
rough_air_spacer	Effective value of the silicon oxide surface roughness (m). **NOTE**: This parameter defines the residual air gap due to the surface roughness when the pull-in condition is reached.
K_touch	Fictitious elastic constant defining the lower contact plane when pull-in occurs (N/m).
min_dist	Minimum distance between two adjacent holes placed on the beam surface (m). **NOTE**: This feature is not used in the beam model, as it mainly refers to the rigid plate (see next pages).
Stat_damp	Static friction coefficient when collision occurs. **NOTE**: It is not used in the beam model, as it mainly refers to the rigid plate (see next pages).
Dyna_damp	Dynamic friction coefficient when collision occurs ($N{*}s\ m^{-1}$). **NOTE**: It is not used in the beam model, as it mainly refers to the rigid plate (see next pages).

The meaning and explanation of each parameter listed above is summarised in table B.4.

When a new technology has to be defined, it is possible either to modify the values of an existing file (like **irst_new.tf**), or to create a new file within the folder **/users/OPUS/MEMS_MODELER_CONTAINER_VerilogA/quantities/tech/**, and then by including it in the **quantities.tf** header file (the latter solution being more straightforward).

NOTE: Every time a value is modified (in the constants or in the technology part), the model must be recompiled (as reported in figures B.5 and B.6) in order to make the change(s) effective.

B.5 The rigid plate suspended electromechanical transducer

Following the same approach of the previous section, the current section focuses on the rigid plate electromechanical transducer model. All the details concerning the mathematical formulation, the physical features and the working principles of the MEMS transducer were previously reported and discussed in section 5.2.1.

Whatever is concerned with the materials' and general properties' definitions was already covered in the previous section on the flexible straight beam. Therefore, such information will not be duplicated in this section, as they still hold validity for the rigid plate model.

B.5.1 Features definition

The name of the subfolder containing the plate model is **plate_Xdof_master** in both the VerilogA and SpectreHDL implementations. The list of all the possible features is contained in the files with the absolute paths: **/users/OPUS/MEMS_MODELER_CONTAINER_VerilogA/plate_Xdof_master/main/main_-plate.rgd** and **/users/OPUS/MEMS_MODELER_CONTAINER_SpectreHDL/plate_Xdof_master/main/main_plate.rgd**.

The content of the file with all the selectable features, for only the case of VerilogA implementation, is reported in the following:

```
// ——————————————————————————————————————————————————— //
//`define Model_Reduction_To_4_Degrees_Of_Freedom
// ——————————————————————————————————————————————————— //
//`define Rotation_Matrix_Modified
`define Offset_Connectivity_Point
`define Contact_Effects
//`define In_Plane_Contact_Effects_Static
//`define In_Plane_Contact_Effects_Dynamic
// ——————————————————————————————————————————————————— //
`define Straight_Lines
`define Oxide_Electrical_Effects
`define Electrical_Effects_Capacitance
`define Electrical_Effects_Force
`define Boundary_Fringing_Capacitance
`define Boundary_Fringing_Electrostatic
`define Lateral_Fringing_Capacitance_Integral
`define Lateral_Fringing_Electrostatic_Integral
`define Manual_Holes_Placement
`define Holes_Capacitance
`define Holes_Electrostatic
`define Holes_Boundary_Fringing_Capacitance
`define Holes_Boundary_Fringing_Electrostatic
`define Holes_Lateral_Fringing_Capacitance_Integral
`define Holes_Lateral_Fringing_Electrostatic_Integral
//`define Electrostatic_Torques
// ——————————————————————————————————————————————————— //
`define Viscous_Damping_Effects
//`define Simplified_Viscous_Damping_Model
//`define Viscous_Damping_Torques_Model
//`define Simplified_Viscous_Damping_Torques_Model
//`define Discrete_Damping_Model
`define Mean_Free_Path_Model
// ——————————————————————————————————————————————————— //
//`define DC_Simulation_Model_Reduction
```

```
`define Torque_Nodes_Static_Damping_Inclusion
`define Torque_Nodes_Dynamic_Damping_Inclusion
`define Center_Of_Mass_Dynamic_Nodes_For_X
`define Center_Of_Mass_Dynamic_Nodes_For_Y
`define Center_Of_Mass_Dynamic_Nodes_For_Z
`define Center_Of_Mass_Dynamic_Nodes_For_Phi_X
`define Center_Of_Mass_Dynamic_Nodes_For_Phi_Y
`define Center_Of_Mass_Dynamic_Nodes_For_Phi_Z
// ———————————————————————————————————————————————— //
```

The meaning and description of all the rigid plate features reported above is listed in table B.5.

The properties mask associated with the rigid plate Cadence symbol is shown in figure B.8. The meaning and description of all the parameters is listed in table B.6 and they correspond to the configuration of the **main_plate.rgd** file shown in the previous pages.

B.6 Complementary components

In this section, all the complementary components included in the library, and necessary to impose mechanical boundary conditions and provide stimuli, are briefly described. Such components, along with the symbols in Cadence, were already introduced and shown in section 5.2.3, to which the reader can refer for more details. The mentioned elements are going to be reviewed here from a more practical and operational point of view. The complementary components can be grouped in two categories, i.e. boundary conditions and stimulus sources, as reported in the following.

B.6.1 Anchoring point—boundary condition component

This device imposes a mechanical constraint to all the linear and angular DoFs, realising an anchoring point. Each of its nodes does not allow any displacement or rotation to the device connected to it. The 6 DoF anchoring element subfolder name within both libraries is **anchor_6_degs**, and the Cadence schematic symbol is shown in figure 5.22 in the previous chapter 5.

B.6.2 Anchoring point with minimum resistance—boundary condition component

When the anchoring point has to be connected to a rigid body (i.e. suspended rigid plate), it is similar to shorting together two voltage generators in the electrical domain. This case cannot be managed by the Spectre simulator and an error occurs. To avoid this condition, an on-purpose anchoring element is available within both libraries. It is called **anchor_6_degs_min_res** and the Cadence schematic symbol is shown in figure 5.23, in the previous chapter 5. The element at stake features a minimum resistance value defined within the model, preventing from the error condition mentioned above. The parameter **r_min** is accessible through the

Table B.5. Description of all the possible features related to the rigid plate suspended electromechanical transducer model.

Parameter name	Meaning
——————**Mechanical model**——————	#########################
Model_Reduction_To_4_Degrees_Of_Freedom	When set to **TRUE**, it reduces the plate mechanical model to 4 Degrees of Freedom (DoFs) (three linear displacements and a rotation allowed only around the vertical *Z* axis). This means that the plate is always parallel to the substrate. When set to **FALSE**, the plate admits any position in space (6 DoFs, i.e. three linear displacements and rotations around the three axes *X*, *Y*, *Z*).
——————**Plate features**——————	#########################
Rotation_Matrix_Modified	It defines a different analytical implementation for the rotation matrix compatible with large deformations [6]. **NOTE**: This feature is not tested and the variable should be always set to **FALSE**.
Offset_Connectivity_Point	When set to **TRUE**, it allows displacing the four interconnection points between the plate and the external world with respect to the four plate corners North-East (NE), South-East (SE), South-West (SW) and North-West (NW) (refer to [8], their appendix A, for more details).
Contact_Effects	When set to **TRUE**, it accounts for the presence of the substrate underneath the plate, i.e. it makes possible the collapse of the plate on the substrate (pull-in). **NOTE**: When set to **FALSE**, none of the following features can be included.
In_Plane_Contact_Effects_Static	When set to **TRUE**, it includes the *XY* in-plane static friction between the plate and the substrate when physical collision occurs. **NOTE**: The accuracy of this feature is not verified and it might lead to convergence issues. Moreover, such an effect is usually not dominant in standard RF-MEMS switch topologies. Therefore, this feature can be kept **FALSE**.
In_Plane_Contact_Effects_Dynamic	When set to **TRUE**, it includes the *XY* in-plane dynamic (sliding) friction between the plate

(Continued)

Table B.5. (*Continued*)

Parameter name	**Meaning**
	and the substrate when physical collision occurs. **NOTE**: The accuracy of this feature is not verified and it might lead to convergence issues. Moreover, such an effect is usually not dominant in standard RF-MEMS switch topologies. Therefore, this feature can be kept **FALSE**.
————Electrostatic effects————	**#########################**
Straight_Lines	When set to **TRUE**, the electric field lines between the plate and the substrate are approximated to straight instead of curved lines. **NOTE**: When the plate mechanical model is reduced to 4 DoFs, the field lines are always straight. Instead, when dealing with the 6 DoF model, it might happen very rarely that the curved lines model causes convergence issues. In this case, it can be set to **TRUE**. The error increase in electrostatic force and capacitance calculations due to straight line approximation is generally rather small.
Oxide_Electrical_Effects	When set to **TRUE**, the electrical effects due to the oxide layer laying on the lower substrate are accounted. The air gap and the oxide layer are treated as two series capacitances (refer to [8], their appendix A, for more details).
Electrical_Effects_Capacitance	When set to **TRUE**, it includes the calculation of the capacitance between the suspended plate and the substrate.
Electrical_Effects_Force	When set to **TRUE**, it includes the calculation of the attractive electrostatic force between the suspended plate and the substrate.
Boundary_Fringing_Capacitance	When set to **TRUE**, the fringing capacitance due to the distortion of the field lines nearby the plate's four edges is included in the capacitance calculation (refer to [8], their appendix A, for more details).
Boundary_Fringing_Electrostatic	When set to **TRUE**, the fringing electrostatic force due to the distortion of the field lines nearby the plate's four edges is included in the attractive force calculation (refer to [8], their appendix A, for more details).

Lateral_Fringing_Capacitance_Integral	When set to **TRUE**, the fringing capacitance due to the field lines starting from the four plate boundary vertical sidewalls is included in the capacitance calculation (refer to [8], their appendix A, for more details). **NOTE**: When the plate is parallel to the substrate, this contribution is null.
Lateral_Fringing_Electrostatic_Integral	When set to **TRUE**, the fringing electrostatic force due to the field lines starting from the four plate boundary vertical sidewalls is included in the attractive force calculation (refer to [8], their appendix A, for more details). **NOTE**: When the plate is parallel to the substrate, this contribution is null.
Manual_Holes_Placement	When set to **TRUE**, the distance between the most peripheral holes' edges and the plate boundaries can be defined by the user (refer to figure 13 in chapter 5). Otherwise, the hole-to-edges clearance is automatically defined.
Holes_Capacitance	When set to **TRUE**, it includes the presence of holes on the plate surface and calculates the actual capacitance value affected by the holes. **NOTE**: When set to **TRUE**, the presence of holes is automatically accounted for in the mechanical plate model (inertia) and in the viscous damping model, as well.
Holes_Electrostatic	When set to **TRUE**, it includes the presence of holes on the plate surface and calculates the actual electrostatic force value affected by the holes. **NOTE**: When set to **TRUE**, the presence of holes is automatically accounted in the mechanical plate model (inertia) and in the viscous damping model as well.
Holes_Boundary_Fringing_Capacitance	When set to **TRUE**, the fringing capacitance due to the distortion of the field lines nearby the holes' edges is included in the capacitance calculation (refer to [8], their appendix A, for more details).
Holes_Boundary_Fringing_Electrostatic	When set to **TRUE**, the fringing electrostatic force due to the distortion of the field lines nearby the holes' edges is included in the attractive force calculation (refer to [8], their appendix A, for more details).

(*Continued*)

Table B.5. (*Continued*)

Parameter name	Meaning
Holes_Lateral_Fringing_Capacitance_Integral	When set to **TRUE**, the fringing capacitance due to the field lines starting from the holes' vertical sidewalls is included in the capacitance calculation (refer to [8], their appendix A, for more details). **NOTE**: When the plate is parallel to the substrate, this contribution is null.
Holes_Lateral_Fringing_Electrostatic_Integral	When set to **TRUE**, the fringing electrostatic force due to the field lines starting from the holes' vertical sidewalls is included in the attractive force calculation (refer to [8], their appendix A, for more details). **NOTE**: When the plate is parallel to the substrate, this contribution is null.
Electrostatic_Torques	When set to **TRUE**, the torque due to the attractive electrostatic force (occurring when the plate is not parallel to the substrate) is calculated (refer to [8], their appendix A, for more details). **NOTE**: The electrostatic torque contribution is usually very small and can be excluded from the model without significant loss in predictive accuracy.
————Viscous damping effect————	**########################**
Viscous_Damping_Effects	When set to **TRUE**, it includes the viscous damping effect in the rigid plate model. **NOTE**: When **Holes_Capacitance** or **Holes_Electrostatic** are set to **TRUE**, the viscous damping model accounts for the holes. Otherwise, the plate is considered without openings on its surface.
Simplified_Viscous_Damping_Model	This parameter addresses a simplified mathematical implementation of the viscous damping model, in case holes are not present on the plate surface. **NOTE**: When the plate is set without holes, it is better to select the simplified model as it allows considerable simulation time saving, without introducing significant loss in predictive accuracy.
Viscous_Damping_Torques_Model	When set to **TRUE**, it includes the calculation of the torque due to the viscous damping effect. **NOTE**: This contribution is not dominant, therefore it can be reasonably neglected in the model definition.

Simplified_Viscous_Damping_Torques_Model	This parameter considers a simplified mathematical implementation of the viscous damping torque calculation. **NOTE**: This contribution is not dominant, therefore it can be reasonably neglected in the model definition.
Discrete_Damping_Model	This parameter considers a discrete implementation of the viscous damping effect, when there are no holes on the plate surface. **NOTE**: Since **Simplified_Viscous_Damping_Model** is a better choice when there are no holes, this feature can be always kept **FALSE**.
Mean_Free_Path_Model	When set to **TRUE**, it includes the calculation of an effective fluid viscosity (according to the model proposed by Veijola in [10]). **NOTE**: Since the so-called slip-flow becomes dominant when the air gap is very small (i.e. same order of magnitude of the gas molecules' mean free path), this feature must be included every time the pull-in is expected to be reached.
————**Convergence aid**————	########################
DC_Simulation_Model_Reduction	When set to **TRUE**, it cuts out of the compilation all the dynamic nodes' definitions (i.e. time derivatives). **NOTE**: It makes the model more robust against convergence issues, even though it can be used only for static DC simulations, as all the dynamics due to inertia and viscous damping are ignored.
Torque_Nodes_Static_Damping_Inclusion	This parameter includes internal nodes for the static XY in-plane friction. It must be set to **TRUE** when **In_Plane_Contact_Effects_Static** is **TRUE**.
Torque_Nodes_Dynamic_Damping_Inclusion	This parameter includes internal nodes for the dynamic XY in-plane friction. It must be set to **TRUE** when **In_Plane_Contact_Effects_Dynamic** is **TRUE**.
Center_Of_Mass_Dynamic_Nodes_For_X	When set to **FALSE**, it does not include all the dynamic nodes (velocity and acceleration) for the X axis linear displacement DoF. **NOTE**: This parameter may be used to increase the convergence robustness, despite

(*Continued*)

Table B.5. (*Continued*)

Parameter name	Meaning
	it cutting off all the dynamic effects (inertia and viscous damping) related to the *X* axis linear displacement DoF.
Center_Of_Mass_Dynamic_Nodes_For_Y	When set to **FALSE**, it does not include all the dynamic nodes (velocity and acceleration) for the *Y* axis linear displacement DoF. **NOTE**: This parameter may be used to increase the convergence robustness, despite it cutting off all the dynamic effects (inertia and viscous damping) related to the *Y* axis linear displacement DoF.
Center_Of_Mass_Dynamic_Nodes_For_Z	When set to **FALSE**, it does not include all the dynamic nodes (velocity and acceleration) for the *Z* axis linear displacement DoF. **NOTE**: This parameter may be used to increase the convergence robustness, despite it cutting off all the dynamic effects (inertia and viscous damping) related to the *Z* axis linear displacement DoF.
Center_Of_Mass_Dynamic_Nodes_For_Phi_X	When set to **FALSE**, it does not include all the dynamic nodes (velocity and acceleration) for the rotation around the *X* axis DoF. **NOTE**: This parameter may be used to increase the convergence robustness, despite it cutting off all the dynamic effects (inertia and viscous damping) related to the rotation around the *X* axis DoF.
Center_Of_Mass_Dynamic_Nodes_For_Phi_Y	When set to **FALSE**, it does not include all the dynamic nodes (velocity and acceleration) for the rotation around the *Y* axis DoF. **NOTE**: This parameter may be used to increase the convergence robustness, despite it cutting off all the dynamic effects (inertia and viscous damping) related to the rotation around the *Y* axis DoF.
Center_Of_Mass_Dynamic_Nodes_For_Phi_Z	When set to **FALSE**, it does not include all the dynamic nodes (velocity and acceleration) for the rotation around the *Z* axis DoF. **NOTE**: This parameter may be used to increase the convergence robustness, despite it cutting off all the dynamic effects (inertia and viscous damping) related to the rotation around the *Z* axis DoF.

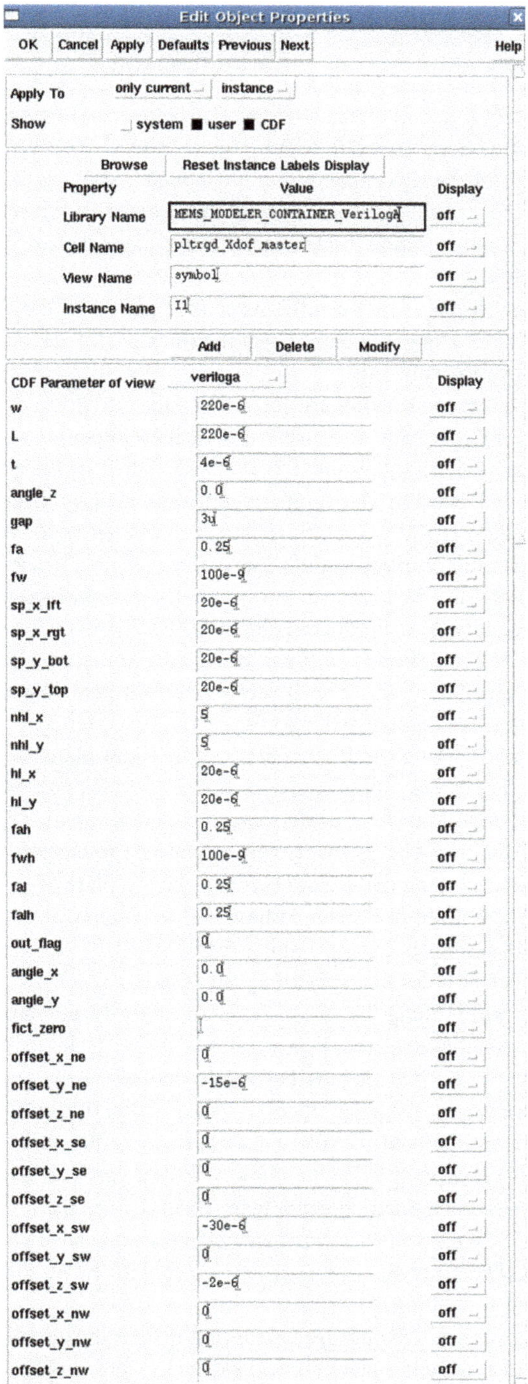

Figure B.8. Cadence window highlighting all the rigid plate properties that can be set by the user.

Table B.6. Description of all the parameters available in the plate schematic properties mask within Cadence.

Parameter	Meaning
w	Width of the rigid plate along the *X* axis (refer to section 5.2.1 for more details). In this case, **w = 220 μm** (see figure B.8).
L	Length of the rigid plate along the *Y* axis (see figure B.8).
t	Thickness of the rigid plate along the *Z* axis (see figure B.8).
angle_z	Plate initial rotation angle around the *Z* axis (expressed in degrees) (see figure B.8).
gap	Vertical distance between the suspended plate and the underlying substrate. **NOTE**: The **gap** includes also the thickness of the insulating layer laying above the substrate.
fa	The parameter **fringing arc (fa)** defines the curvature of the fringing field lines nearby the plate edges (refer to [8], their appendix A, for more details). It varies within the interval [0,1]. The larger it is, the more the field lines are curved. **NOTE**: The **fa** parameter was tuned by means of Finite Element Method (FEM) simulations and it can be set to 0.25 [8, 11].
fw	The parameter **fringing width (fw)** defines the area nearby the plate edges within which the fringing effect is supposed to be confined (refer to [8], their appendix A, for more details). **NOTE**: The **fw** parameter was tuned with FEM simulations and it can be set to 100 nm [8, 11].
sp_x_lft	Clearance (along the *X* axis) between the left plate edge and the left edge of the first holes column (refer to figure 13 in chapter 5).
sp_x_rgt	Clearance (along the *X* axis) between the right plate edge and the right edge of the last holes column (refer to figure 13 in chapter 5).
sp_y_bot	Clearance (along the *Y* axis) between the bottom plate edge and the bottom edge of the lowest holes row (refer to figure 13 in chapter 5).
sp_y_top	Clearance (along the *Y* axis) between the top plate edge and the top edge of the highest holes row (refer to figure 13 in chapter 5).
nhl_x	Number of holes (integer) along the *X* axis.
nhl_y	Number of holes (integer) along the *Y* axis.
hl_x	Holes width along the *X* axis (refer to figure 13 in chapter 5).
hl_y	Holes length along the *Y* axis (refer to figure 13 in chapter 5).
fah	The parameter **fringing arc holes (fah)** defines the curvature of the fringing field lines nearby the holes' edges. It varies within the interval]0,1]. The larger it is, the more curved the field lines are (refer to **fa**). **NOTE**: The **fah** parameter was tuned with FEM simulations and in can be kept to 0.25 [8, 11].
fwh	The parameter **fringing width holes (fwh)** defines the area nearby the holes' edges, within which the fringing effect is supposed to be confined (refer to **fw**). **NOTE**: The **fwh** parameter was tuned with FEM simulations and it can be kept to 100 nm [8, 11].
fal	The parameter **fringing arc lateral (fal)** defines the curvature of the fringing field lines nearby the plate vertical sidewalls (refer to [8], their appendix A, for more details). It swings within the interval]0,1]. The larger it is, the more curved the field lines are. **NOTE**: The **fal** parameter was tuned with FEM simulations and it can be kept to 0.25 [8, 11].

falh	The parameter **fringing arc lateral holes (falh)** defines the curvature of the fringing field lines nearby the vertical sidewalls of holes. It varies within the interval]0,1]. The larger it is, the more curved the field lines are (refer to **fal**). **NOTE**: The **falh** parameter was tuned with FEM simulations and it can be kept to 0.25 [8, 11].
out_flag	Boolean flag. If set to **1**, real-time information about the electrostatic force and capacitance values are displayed during simulations.
angle_x	Plate initial rotation angle around the *X* axis (expressed in degrees) (refer to the images in section 5.2.1 for more details).
angle_y	Plate initial rotation angle around the *Y* axis (expressed in degrees) (refer to the images in section 5.2.1 for more details).
fict_zero	Value considered as fictitious zero in order to avoid division by zero. **NOTE**: When nothing is specified in the mask, it is set to 1e-6 by default. This value usually does not need to be modified.
offset_x_ne	Defines the displacement of the North-East (NE) node from the top-right plate corner along the *X* axis. The parameter is not available when **Offset_Connectivity_Point** is **FALSE** (see table B.5). **NOTE**: If this parameter has a non-zero value, the following one (i.e. **offset_y_ne**) must be zero.
offset_y_ne	Defines the displacement of the North-East (NE) node from the top-right plate corner along the *Y* axis. The parameter is not available when **Offset_Connectivity_Point** is **FALSE** (see table B.5). **NOTE**: If this parameter has a non-zero value, the previous one (i.e. **offset_x_ne**) must be zero.
offset_z_ne	Defines the displacement of the North-East (NE) node from the top-right plate corner along the *Z* axis. The parameter is not available when **Offset_Connectivity_Point** is **FALSE** (see table B.5).
offset_x_se	Defines the displacement of the South-East (SE) node from the bottom-right plate corner along the *X* axis. The parameter is not available when **Offset_Connectivity_Point** is **FALSE** (see table B.5). **NOTE**: If this parameter has a non-zero value, the following one (i.e. **offset_y_se**) must be zero.
offset_y_se	Defines the displacement of the South-East (SE) node from the bottom-right plate corner along the *Y* axis. The parameter is not available when **Offset_Connectivity_Point** is **FALSE** (see table B.5). **NOTE**: If this parameter has a non-zero value, the previous one (i.e. **offset_x_se**) must be zero.
offset_z_se	Defines the displacement of the South-East (SE) node from the bottom-right plate corner along the *Z* axis. The parameter is not available when **Offset_Connectivity_Point** is **FALSE** (see table B.5).
offset_x_sw	Defines the displacement of the South-West (SW) node from the bottom-left plate corner along the *X* axis. The parameter is not available when **Offset_Connectivity_Point** is **FALSE** (see table B.5). **NOTE**: If this parameter has a non-zero value, the following one (i.e. **offset_y_sw**) must be zero.
offset_y_sw	Defines the displacement of the South-West (SW) node from the bottom-left plate corner along the *Y* axis. The parameter is not available when **Offset_Connectivity_Point** is **FALSE** (see table B.5). **NOTE**: If this parameter has a non-zero value, the previous one (i.e. **offset_x_sw**) must be zero.

(*Continued*)

Table B.6. (*Continued*)

Parameter	Meaning
offset_z_sw	Defines the displacement of the South-West (SW) node from the bottom-left plate corner along the *Z* axis. The parameter is not available when **Offset_Connectivity_Point** is **FALSE** (see table B.5).
offset_x_nw	Defines the displacement of the North-West (NW) node from the top-left plate corner along the *X* axis. The parameter is not available when **Offset_Connectivity_Point** is **FALSE** (see table B.5). **NOTE**: If this parameter has a non-zero value, the following one (i.e. **offset_y_nw**) must be zero.
offset_y_nw	Defines the displacement of the North-West (NW) node from the top-left plate corner along the *Y* axis. The parameter is not available when **Offset_Connectivity_Point** is **FALSE** (see table B.5). **NOTE**: If this parameter has a non-zero value, the previous one (i.e. **offset_x_nw**) must be zero.
offset_z_nw	Defines the displacement of the North-West (NW) node from the top-left plate corner along the *Z* axis. The parameter is not available when **Offset_Connectivity_Point** is **FALSE** (see table B.5).

properties window within the schematic editor. When nothing is specified (i.e. the field is left blank), its default value is set to 1e-6 Ω. If necessary, this value can be modified.

B.6.3 Minimum resistance block—boundary condition component

Recalling what was just discussed, a similar short circuit condition is detected by Spectre when two rigid components (i.e. rigid plates) are connected together within the same schematic. In order to avoid this issue, an element defining a minimum resistance between the input and output nodes is available within the libraries. Its subfolder name is **min_res_l_a_6_pin** and the Cadence schematic symbol is shown in figure 5.24 in the previous chapter 5. Such an element must be instanced between the two nodes of the rigid bodies to be connected together. The parameter **r_min** (minimum resistance) is accessible in the properties window within the schematic editor. When nothing is specified (i.e. the field is left blank), its default value is 1e-6 Ω. If necessary it can be modified. Depending on the number of nodes to be connected together, the minimum resistance element is available with different combinations other than the one in figure 5.24. The additional minimum resistance elements are available with one, two or three linear node(s), named, respectively, as **min_res_lin_1_pin**, **min_res_lin_2_pin** and **min_res_lin_3_pin**, as well as with one, two or three angular node(s), named, respectively, as **min_res_ang_1_pin**, **min_re-s_ang_2_pin** and **min_res_ang_3_pin**.

B.6.4 Unconnected nodes termination—boundary condition component

Given the schematic implementation corresponding to a certain RF-MEMS device to be simulated, it might happen that one or more nodes remain floating (i.e. unconnected). This does not represent an actual issue impairing the analysis. However, in such a circumstance Spectre generates a warning message and shorts unconnected nodes to ground with a minimum conductance. In order to avoid the warning messages, a module defining an unconnected termination to which the floating nodes must be wired is available in the library. Its name within both libraries is **unc_l_a_6_pin** and the Cadence schematic symbol is shown in figure 5.25 in the previous chapter 5. Depending on the number of nodes to be terminated, the floating element is available with different combinations other than the one in figure 5.25. Such additional implementations of the unconnected nodes element are available with one, two or three linear node(s), named, respectively, as **unc_lin_1_pin**, **unc_lin_2_pin** and **unc_lin_3_pin**, as well as with one, two or three angular node (s), named, respectively, as **unc_ang_1_pin**, **unc_ang_2_pin** and **unc_ang_3_pin**.

B.6.5 Force generator—stimulus source component

This module generates a force to be applied to a certain MEMS structure. The name of the subfolder in both libraries is **force_ang_6_full_degs**. The Cadence schematic symbol is shown in figure 5.26 in the previous chapter 5. Within the Cadence schematic properties interface, four parameters are available for this component, as reported in the following table B.7.

B.6.6 Force generator (AC simulations)—stimulus source component

This module generates a force to be applied to a certain MEMS structure for small signal (AC) simulations within Spectre. Its name within both libraries is **force_-ac_5_degs** and it allows only 5 DoFs (i.e. torques around the X axis cannot be imposed). Four parameters are available within the Cadence schematic properties window, as described in table B.8.

B.6.7 Force generator (pulse)—stimulus source component

The module here at stake generates a force pulse to be applied to a certain MEMS within Spectre. Its name within both libraries is **force_ang_pulse_6_degs**. Seven parameters are available within the Cadence schematic properties window, as described in table B.9.

Table B.7. Description of all the available parameters for the **force_ang_6_full_degs** model.

Parameter	Meaning
angle_y	This parameter defines the rotation (in degrees) of the force and torque (refer to the following parameters) around the Y axis.
angle_z	This parameter defines the rotation (in degrees) of the force and torque (refer to the following parameters) around the Z axis.
torque_x	This parameter defines the torque modulus (N*m) around the X axis.
force_mod	This parameter defines the force modulus (N) to be applied.

Table B.8. Description of all the parameters available for the **force_ac_5_degs** model.

Parameter	Meaning
forza	This parameter defines the force modulus (N) to be applied.
phase	This parameter defines the phase rotation to be applied to the force.
angle_y	This parameter defines the rotation (in degrees) of the force around the *Y* axis.
angle_z	This parameter defines the rotation (in degrees) of the force around the *Z* axis.

Table B.9. Description of all the parameters available for the **force_ang_pulse_6_degs** model.

Parameter	Meaning
angle_y	This parameter defines the rotation (in degrees) of the force around the *Y* axis.
angle_z	This parameter defines the rotation (in degrees) of the force around the *Z* axis.
t_delay	This parameter defines the delay (seconds) of the pulse.
t_rise	This parameter defines the rise time (seconds) of the OFF to ON edge.
t_on	This parameter defines the duration of the pulse ON value (seconds).
t_fall	This parameter defines the fall time (seconds) of the ON to OFF edge.
force_mod	This parameter defines the force modulus (N) to be applied during the pulse ON time. The pulse low (OFF) force value is always zero.

Table B.10. Description of all the parameters available for the **displacements_ang_6_full_degs** model.

Parameter	Meaning
displacement	This parameter defines the displacement (m) to be applied to the MEMS structure.
rot_x	This parameter defines the angular rotation (degrees) to be applied to the MEMS structure around the *X* axis.
angle_y	This parameter defines the rotation (in degrees) of the linear and angular displacement around the *Y* axis.
angle_z	This parameter defines the rotation (in degrees) of the linear and angular displacement around the *Z* axis.

B.6.8 Displacement—stimulus source component

This module generates a displacement to be applied to a certain MEMS structure. The name of the model available in the libraries is **displacements_ang_6_full_degs**, while the Cadence schematic symbol is shown in figure 5.27 in the previous chapter 5. Four parameters are available in the Cadence schematic properties window, as listed in table B.10.

B.7 Conclusion

The target of this appendix was that of providing the reader with the necessary knowledge and information in order to utilise the MEMS compact model library, discussed in the previous chapter 5, concerning the mathematical models within the Cadence Integrated Circuits (ICs) development environment. The appendix was arranged as a user guide, while the software library is made available for download together with this book, and is implemented in two different Hardware Description Language (HDL)-based codes, i.e. VerilogA and SpectreHDL, within the Cadence environment (www.cadence.com), for the Spectre Circuit Simulator. The tool can be ported to other simulation environments, i.e. other than Cadence, as long as HDL codes are supported.

References

[1] *Verilog-A—Language Reference Manual* https://siue.edu/~gengel/ece585WebStuff/OVI_VerilogA.pdf)

[2] *Affirma VerilogA Language Reference*, December, 1999 (Cadence Design Systems, Inc. Product Version 4.4.5)

[3] *SpectreHDL Reference* December, 1999 (Cadence Design Systems, Inc. Product Version 4.4.5)

[4] *Spectre Simulation Platform* https://cadence.com/en_US/home/tools/custom-ic-analog-rf-design/circuit-simulation/spectre-simulation-platform.html)

[5] *Affirma Spectre Circuit Simulator User Guide* June, 2000 (Cadence Design Systems, Inc. Product Version 4.4.6)

[6] Jing Q 2003 Modeling and for design of suspended MEMS, *PhD Dissertation in Electrical and Computer Engineering* Carnegie Mellon University https://research.ece.cmu.edu/~mems/pubs/pdfs/ece/PhD_thesis/0117_jing-2003.pdf)

[7] Przemieniecki J S 1968 *Theory of Matrix Structural Analysis* (New York: McGraw-Hill)

[8] Iannacci J 2013 *Practical Guide to RF-MEMS* 1st edn (Weinheim: Wiley) p 372

[9] Cho Y-H, Pisano A P and Howe R T 1994 Viscous damping model for laterally oscillating microstructures *J. Microelectromech. Syst.* **3** 81–7

[10] Veijola T, Kuisma H and Lahdenpera J 1997 Model for gas film damping in a silicon accelerometer *Proc. of the Int. Conf. on Solid-State Sensors Actuators* **4** 1097–100

[11] Iannacci J, Gaddi R and Gnudi A 2010 Experimental validation of mixed electromechanical and electromagnetic modeling of RF-MEMS devices within a standard IC simulation environment *J. Microelectromech. Syst.* **19** 526–37

www.ingramcontent.com/pod-product-compliance
Ingram Content Group UK Ltd.
Pitfield, Milton Keynes, MK11 3LW, UK
UKHW051858100926
8556IPUK00005B/686

9 780750 3419